POWER PIPING

THE COMPLETE GUIDE TO ASME B31.1

by
Charles Becht IV

ASME PRESS

Library of Congress Cataloging-in-Publication Data

Becht, Charles, IV.
    Power piping : the complete guide to ASME B31.1 / by Charles Becht IV.
    pages cm
    ISBN 978-0-7918-6014-4
    1. Piping–Standards—United States. I. Title.
    TJ930.B346 2013
    621.8'672021873–dc23

                                    2013001888

ABOUT THE AUTHOR

Dr. Becht is a recognized authority in pressure vessels, piping, expansion joints, and elevated temperature design. He is President of Becht Engineering Co. Inc, a consulting engineering company that provides both process and equipment engineering services as well as project and turnaround services for the process and power industries; President of Becht Engineering Canada Ltd.; CEO of Helidex, LLC; and Director of Sonomatic Ltd. (also dba Becht Sonomatic in North America) an NDE company that provides advanced ultrasonic imaging equipment and services. He has performed numerous expert troubleshooting and failure investigations, design reviews and construction inspections for capital projects into the billion dollar range, and consulting to manufacturers on design, development and code compliance for new and existing equipment. He was previously with Energy Systems Group, Rockwell International and Exxon Research and Engineering where he was a pressure equipment specialist.

Dr. Becht is a member of the ASME B31.3, Process Piping Committee (past Chair); the Post Construction Subcommittee on Repair and Testing (PCC) (founding chair), the Post Construction Standards Committee (past Chair); Post Construction Executive Committee (past Chair); B&PV Code Subgroup on Elevated Temperature Design (past Chair); B31 Code for Pressure Piping Standards Committee; B31 Mechanical Design Committee; and is a past member of the Board on Pressure Technology Codes and Standards; the B&PV Code Subcommittee on Design; the B&PV Code Subcommittee on Transport Tanks; the B31 Executive Committee; and the B&PV Code TG on Class 1 Expansion Joints for liquid metal service. He is a member of ASTM Committee F-17, Plastic Piping Systems Main Committee; and the ASME PVP Division, Design and Analysis Committee.

He received a PhD from Memorial University in Mechanical Engineering (dissertation: Behavior of Bellows), a MS from Stanford University in Structural Engineering and BSCE from Union College, New York. Chuck is a licensed professional engineer in 16 states and provinces, an ASME Fellow since 1996, recipient of the ASME Dedicated Service Award in 2001, recipient of the PVP Medal in 2009 and has more than 60 publications and six patents.

CONTENTS

LIST OF FIGURES

LIST OF TABLES

BACKGROUND AND GENERAL INFORMATION

This book is based on the 2012 edition of ASME B31.1, Power Piping Code. As changes, some very significant, are made to the Code with every new edition, the reader should refer to the most recent edition of the Code for specific requirements. The purpose of this book is to provide background information and not the specific, current Code rules.

References herein to ASME BPVC Sections I, II, III, V, VIII, and IX are references to Sections of the ASME Boiler and Pressure Vessel Code. References to a paragraph are generally references to a paragraph in ASME B31.1 or to a paragraph in this book.

The equations that are numbered in this book use the same numbers as are used in ASME B31.1. Equations that are not numbered are either not in ASME B31.1 or are not numbered therein.

1.1 HISTORY OF B31.1

In 1926, the American Standards Institute initiated Project B31 to develop a piping code. The ASME was the sole administrative sponsor. The first publication of this document, American Tentative Standard Code for Pressure Piping, occurred in 1935. From 1942 through 1955, the Code was published as the American Standard Code for Pressure Piping, ASA B31.1. It consisted of separate sections for different industries.

These separate sections were split off, starting in 1955, with the Gas Transmission and Distribution Piping Systems, ASA B31.8. ASA B31.3, Petroleum Refinery Piping Code, was first published in 1959. A number of separate documents have been prepared, most of which have been published, and some of which have been withdrawn. The various designations are as follows:

(1) B31.1, Power Piping
(2) B31.2, Fuel Gas Piping (withdrawn in 1988)
(3) B31.3, Process Piping
(4) B31.4, Pipeline Transportation Systems for Liquid Hydro-Carbons and Other Liquids
(5) B31.5, Refrigeration Piping
(6) B31.6, Chemical Plant Piping (never published; merged into B31.3)
(7) B31.7, Nuclear Piping (moved to ASME BPVC, Section III)
(8) B31.8, Gas Transmission and Distribution Piping Systems
(9) B31.9, Building Services Piping
(10) B31.10, Cryogenic Piping (never published; merged into B31.3)
(11) B31.11, Slurry Piping
(12) B31.12, Hydrogen Piping and Pipelines

With respect to the initials that appear in front of B31.1, these have been ASA, ANSI, and ASME. It is currently correct to refer to the Code as ASME B31.1. The initial designation, ASA, referred to the American Standards Association. This organization later became the United States of America Standards Institute and then the American National Standards Institute (ANSI) between 1967 and 1969; thus, ASA was changed to ANSI. In 1978, the B31 Code Committees were reorganized as a committees operating under ASME procedures that are accredited by ANSI. Therefore, the initials ASME now appear in front of B31.1. These changes in acronyms have not changed the committee structure or the Code itself.

1.2   SCOPE OF B31.1

The B31.1 Code for Power Piping is generally thought of as a Code for addressing piping systems within electrical power-generating plants. The original 1935 B31.1 Code for Pressure Piping was written to address all pressure piping. Specific sections within the original B31.1 Code addressed piping for various industries. These sections were split off into individual B31 series Codes starting in 1955 and as they were split off, specific rules for those industries were no longer included in B31.1. As it exists at this writing, the B31.1 Code for Power Piping includes rules for addressing piping within electric power-generating plants, industrial and institutional plants, geothermal heating systems, and central and district heating and cooling systems.

Through the 1998 edition, the B31.1 Code defined "Power Piping" systems as (with exceptions) all piping systems and their component parts within the plants mentioned above to include steam, water, oil, gas, and air services. The exceptions were the systems that were explicitly excluded by para. 100.1.3 as listed below:

(a) Piping specifically covered by other sections of the B31 Code for Pressure Piping

(b) Pressure Vessels (e.g., economizers, heaters, etc.) and other components covered by sections of the ASME Boiler and Pressure Vessel Code (note that the connecting piping is covered by B31.1)

(c) Building heating and distribution steam piping designed for 15 psig or less, or hot water heating systems designed for 30 psig or less

(d) Roof and floor drains, plumbing, sewers, and sprinkler systems, and other fire protection systems

(e) Piping for hydraulic or pneumatic tools and their components downstream of the first stop or block valve off the system distribution header

(f) Piping for marine or other installations under Federal control

(g) Piping for nuclear installations covered by Section III of the ASME Boiler and Pressure Vessel Code

(h) Towers, building frames, tanks, mechanical equipment, instruments, and foundations

(i) Building services piping within the property limits or buildings or buildings of industrial and institutional facilities, which is within the scope of ASME B31.9 except that piping beyond the limits of material, size, temperature, pressure, and service specified in ASME B31.9 shall conform to the requirement of ASME B31.1

(j) Fuel gas piping inside industrial and institutional buildings, which is within the scope of ANSI/NFPA Z223.1, National Fuel Gas Code

(k) Pulverized fossil fuel piping, which is within the scope of NFPA 85F

Note that through the 1998 edition of B31.1, for fuel gas or fuel oil brought to the plant site from a distribution system, the piping upstream of the meters was excluded from the scope of B31.1. Fuel gas or fuel oil downstream of the meters and into the plant was included in the scope of B31.1. Plant gas and oil systems other than fuel systems, air systems, and hydraulic fluid systems were included in the scope of B31.1.

In the 2012 edition, packaged equipment piping was introduced. Packaged equipment piping included as part of a shop-assembled packaged equipment assembly that is constructed to another B31 Code section is exempted, with owner's approval.

A number of these explicit definitions of scope were removed from B31.1 (specifically a, d, g, i, j, and k) when the ASME B31 Standards Committee directed that the B31 Codes be revised to permit the owner to select the piping code most appropriate to their piping installation; this change is incorporated in the 1999 addenda. The Introduction to ASME B31.1 (as well as the Introductions to the other B31 Codes) now states the following:

"It is the owner's responsibility to select the Code Section which most nearly applies to a proposed piping installation. Factors to be considered by the owner include: limitations of the Code Section; jurisdictional requirements; and the applicability of other Codes and Standards. All applicable requirements of the selected Code Section shall be met."

While ASME B31 now assigns responsibility to the owner for selecting the Code Section that the owner considers the most appropriate to the piping installation, the ASME B31.1 Section Committee has generally considered industrial and institutional piping, other than process piping, to be within the scope of ASME B31.1. In process facilities, nearly all piping, including utilities, generally, are constructed in accordance with ASME B31.3. In other industrial and institutional facilities, ASME B31.9 should generally be the Code of choice unless the system is not within the coverage limitations of ASME B31.9, in which case, B31.1 would normally be the most applicable Code. These B31.9 limits include:

(1) Maximum size and thickness limitations, depending on material:
    (a) Carbon steel: NPS 48 (DN 1200) and 0.50 in. (12.5 mm)
    (b) Stainless steel: NPS 24 (DN 600) and 0.50 in. (12.5 mm)
    (c) Aluminum: NPS 12 (DN 300)
    (d) Brass and copper: NPS 12 (DN 300) [12.125 in. OD (308 mm) for copper tubing]
    (e) Thermoplastics: NPS 24 (DN 600)
    (f) Ductile iron: NPS 48 (DN 1200)
    (g) Reinforced thermosetting resin: 24 in. (600 mm) nominal

(2) Maximum pressure limits:
    (a) Boiler external piping for steam boilers: 15 psig (105 kPa)
    (b) Boiler external piping for water heating units: 160 psig (1100 kPa)
    (c) Steam and condensate: 150 psig (1035 kPa)
    (d) Liquids: 350 psig (2415 kPa)
    (e) Vacuum: 1 atm external pressure
    (f) Compressed air and gas: 150 psig (1035 kPa)

(3) Maximum temperature limits:
    (a) Boiler external piping for water heating units: 250°F (121°C)
    (b) Steam and condensate: 366°F (186°C)
    (c) Other gases and vapors: 200°F (93°C)
    (d) Other nonflammable liquids: 250°F (121°C)

Note that within the ASME B31.9 Code the minimum temperature for piping is 0°F (–18°C). Also note that piping for toxic and flammable gases and toxic liquids are excluded from the scope of ASME B31.9.

High pressure and/or temperature steam and water piping within industrial and institutional facilities should generally be designed and constructed in accordance with ASME B31.1.

The steam–water loop piping associated with power plant boilers has three general types: boiler proper piping, boiler external piping (BEP), and non-boiler external piping (NBEP). Boiler proper piping is internal to the boiler and is entirely covered by Section I of the Boiler and Pressure Vessel Code. Boiler proper piping is actually part of the boiler (e.g., downcomers, risers, transfer piping, and piping between the steam drum and an attached superheater). It is entirely within the scope of ASME BPVC, Section I and is not addressed by ASME B31.1. A discussion of boiler piping classification and the history behind it is provided by Bernstein and Yoder (1998) and Mackay and Pillow (2011).

Boiler external piping includes piping that is part of the boiler, but which is external to the boiler. Boiler external piping (BEP) begins at the boiler where the boiler proper ends (boiler terminal points):

(1) at the first circumferential weld joint for a welding end connection, or
(2) at the face of the first flange in bolted flange connections, or
(3) at the first threaded joint in that type of connection.

The BEP extends from these boiler terminal points up to and including the valves required by ASME B31.1 para. 122.1. This piping is considered part of the boiler, and thus within the scope of Section I; however, the rules covering the design and construction of this piping are provided in B31.1.

Non-boiler external piping comprises all the piping that is covered by the ASME B31.1 Code except the piping described as boiler external piping. For this piping, the rules fall entirely within ASME B31.1. Figures 1.1 through 1.3 illustrate the jurisdictional limits of boiler proper, boiler external, and non-boiler external piping.

Because the ASME B31.1 Code is written for a very specific application—power plant piping—very detailed piping system-specific rules are provided. This differs from, for example, the ASME B31.3 Code, where rules are written in respect to service conditions (e.g., pressure, temperature, flammable, and toxic) rather than (as is the case with ASME B31.1) in respect to specific systems (e.g., steam, feedwater, drains, blowoff, and blowdown).

1.3   WHAT IS PIPING?

ASME B31.1 covers power piping, but what is within the scope of piping? Piping is defined in para. 110.1.1 to include "pipe, flanges, bolting, gaskets, valves, pressure-relieving valves/devices, fittings, and other pressure containing portions of other piping components, whether manufactured in accordance with Standards listed in Table 126.1 or specially designed. It also includes hangers and supports and other equipment items necessary to prevent overstressing the pressure containing components."

Pipe supporting elements are defined in para. 100.1.2 to include "hangers, supports, and structural attachments." Hangers and supports are defined to "include elements that transfer the load from the pipe or structural attachment to the supporting structure or equipment." Examples such as hanger rods, spring hangers, sway braces, and guides are given.

The supporting structure itself is not within the scope of ASME B31.1; its design and construction is governed by civil/structural codes and standards.

1.4   INTENT

The ASME B31.1 Code provides the minimum requirements for safety. It is not a design handbook; furthermore, it is for design of new piping. However, it is used for guidance in the repair, replacement, or modification of existing piping. See B31.1 Non-mandatory Appendix V, Recommended Practice for Operation, Maintenance, and Modification of Power Piping Systems, para. V-8.1, which states the following:

"Piping and piping components which are replaced, modified, or added to existing piping systems are to conform to the edition and addenda of the Code used for design and construction of the original systems, or to later Code editions or addenda as determined by the Operating Company. Any additional piping systems installed in existing plants shall be considered as new piping and shall conform to the latest issue of the Code."

Also see B31.1, Chapter VII, Operation and Maintenance, which was added in the 2007 edition.

Further clarification on the issue of using a more recent edition of the Code for replacement, modification, or addition is provided in Interpretation 26-1, Question (2).

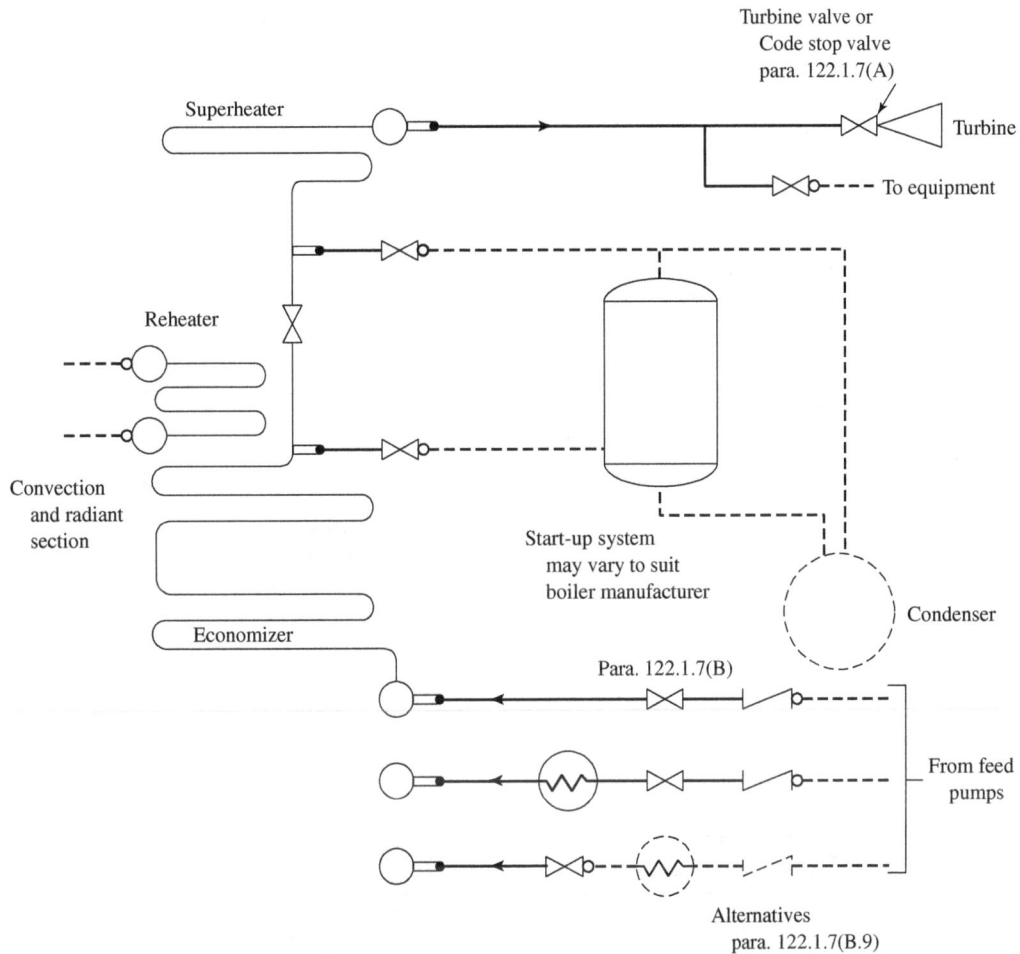

Superheater

Turbine valve or
Code stop valve
para. 122.1.7(A)

Turbine

To equipment

Reheater

Convection
and radiant
section

Start-up system
may vary to suit
boiler manufacturer

Condenser

Economizer

Para. 122.1.7(B)

From feed
pumps

Alternatives
para. 122.1.7(B.9)

Administrative Jurisdiction and Technical Responsibility

——— Boiler Proper — The ASME Boiler and Pressure Vessel Code (ASME BPVC) has total administrative jurisdiction and technical responsibility. Refer to ASME BPVC Section I Preamble.

●——— Boiler External Piping and Joint (BEP) — The ASME BPVC has total administrative jurisdiction (mandatory certification by Code Symbol stamping, ASME Data Forms, and Authorized Inspection) of BEP. The ASME Section Committee B31.1 has been assigned technical responsibility. Refer to ASME BPVC Section I Preamble, fifth, sixth, and seventh paragraphs and ASME B31.1 Scope, para. 100.1.2(A). Applicable ASME B31.1 Editions and Addenda are referenced in ASME BPVC Section I, PG-58.3.

○- - - - Nonboiler External Piping and Joint (NBEP) — The ASME Code Committee for Pressure Piping, B31, has total administrative and technical responsibility.

FIG. 1.1
CODE JURISDICTION LIMITS FOR PIPING – AN EXAMPLE OF FORCED FLOW STEAM GENERATORS WITH NO FIXED STEAM AND WATER LINE (ASME B31.1, FIG. 100.1.2 (A.1))

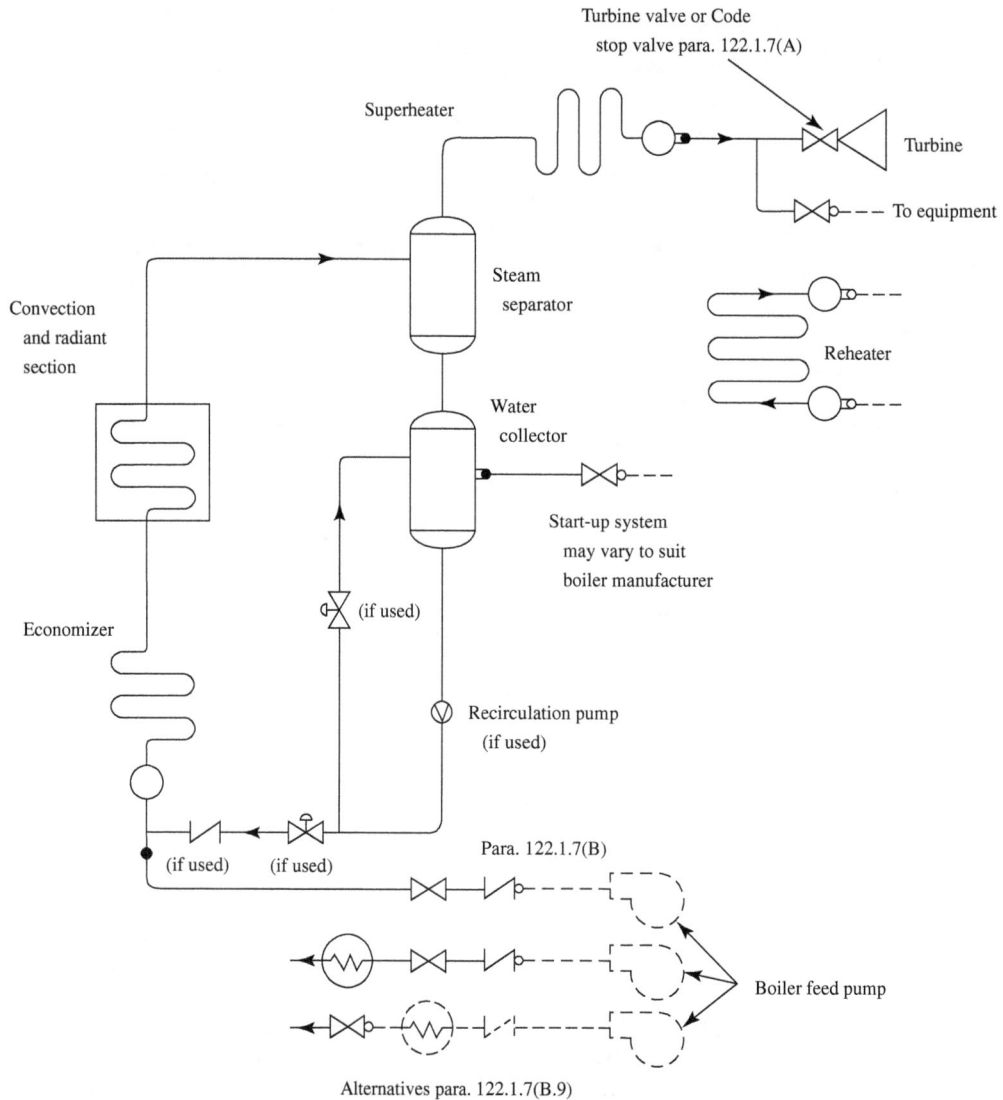

Turbine valve or Code
stop valve para. 122.1.7(A)

Superheater

Turbine

To equipment

Reheater

Convection
and radiant
section

Steam
separator

Water
collector

Start-up system
may vary to suit
boiler manufacturer

Economizer

(if used)

Recirculation pump
(if used)

Para. 122.1.7(B)

(if used)   (if used)

Boiler feed pump

Alternatives para. 122.1.7(B.9)

Administrative Jurisdiction and Technical Responsibility

—————— Boiler Proper – The ASME Boiler and Pressure Vessel Code (ASME BPVC) has total
administrative jurisdiction and technical responsibility. Refer to ASME BPVC Section I Preamble.

●—————— Boiler External Piping and Joint (BEP) – The ASME BPVC has total administrative jurisdiction
(mandatory certification by Code Symbol stamping, ASME Data Forms, and Authorized
Inspection) of BEP. The ASME Section Committee B31.1 has been assigned technical
responsibility. Refer to ASME BPVC Section I Preamble, fifth, sixth, and seventh paragraphs
and ASME B31.1 Scope, para. 100.1.2(A). Applicable ASME B31.1 Editions and Addenda are
referenced in ASME BPVC Section I, PG-58.3.

○ – – – – Nonboiler External Piping and Joint (NBEP) – The ASME Code Committee for Pressure Piping,
B31, has total administrative and technical responsibility.

FIG. 1.2
CODE JURISDICTIONAL LIMITS FOR PIPING – AN EXAMPLE OF STEAM SEPARATOR
TYPE FORCED FLOW STEAM GENERATORS WITH NO FIXED STEAM AND WATER LINE
(ASME B31.1, FIG. 100.1.2(A.2))

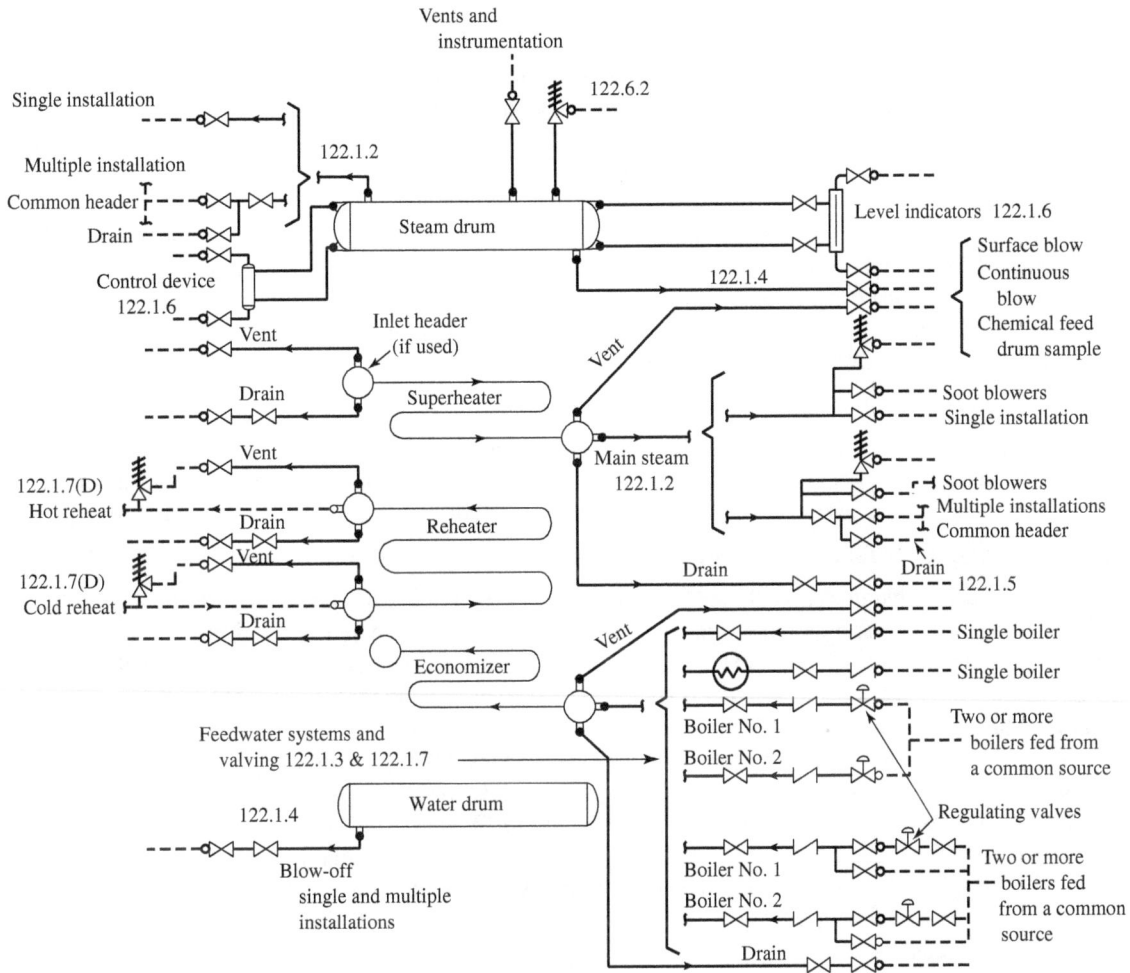

Administrative Jurisdiction and Technical Responsibility

———— Boiler Proper — The ASME Boiler and Pressure Vessel Code (ASME BPVC) has total administrative jurisdiction and technical responsibility. Refer to ASME BPVC Section I Preamble.

●———— Boiler External Piping and Joint (BEP) — The ASME BPVC has total administrative jurisdiction (mandatory certification by Code Symbol stamping, ASME Data Forms, and Authorized Inspection) of BEP. The ASME Section Committee B31.1 has been assigned technical responsibility. Refer to ASME BPVC Section I Preamble and ASME B31.1 Scope, para. 100.1.2(A). Applicable ASME B31.1 Editions and Addenda are referenced in ASME BPVC Section I, PG-58.3.

o— — — — Nonboiler External Piping and Joint (NBEP) — The ASME Code Committee for Pressure Piping, B31, has total administrative jurisdiction and technical responsibility.

FIG. 1.3
CODE JURISDICTIONAL LIMITS FOR PIPING – DRUM TYPE BOILERS
(ASME B31.1, FIG. 100.1.2(A.1))

Question (2): If a Code edition or addenda later than the original construction edition (and applicable addenda) is used, is a reconciliation of the differences required?

Reply (2): No. However, the Committee recommends that the impact of the applicable provisions of the later edition or addenda be reconciled with the original Code edition and applicable addenda.

ASME B31.1 is intended to parallel the ASME BPVC, Section I, Power Boilers, to the extent that it is applicable to power piping. Some of the philosophy of the Code is discussed in the Foreword.

The Foreword states that the Code is more conservative than some other piping Codes; however, conservatism consists of many aspects, including allowable stress, fabrication, examination, and testing. When comparing ASME B31.1 with ASME B31.3, one will find that ASME B31.1 is more proscriptive and, depending on the circumstances, more or less conservative. For example, wall thickness of ASME B31.1 will generally be the same or greater. Degree of examination will be more or less, depending on the service. Hydrostatic test pressure will be lower, but pneumatic test pressure will be higher.

The Foreword also contains the following additional paragraph:

"The Code never intentionally puts a ceiling limit on conservatism. A designer is free to specify more rigid requirements as he feels they may be justified. Conversely, a designer who is capable of a more rigorous analysis than is specified in the Code may justify a less conservative design, and still satisfy the basic intent of the Code."

In the Introduction, the following paragraph is provided:

"The specific design requirements of the Code usually revolve around a simplified engineering approach to a subject. It is intended that a designer capable of applying more complete and rigorous analysis to special or unusual problems shall have latitude in the development of such designs and the evaluation of complex or combined stresses. In such cases, the designer is responsible for demonstrating the validity of his approach."

Thus, while ASME B31.1 is generally very proscriptive, it provides the latitude for good engineering practice when appropriate to the situation. Note that designers are essentially required to demonstrate the validity of their approach to the owner's satisfaction and, for boiler external piping, to the Authorized Inspector's satisfaction. This is addressed in B31.1 Interpretations 11 to 13, Question (1).

Question (1): To whom should a designer justify a less conservative design by more rigorous analysis to satisfy the basic intent of the Code as allowed in the Foreword and Introduction?

Reply (1): The owner of a piping installation has overall responsibility for compliance with the B31.1 Code, and for establishing the requirements for design, construction, examination, inspection, and testing. For boiler external piping, the requirements of para. 136.3 shall also be satisfied. A designer capable of more rigorous design analysis than is specified in the B31.1 Code may justify less conservative designs to the owner or his agent and still satisfy the intent of the Code. The designer is cautioned that applicable jurisdictional requirements at the point of installation may have to be satisfied.

1.5 RESPONSIBILITIES

1.5.1 Owner

The owner's first responsibility is to determine which Code Section should be used. The owner is also responsible for imposing requirements supplementary to those of the selected Code Section, if necessary, to ensure safe piping for the proposed installation. These responsibilities are included in the Introduction.

The owner is also responsible for inspection of non-boiler external piping to ensure compliance with the engineering design and with the material, fabrication, assembly, examination, and test requirements of ASME B31.1.

1.5.2 Designer

While not specifically stated in ASME B31.1, the designer is responsible to the owner for assurance that the engineering design of piping complies with the requirements of the Code and with any additional requirements established by the owner.

1.5.3   Manufacturer, Fabricator, and Erector

While not specifically stated in ASME B31.1, the manufacturer, fabricator, and erector of piping are responsible for providing materials, components, and workmanship in compliance with the requirements of the Code and of the engineering design.

1.5.4   Inspector

The inspector is responsible to the owner, for non-boiler external piping, to ensure compliance with the engineering design and with the material, fabrication, assembly, examination, and test requirements of the Code. An Authorized Inspector, which is a third party, is required for boiler external piping. The manufacturer or assembler is required to arrange for the services of the Authorized Inspector. The Authorized Inspector's duties are described in Section 14.1 herein. The qualifications of the Authorized Inspector are specified in ASME BPVC, Section I, PG-91, as follows:

An Inspector employed by an ASME accredited Authorized Inspection Agency, that is, the inspection organization of a state or municipality, of the United States, a Canadian province, or of an insurance company authorized to write boiler and pressure vessel insurance. They are required to have been qualified by written examination under the rules of any state of the United States or province of Canada, which has adapted the Code (ASME BPVC, Section I).

1.6   HOW IS B31.1 DEVELOPED AND MAINTAINED?

ASME B31.1 is a consensus document. It is written by a committee that is intended to contain balanced representation from a variety of interests. Membership includes the following:

(1) Manufacturers
(2) Owners/operators
(3) Designers/constructors
(4) Regulatory agents
(5) Insurers/inspectors
(6) General interest parties

The members of the committee are not intended to be representatives of specific organizations; their membership is considered based on qualifications of the individual and desire for balanced representation of various interest groups. B31.1 is written as a consensus Code and is intended to reflect industry practice. This differs from a regulatory approach in which rules may be written by a government body.

Changes to the Code are prepared by the B31.1 Section Committee. Within the Section Committee, responsibility for specific portions of the Code is split among Subgroups. These are the following:

(1) Subgroup on general requirements
(2) Subgroup on materials
(3) Subgroup on design
(4) Subgroup on fabrication and examination
(5) Subgroup on operations and maintenance
(6) Task group on special assignments

To make a change to the Code, the responsible Subgroup prepares documentation of the change, which is then sent out as a ballot to the entire Section Committee to vote on. Anyone who votes against the change (votes negatively) must state their reason for doing so, which is shared with the entire Section Committee. The responsible Subgroup usually makes an effort to resolve any negatives. A two-thirds majority is required to approve an item. Any changes to the Code are forwarded to the B31 Standards Committee along with the written reasons for any negative votes. In this fashion, the Standards Committee is given the opportunity to see any opposing viewpoints. If anyone on the B31 Standards Committee votes negatively on the change, on first consideration, the item is returned to the Section Committee with written reasons for the negative. The Section Committee must consider and respond to any negatives and comments, either by withdrawing or modifying the proposed change or by providing explanations that respond to the negatives or comments. If the item is returned to the Standards Committee for second consideration, it requires a two-thirds approval to pass.

Once an item is passed by the Standards Committee, it is forwarded to the Board on Pressure Technology Codes and Standards, which is the final level at which the item is voted on within ASME. Board member are given the opportunity to offer technical comments when the Standards Committee votes. When the Board votes, it is a vote as to whether procedures have been properly followed. Any negative vote by the Board returns the ballot to the Section Committee.

While the Board on Pressure Technology Codes and Standards reports to the Council on Standards and Certification, the Council does not vote on changes to the Code.

The final step is a public review process. Availability of document drafts is announced in two publications: ANSI's Standards Action and ASME's Mechanical Engineering. Copies of the proposed changes are also forwarded to the B31 Conference Group for review. Any comments from the public or the Group are considered by the Section Committee.

While there are a lot of steps in the process, an item can be published as a change to the Code within 1 year of approval by the Section Committee, assuming it is passed on first consideration by the higher committees. The procedures provide for careful consideration and public review of any change to the Code.

1.7 CODE EDITIONS AND ADDENDA

A new edition of the Code is issued every 2 years. Prior to the 2010 edition, a new edition was issued every 3 years, with addenda issued each year between editions. New editions (and previously addenda) include the following:

    (1) technical changes that have been approved by ballot;
    (2) editorial changes, which clarify the Code but do not change technical requirements; and
    (3) errata items

Until 1998, three addenda were issued between new editions, with one addendum being issued in the same year as that in which the new edition was published. All technical changes were made in addenda, and only editorial changes and errata were included in any new edition. In 1998, this was changed to two addenda with technical changes included in the new edition. Then, in 2010, addenda were eliminated, and the code was put on a 2-year cycle for new editions, rather than the prior 3-year cycle.

This chapter is prepared based on the 2012 edition. The next new edition is planned to be in 2014. Significant changes can occur in each new edition. An engineer whose practice includes power piping should keep current Codes. ASME sells new editions of the B31.1 Code.

1.8 HOW DO I GET ANSWERS TO QUESTIONS ABOUT THE CODE?

The B31.1 Section Committee responds to all questions about the Code via the inquiry process. Instructions for writing a request for an interpretation are provided in Appendix H. The Committee will provide a strict interpretation of the existing rules. However, as a matter of policy, the Committee will not approve, certify, rate, or endorse any proprietary device, nor will it act as a consultant on specific engineering problems or the general understanding or application of Code rules. Furthermore, it will not provide explanations for the background or reasons for Code rules. If you need any of the above, you should engage in research or education, read this book, and/or hire a consultant, as appropriate.

The Section Committee will answer any request for interpretation with a literal interpretation of the Code. It will not create rules that do not exist in the Code and will state that the Code does not address an item if it is not specifically covered by rules written into the Code. An exception to this is an intent interpretation. On occasion, it is determined that the Code wording is unclear; in that case, an intent interpretation can be issued together with a Code change to clarify the wording in the Code. The intent interpretation is not released until the Code change is approved.

Inquiries are assigned to a committee member who develops a proposed question and reply between meetings. Although the procedures permit these to be considered between meetings, the practice is for the Section Committee as a whole to consider and approve interpretations at the Section Committee meetings. The approved question and reply are then forwarded to the inquirer by the ASME staff. Note that the inquiry may not be considered at the next meeting after it is received (the person responsible for handling the inquiry may not have prepared a response yet).

Interpretations are posted on the ASME B31.1 website for the benefit of all Code users.

1.9   HOW CAN I CHANGE THE CODE?

The simplest means for trying to change the Code is to write a letter suggesting a change. Any requests for revision to the Code are considered by the Code Committee.

To be even more effective, the individual should come to the meeting at which the item will be discussed. ASME B31.1 Section Committee meetings are open to the public, and participation of interested parties is generally welcomed. Having a person explain the change and the need for it is generally more effective than a letter alone. If you become an active participant and have appropriate professional and technical qualifications, you could be invited to become a member.

Your request for a Code change may be passed to one of three technical committees under ASME B31. These are the Fabrication and Examination Technical Committee, the Materials Technical Committee and the Mechanical Design Technical Committee, which are technical committees intended to provide technical advice to and consistency among the various Code Sections.

ORGANIZATION OF B31.1

2.1 BOILER EXTERNAL PIPING AND NON-BOILER EXTERNAL PIPING

The Code has separate requirements for boiler external and non-boiler external piping. Boiler external piping is actually within the scope of ASME BPVC, Section I. ASME BPVC, Section I refers to ASME B31.1 for technical requirements. Non-boiler external piping falls entirely within the scope of ASME B31.1. Thus, boiler external piping is treated as part of the boiler and subject to the Boiler and Pressure Vessel Code, whereas non-boiler external piping is not.

Boiler external piping is considered to start at the first weld for welded pipe, flange-face for flanged piping, or threaded joint for threaded piping outside of the boiler. It extends to the valve or valves required by ASME BPVC, Section I (and B31.1 para. 122). Both the joint with the boiler proper piping and the valve(s) at the end of the piping fall within the scope of boiler external piping.

2.2 CODE ORGANIZATION

Since the systems in a power plant are well defined, requirements are given for specific piping systems. Specific requirements for a piping system, including the basis for determining the design pressure and temperature for specific systems, can be found in Chapter II, Part 6 (para. 122). The following systems are covered:

(1) boiler external piping including steam, feedwater, blowoff, and drain piping;
(2) instrument, control, and sampling piping;
(3) spray-type desuperheater piping for use on steam generators and reheat piping;
(4) piping downstream of pressure-reducing valves;
(5) pressure-relief piping;
(6) piping for flammable and combustible liquids;
(7) piping for flammable gases, toxic gases or liquids, or non-flammable nontoxic gases;
(8) piping for corrosive liquids and gases;
(9) temporary piping systems;
(10) steam-trap piping;
(11) pump-discharge piping; and
(12) district heating and steam distribution systems.

The Code consists of six chapters and 14 appendices. Appendices with a letter designation are mandatory; those with a Roman numeral designation are non-mandatory.

The paragraphs in the Code follow a specific numbering scheme. All paragraphs in the Code are in the 100 range. The 100-series paragraphs are the ASME B31.1 Code Section of the ASME B31 Code for Pressure Piping.

2.3   NON-MANDATORY APPENDICES

ASME B31.1 contains several non-mandatory appendices. These are described below, but are not covered in detail, except as otherwise noted.

Appendix II: Non-mandatory Rules for the Design of Safety Valve Installations provides very useful guidance for the design of safety-relief-valve installations. In addition to general guidance on layout, it provides specific procedures for calculating the dynamic loads that occur when these devices operate.

Appendix III: Rules for Nonmetallic Piping provides rules for some of the services in which nonmetallic piping is permitted by ASME B31.1. It does not cover all potential non-metallic piping system applications within the scope of ASME B31.1. Appendix III is discussed in greater detail in Chapter 16.

Appendix IV: Corrosion Control for ASME B31.1 Power Piping Systems contains guidelines for corrosion control both in the operation of existing piping systems and the design of new piping systems. Though non-mandatory, Appendix IV is considered to contain minimum "requirements." It includes discussions of external corrosion of buried pipe, internal corrosion, external corrosion of piping exposed to the atmosphere, and erosion–corrosion.

Appendix V: Recommended Practice for Operation, Maintenance, and Modification of Power Piping Systems provides minimum recommended practices for maintenance and operation of power piping. It includes recommendations for procedures; documentation; records; personnel; maintenance; failure investigation and restoration; piping position history and hanger/support inspection; corrosion and/or erosion; piping addition and replacement; safety, safety-relief, and relief valves; considerations for dynamic load and high-temperature creep; and rerating.

Appendix VI: Approval of New Materials offers guidance regarding information generally required to be submitted to the ASME B31.1 Section Committee for the approval of new materials.

Appendix VII: Procedures for the Design of Restrained Underground Piping provides methods to evaluate the stresses in hot underground piping where the thermal expansion of the piping is restrained by the soil. It includes not only the axial compression of fully restrained piping but also the calculation of bending stresses that occur at changes of direction, where the piping is only partially restrained by the soil. Note that there are other procedures for such evaluations that are more amenable to computer analysis of piping, such as those published by the American Lifelines Alliance (2005).

CHAPTER

3

DESIGN CONDITIONS
AND CRITERIA

3.1 DESIGN CONDITIONS

Design conditions in ASME B31.1 are specifically intended for pressure design. The design pressure and temperature are the most severe coincident conditions that result in the greatest pipe wall thickness or highest required pressure class or other component rating. Design conditions are not intended to be a combination of the highest potential pressure and the highest potential temperature unless such conditions occur at the same time.

While it is possible for one operating condition to govern the design of one component in a piping system (and be the design condition for that component) and another to govern the design of another component, this is a relatively rare event. If this case was encountered, the two different components in a piping system would have different design conditions.

3.1.1 Design Pressure

In determining the design pressure, all conditions of internal pressure must be considered. These include thermal expansion of trapped fluids, surge, and failure of control devices. The determination of design pressure can be significantly affected by the means used to protect the pipe from overpressure. An example is the piping downstream of a pressure-reducing valve. As per para. 122.5, this piping must either be provided with a pressure-relief device or the piping must be designed for the same pressure as the upstream piping.

In general, piping systems are permitted to be used without protection of safety-relief valves. However, in the event that none are provided on the pipe (or attached equipment that would also protect the pipe), the piping system must be designed to safely contain the maximum pressure that can occur in the piping system, including consideration of failure of any and all control devices.

ASME B31.1 dictates how the design pressure is determined in para. 122 for specific systems. For example, for boiler external feedwater piping, the design pressure is required to exceed the boiler design pressure by 25% or 225 psi (1,550 kPa), whichever is less. These requirements are based on system-specific experience. For example, the aforementioned 25% higher pressure is required because this piping is considered to be in shock service and subject to surge pressure from pump transients.

While short-term conditions, such as surge must be considered, they do not necessarily become the design pressure. The Code permits short-term pressure and temperature variations as per para. 102.2.4. If the event being considered complies with the Code requirements of para. 102.2.4, the allowable stress and/or component pressure rating may be exceeded for a short time, as discussed in Section 3.5. While this is often considered to be an allowable variation above the design condition, the variation limitations are related to the maximum allowable working pressure of the piping, not the design conditions, which could be lower than the maximum allowable pressure at temperature.

3.1.2   Design Temperature

It is the metal temperature that is of interest in establishing the design temperature. The design temperature is assumed to be the same as the fluid temperature, unless calculations or tests support use of other temperatures. If a lower temperature is determined by such means, the design metal temperature is not permitted to be less than the average of the fluid temperature and the outside surface temperature.

Boilers are fired equipment and therefore subject to possible overtemperature conditions. Paragraph 101.3.2(C) requires that steam, feedwater, and hot-water piping leading from fired equipment have the design temperature based on the expected continuous operating condition plus the equipment manufacturer's guaranteed maximum temperature tolerance. Short-term operation at temperatures in excess of that condition fall within the scope of para. 102.2.4 covering permitted variations.

ASME B31.1 does not have a design minimum temperature for piping, as it does not contain impact test requirements. This is perhaps because power piping generally does not run cold. Certainly, operation of water systems below freezing is not a realistic condition to consider.

3.2   ALLOWABLE STRESS

The Code provides allowable stresses for metallic piping in Appendix A. These are, as of addendum a to the 2004 edition, the lowest of the following with certain exceptions:

(1) 1/3.5 times the specified minimum tensile strength (which is at room temperature);
(2) 1/3.5 times the tensile strength at temperature (times 1.1);
(3) two-thirds specified minimum yield strength (which is at room temperature);
(4) two-thirds "minimum" yield strength at temperature;
(5) average stress for a minimum creep rate of 0.01%/1000 hr.;
(6) two-thirds average stress for creep rupture in 100,000 hr.; and
(7) 80% minimum stress for a creep rupture in 100,000 hr.

Specified values are the minimum required in the material specifications. The "minimum" at temperature is determined by multiplying the specified (room temperature) values by the ratio of the average strength at temperature to that at room temperature. The allowable stresses listed in the Code are determined by the ASME Boiler and Pressure Vessel Code Subcommittee II and are based on trend curves that show the effect of temperature on yield and tensile strengths (the trend curve provides the aforementioned ratio). An additional factor of 1.1 is used with the tensile strength at temperature.

An exception to the above criteria is made for austenitic stainless steel and nickel alloys with similar stress–strain behavior, which can be as high as 90% of the yield strength at temperature. This is not due to a desire to be less conservative, but is a recognition of the differences between the behaviors of these alloys. The quoted yield strength is determined by drawing a line parallel to the elastic loading curve, but with a 0.2% offset in strain. The yield strength is the intercept of this line with the stress–strain curve. Such an evaluation provides a good yield strength value of carbon steel and alloys with similar behavior, but it does not represent the strength of austenitic stainless steel, which has considerable hardening and additional strength beyond this value. However, the additional strength is achieved with the penalty of additional deformation. Thus, the higher allowable stresses relative to yield are only applicable to components that are not deformation sensitive. Thus, while one might use the higher allowable stress for pipe, it should not be used for flange design.

The allowable stress for Section I of the ASME Boiler and Pressure Vessel Code was revised to change the factor on tensile strength from 1/4 to 1/3.5 in 1999. B31.1 Code Case 173 was issued in 2001 to permit use of the higher allowable stresses, while new allowable stress tables were under preparation for B31.1. The new allowable stress tables were issued with addenda 2005a (issued in 2006) to the 2004 edition.

The increase in allowable stress for ASME BPVC, Section I was not applied to bolting. Bolting remains at one-fourth tensile strength.

For cast iron materials, the behavior is brittle, and the allowable stress differs accordingly. For cast iron, the basic allowable stress is the lower of one-tenth of the specified minimum tensile strength (at room temperature) and one-tenth of the "minimum" strength at temperature, also based on the trend of average material strength with temperature. For ductile iron, a factor of one-fifth is used rather than a factor of one-tenth, and the stress is also limited to two-thirds times the yield strength. These are in accordance with ASME BPVC, Section VIII, Division 1, Appendix P, and Tables UCI-23 and UCD-23.

3.3   WELD JOINT EFFICIENCY AND CASTING QUALITY FACTORS

Weld joint efficiency factors for straight seam and spiral seam welded pipe are used in pressure design. The weld joint efficiency factors are based on the assumption of full penetration welds. The factors vary from 0.6 to 1.0, with furnace butt weld pipe having the lowest factor.

Electric resistance welded pipe has a quality factor of 0.85. This cannot be improved by additional examination.

The quality factor for electric fusion welded pipe varies from 0.80 to 1.0, depending upon whether it is a single- or double-sided weld and the degree of radiographic (RT) or ultrasonic (UT) examination (either as required by the specification or with 100% RT or UT examination).

The factors are provided in Table 102.4.3 of the Code (Table 3.1). The weld joint efficiency is included in the allowable stress values provided in B31.1 Appendix A.

Quality factors are applied to the allowable stress used in the design of cast components. A quality factor of 0.80 is included in the allowable stress values for castings that are provided in Appendix A of ASME B31.1. A quality factor as high as 1.0 may be used for cast steel components if the requirements of para. 102.4.6(B) are satisfied; this paragraph includes requirements for examination and repair of steel castings.

Note that casting quality factors are not applied to the pressure-temperature ratings of components listed in Table 126.1 or, if allowable stresses for cast components were included in such listed standards, such allowable stress values.

3.4   WELD JOINT STRENGTH REDUCTION FACTORS

Weld joint strength reduction factors, W, for weldments at elevated temperatures were introduced as para. 102.4.7 in 2008 with addenda a to ASME B31.1. They were added because weldment creep rupture strength had been determined to be lower than the base metal creep rupture strength in some circumstances.

The factor is used when calculating the required thickness of longitudinal and spiral welded pipe and fittings in pressure design. It is not required for evaluating stresses at circumferential weld locations; the Code states the designer is responsible for evaluating whether to apply weld joint strength reduction factors to welds other than longitudinal and spiral welds (e.g., circumferential welds). While it is generally agreed that it is not appropriate to apply the strength reduction factor to stresses due to displacement loading, there was no general agreement as to whether to apply the factor for stresses due to sustained loads at circumferential weld joints. As a result, the Code has left it up to the designer.

The factor does not apply to the following conditions:

1. It is not used to reduce the allowable displacement stress range, S_A, because these stresses are not sustained. The displacement stresses relax over time.
2. It is not used for evaluating stresses due to occasional loads, as such loads have short durations.

TABLE 3.1
LONGITUDINAL WELD JOINT EFFICIENCY FACTORS (ASME B31.1, TABLE 102.4.3)

No.	Type of Joint		Type of Seam	Examination	Factor E
1	Furnace butt weld, continuous weld		Straight	As required by listed specification	0.60 [Note (1)]
2	Electric resistance weld		Straight or spiral	As required by listed specification	0.85 [Note (1)]
3	Electric fusion weld				
	(a) Single butt weld (without filler metal)		Straight or spiral	As required by listed specification	0.85
				Additionally 100% RT or UT	1.00 [Note (2)]
	(b) Single butt weld (with filler metal)		Straight or spiral	As required by listed specification	0.80
				Additionally 100% RT or UT	1.00 [Note (2)]
	(c) Double butt weld (without filler metal)		Straight or spiral	As required by listed specification	0.90
				Additionally 100% RT or UT	1.00 [Note (2)]
	(d) Double butt weld (with filler metal)		Straight or spiral	As required by listed specification	0.90
				Additionally 100% RT or UT	1.00 [Note (2)]
4	API 5L	Submerged arc weld (SAW) Gas metal arc weld (GMAW) Combined GMAW, SAW 	Straight with one or two seams Spiral	As required by specification Additionally 100% RT or UT	0.90 1.00 [Note (2)]

NOTES:
(1) It is not permitted to increase the longitudinal weld joint efficiency factor by additional examination for joint 1 or 2.
(2) RT (radiographic examination) shall be in accordance with the requirements of para. 136.4.5 or the material specification, as applicable. UT (ultrasonic examination) shall be in accordance with the requirements of para. 136.4.6 or the material specification, as applicable.

3. It is not used when considering the allowable stress for permissible variations, as provided in para. 102.2.4, as such loads have short durations.
4. Equation (15) for evaluating stress due to sustained loads does not include weld joint strength reduction factors indicating that their use is not required; although para. 102.4.7 indicates that it is up to the designer.

The material-specific factors are provided in Table 102.4.7 (Table 3.2). Use of longitudinally (or spiral) welded materials in the creep range is not permitted if they are not listed in the table. When the material is in the creep range is determined from the allowable stress tables in Appendix A. When creep properties determine the allowable stress, the allowable stress is shown in italics. The start of the creep range is specified

TABLE 3.2
WELD JOINT STRENGTH REDUCTION FACTORS, (ASME B31.1, TABLE 102.4.7)

Steel Group	Weld Strength Reduction Factor for Temperature, °F (°C) [Notes (1)–(6)]										
	700 (371)	750 (399)	800 (427)	850 (454)	900 (482)	950 (510)	1,000 (538)	1,050 (566)	1,100 (593)	1,150 (621)	1,200 (649)
Carbon (Norm.) [Notes (7), (8)]	1.00	0.95	0.91	NP	NP	NP	NP	NP	NP	NP	NP
Carbon (Sub Crit) [Notes (8), (9)]	1.00	0.95	0.91	NP	NP	NP	NP	NP	NP	NP	NP
CrMo [Notes (8), (10), (11)]	...	...	1.00	0.95	0.91	0.86	0.82	0.77	0.73	0.68	0.64
CSEF (N+T) [Notes (8), (12), (13)]	...	...	...	...	...	1.00	0.95	0.91	0.86	0.82	0.77
CSEF (Sub Crit) [Notes (8), (9)]	...	...	...	...	1.00	0.73	0.68	0.64	0.59	0.55	0.50
Austenitic stainless (incl. 800H & 800HT) [Notes (14), (15)]	...	...	...	...	...	1.00	0.95	0.91	0.86	0.82	0.77
Autogenously welded austenitic stainless [Note (16)]	...	...	...	...	...	1.00	1.00	1.00	1.00	1.00	1.00

NOTES:

(1) NP = not permitted.

(2) Longitudinal welds in pipe for materials not covered in this Table operating in the creep regime are not permitted. For the purposes of this Table, the start of the creep range is the highest temperature where the nonitalicized stress values end in Mandatory Appendix A for the base material involved.

(3) All weld filler metal shall be a minimum of 0.05% C for CrMo and CSEF materials, and 0.04% C for austenitic stainless in this Table.

(4) Materials designed for temperatures below the creep range [see Note (2)] may be used without consideration of the WSRF or the rules of this Table. All other Code rules apply.

(5) Longitudinal seam welds in CrMo and CSEF materials shall be subjected to, and pass, a 100% volumetric examination (RT or UT). For materials other than CrMo and CSEF, see para. 123.4(B).

(6) At temperatures below those where WSRFs are tabulated, a value of 1.0 shall be used for the factor W where required by the rules of this Section. However, the additional rules of this Table and Notes do not apply.

(7) Norm. = normalizing postweld heat treatment (PWHT) is required.

(8) Basicity index of SAW flux ≥ 1.0.

(9) Sub Crit = subcritical PWHT is required. No exemptions from PWHT are permitted. The PWHT time and temperature shall meet the requirements of Table 132; the alternate PWHT requirements of Table 132.1 are not permitted.

(10) The CrMo steels include $\frac{1}{2}$Cr–$\frac{1}{2}$Mo, 1Cr–$\frac{1}{2}$Mo, 1$\frac{1}{4}$Cr–$\frac{1}{2}$Mo–Si, 2$\frac{1}{4}$Cr–1Mo, 3Cr–1Mo, and 5Cr–$\frac{1}{2}$Mo. Longitudinal welds shall either be normalized, normalized and tempered, or subjected to proper subcritical PWHT for the alloy.

(11) Longitudinal seam fusion welded construction is not permitted for C–$\frac{1}{2}$Mo steel for operation in the creep range [see Notes (2) and (4)].

(12) The CSEF (creep strength enhanced ferritic) steels include Grades 91, 92, 911, 122, and 23.

(13) N+T = normalizing + tempering PWHT.

(14) WSRFs have been assigned for austenitic stainless (including 800H and 800HT) longitudinally welded pipe up to 1,500°F as follows:

Temperature, °F	Temperature, °C	Weld Strength Reduction Factor
1,250	677	0.73
1,300	704	0.68
1,350	732	0.64
1,400	760	0.59
1,450	788	0.55
1,500	816	0.5

(15) Certain heats of the austenitic stainless steels, particularly for those grades whose creep strength is enhanced by the precipitation of temper-resistant carbides and carbo-nitrides, can suffer from an embrittlement condition in the weld heat affected zone that can lead to premature failure of welded components operating at elevated temperatures. A solution annealing heat treatment of the weld area mitigates this susceptibility.

(16) Autogenous SS welded pipe (without weld filler metal) has been assigned a WSRF up to 1,500°F of 1.00, provided that the product is solution annealed after welding and receives nondestructive electric examination, in accordance with the material specification.

to be the temperature at which the allowable stress value is given, immediately lower than when the values start being shown in italics.

A few highlights of the table are the following:

1. Factors are provided for carbon steel, CrMo (through 5Cr-1/2Mo), CSEF, and austenitic stainless including 800H and 800HT.

2. CSEF material weldment performance is highly dependent upon the heat treatment.
3. Autogenous (i.e., no filler metal welds) austenitic stainless welds are assigned a *W* factor of 1.0 up to 1,500°F (816°C) if they are solution annealed after welding and receive nondestructive electric examination in accordance with the material specification.

In addition to changes to the weld joint factors, additional changes were made to the fabrication and examination rules for welds in elevated temperature piping.

1. Longitudinal seam welds for CrMo and CSEF materials (in the creep range) are required to be examined by 100% radiography or 100% ultrasonic examination. This is not required for other materials; however, the appropriate joint efficiency factor, which depends upon the examination performed, is multiplied by the weld joint strength reduction factor in the pressure design rules.
2. Weld metal requirements for CSEF materials are provided in note 3 of Table 102.4.7.
3. Required heat treatment conditions for use of the factors in Table 102.4.7 are specified for some materials therein.

3.5   ALLOWANCES FOR TEMPERATURE AND PRESSURE VARIATIONS

While the Code does not use the term "maximum allowable working pressure," the concept is useful in the discussion of the allowances for variations. Pressure design of piping systems is based on the design conditions. However, since piping systems are an assembly of standardized parts, there is quite often significant pressure capacity in the piping beyond the design conditions of the system. The allowances for variations are relative to the maximum permissible pressure for the system. The allowances for variations are not used in sustained (longitudinal), occasional (wind, earthquake), nor displacement (thermal expansion) stress evaluations. They are only used in pressure design.

Increases in pressure and temperature above the design conditions are permitted by para. 102.2.4 for short-term events as long as several conditions are satisfied, one of which is that this maximum allowable working pressure is not exceeded by more than some percentage. Thus, the variation can be much higher than the design conditions, yet remains permissible.

ASME B31.1 does not allow use of the variations provision of the Code to override limitations of component standards or those given by manufacturers of components.

The circumferential pressure stress may exceed the allowable stress provided by ASME B31.1, Appendix A, by the following:

(1) 15% if the event duration occurs for no more than 8 hr at any one time and no more than 800 hour/year; or
(2) 20% if the event duration occurs for no more than 1 hr at any one time and no more than 80 hour/year. Use of the allowances for variations for piping containing toxic fluid is prohibited [see para. 122.8.2(F)].

3.6   OVERPRESSURE PROTECTION

As discussed in the prior section on design conditions, the piping system must either be designed to safely contain the maximum possible pressure, considering such factors as failure of control devices and dynamic events such as surge, or be provided with overpressure protection such as a safety-relief valve. Specific examples are provided in the Systems (Part 6) part of Chapter II for pressure-reducing valves (para. 122.5) and pump discharge piping (para. 122.13), as well as elsewhere in specific system discussions.

For example, if a 600-psi system goes through a pressure letdown valve (irrespective of fail-closed features or other safeguards) to a 300-psi system, if no safety-relief devices are provided, the 300-psi system is required to be designed to safely contain 600 psi.

If a pressure-relieving device is used, ASME B31.1 refers to ASME BPVC, Section I for boiler external piping and non-boiler external piping reheat systems, and to ASME BPVC, Section VIII, Division 1, for non-boiler external piping. See para. 5.2 herein.

Block valves are prohibited from the inlet lines to pressure-relieving safety devices, and diverter or changeover valves for redundant protective devices are permitted under certain conditions (para. 122.6.1). Block valves are also prohibited from use in pressure-relieving device discharge piping (para. 122.6.2).

PRESSURE DESIGN

4.1 METHODS FOR INTERNAL PRESSURE DESIGN

The ASME B31.1 Code provides four basic methods for design of components for internal pressure, as described in para. 102.2.

(1) Components in accordance with standards listed in Table 126.1, for which pressure ratings are provided in the standard, such as ASME B16.5 for flanges, are considered suitable by ASME B31.1 for the pressure rating specified in the standard. Note that the other methods of pressure design provided in ASME B31.1 can be used to determine pressure ratings above the maximum temperature provided in the standard if the standard does not specifically prohibit that.

(2) Some listed standards, such as ASME B16.9 for pipe fittings, state that the fitting has the same pressure rating as matching seamless pipe. If these standards are listed in Table 126.1, the components are considered to have the same allowable pressure as seamless pipe of the same nominal thickness. Note that design calculations are not usually performed for these components, design calculations are performed for the straight pipe, and matching fittings are simply selected.

(3) Design equations for some components such as straight pipe and branch connections are provided in para. 104 of ASME B31.1. These can be used to determine the required wall thickness with respect to internal pressure of components. Also, some specific branch connection designs are assumed to be acceptable.

(4) Specially designed components that are not covered by the standards listed in Table 126.1 and for which design formulas and procedures are not given in ASME B31.1 may be designed for pressure in accordance with para. 104.7.2. This paragraph provides accepted methods, such as burst testing and finite element analysis, to determine the pressure capacity of these components.

The equations in the Code provide the minimum thickness required to limit the membrane and, in some cases, bending stresses in the piping component to the appropriate allowable stress. The pressure design rules in the Code are based on maximum normal stress or maximum principal stress versus maximum shear stress or von Mises stress intensity. When the rules were developed in the 1940s, it was understood that stress intensity provided a better assessment of yielding, but it was felt that the maximum principal stress theory could generally provide a better measure of pressure capacity in situations where local yielding could simply lead to stress redistribution (Rossheim and Markl, 1960).

Mechanical and corrosion/erosion allowances must be added to this thickness. Finally, the nominal thickness selected must be such that the minimum thickness that may be provided, per specifications and considering mill tolerance, is at least equal to the required minimum thickness. Mechanical allowances include physical reductions in wall thickness such as from threading and grooving the pipe. Corrosion and erosion allowances are based on the anticipated corrosion and/or erosion over the lifespan of the pipe. Such allowances are derived from estimates, experience, or references such as NACE publications. These allowances are added to the pressure design thickness to determine the minimum required thickness of the pipe or component when it is new.

For threaded components, the nominal thread depth (dimension *h* of ASME B1.20.1, or equivalent) is used for the mechanical allowance. For machined surfaces or grooves, where the tolerance is not specified, the tolerance is required to be assumed as 1/64 in. (0.40 mm) in addition to the depth of the cut.

Mill tolerances are provided in specifications. The most common tolerance on wall thickness of straight pipe is 12.5%. This means that the wall thickness at any given location around the circumference of the pipe must not be less than 87.5% of the nominal wall thickness. Note that the tolerance on pipe weight is typically tighter, so that volume of metal and its weight may be there but a thin region would control design for hoop stress from internal pressure.

Note that the appropriate specification for the pipe must be consulted to determine the specified mill tolerance. For example, plate typically has an undertolerance of 0.01 in. (0.25 mm). However, pipe formed from plate does not have this undertolerance; it can be much greater. The pipe specification, which can permit a greater undertolerance, governs for the pipe. The manufacturer of pipe can order plate that is thinner than the nominal wall thickness for manufacturing the pipe, as long as the pipe specification mill tolerances are satisfied.

4.2   PRESSURE DESIGN OF STRAIGHT PIPE FOR INTERNAL PRESSURE

Equations for pressure design of straight pipe are provided in para. 104.1. The minimum thickness of the pipe selected, considering manufacturer's minus tolerance, must be at least equal to t_m, as calculated using Eq. (7) or Eq. (8). Equation (7) is provided below. Equation (8) is based on inside diameter, *d*, rather than outside diameter.

$$t_m = \frac{PD_O}{2(SE + Py)} + A \tag{7}$$

where:

- A = additional thickness
- D_O = pipe outside diameter (not nominal diameter)
- P = internal design gage pressure
- SE = maximum allowable stress in material from internal pressure and joint efficiency (or casting quality factor, *SF*) at design temperature from Appendix A
- t_m = minimum required thickness including additional thickness, *A*
- y = coefficient provided in Table 104.1.2(A) of the Code and Table 4.1 herein

The additional thickness, *A*, is to compensate for material removed in threading and grooving; to allow for corrosion and/or erosion; and to accommodate other variations, as described in para. 104.1.2, such as local stresses from pipe support attachments.

When Eqs. (7) or (8) is used for a casting, SF (basic material allowable stress, S, multiplied by casting quality factor, *F*), is used rather than SE.

In the 2008 edition, the title of para. 104.1.2 was changed to indicate that the above equation was applicable for temperatures below the creep range. Para. 104.1.4 was added for longitudinal and spiral-welded pipe operating in the creep range. Equations (11) and (12), for use when operating in the creep range, are the same as Eqs. (7) and (8), except that the term SE is replaced with SEW, to include the weld joint strength reduction factor. Equation (11) is shown below.

$$t_m = \frac{PD_O}{2(SEW + Py)} + A \tag{11}$$

TABLE 4.1
VALUES OF y (ASME B31.1, TABLE 104.1.2 (A))

	Temperature, °F (°C)							
Material	900 (482) and Below	950 (510)	1,000 (538)	1,050 (566)	1,100 (593)	1,150 (621)	1,200 (649)	1,250 (677) and Above
Ferritic steels	0.4	0.5	0.7	0.7	0.7	0.7	0.7	0.7
Austenitic steels	0.4	0.4	0.4	0.4	0.5	0.7	0.7	0.7
Nickel alloys UNS Nos. N06617, N08800, N08810, N08825	0.4	0.4	0.4	0.4	0.4	0.4	0.5	0.7

GENERAL NOTES:

(a) The value of y may be interpolated between the 50°F (27.8°C) values shown in the Table. For cast iron and nonferrous materials, y equals 0.

(b) For pipe with a D_o/t_m ratio less than 6, the value of y for ferritic and austenitic steels designed for temperatures of 900°F (480°C) and below shall be taken as:

$$y = \frac{d}{d + D_o}$$

Note that Eqs. (7) and (11) are based on the outside, rather than the inside diameter, which is used in pressure vessel codes. This is for a very good reason: the fact that the outside diameter of pipe is independent of wall thickness, that is, an NPS 6 pipe will have an outside diameter of 6.625 in. regardless of the wall thickness. Therefore, the wall thickness can be directly calculated when the outside diameter is used in the equation.

The foregoing equation is an empirical approximation of the more accurate and complex Lame equation. The hoop or circumferential stress is higher toward the inside of the pipe than toward the outside. This stress distribution is illustrated in Fig. 4.1. The Lame equation can be used to calculate the stress as a function of location through the wall thickness. Equation (7) is the Boardman equation (Boardman, 1943). While it has no theoretical basis, it provides a good match to the more accurate and complex Lame equation for a wide range of diameter-to-thickness ratios. It becomes increasingly conservative for lower D/t ratios (thicker pipe).

The Lame equation for hoop stress on the inside surface of pipe is given in the following equation. Note that for internal pressure, the stress is higher on the inside than the outside. This is because the strain in the longitudinal direction of the pipe must be constant through the thickness, so that any longitudinal strain caused by the compressive radial stress (from Poisson's effects and considering that the radial stress on the inside surface is equal to the surface traction of internal pressure) must be offset by a corresponding increase in hoop tensile stress to cause an offsetting Poisson's effect on longitudinal strain.

$$\sigma_h = P\left[\frac{0.5(D_O/t)^2 - (D_O/t) + 1}{(D_O/t) - 1}\right]$$

where:

σ_h = hoop stress

The Boardman empirical representation of this simply bases the calculation of pressure stress on some intermediate diameter between the inside and outside diameters of the pipe, as follows:

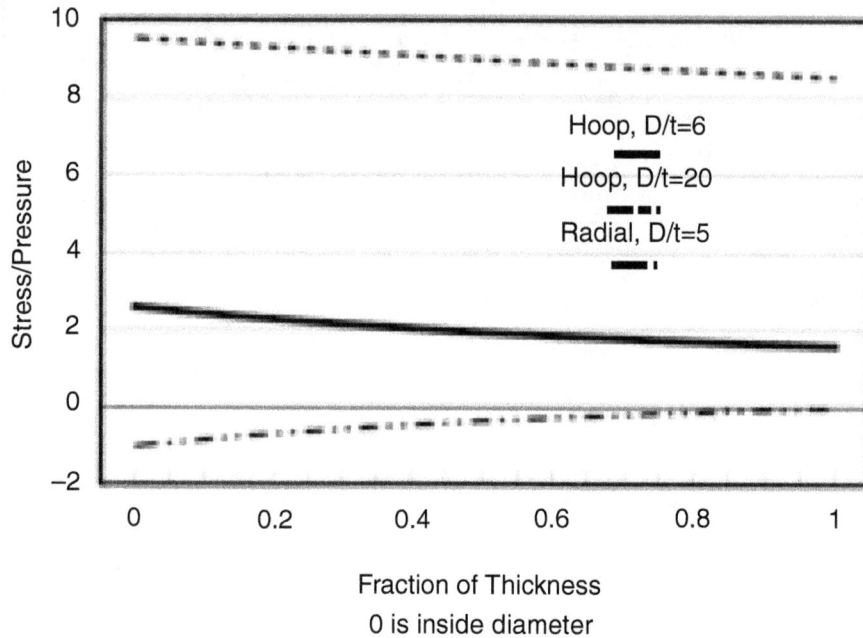

FIG. 4.1
STRESS DISTRIBUTION THROUGH PIPE WALL THICKNESS DUE TO INTERNAL PRESSURE

$$\sigma_h = P\left[\frac{D_O - 2yt}{2t}\right]$$

where:

$y = 0.4$

Simple rearrangement of the above equation, and substituting SE for σ_h, leads to the Code Eq. (7). Furthermore, inside diameter-based formulae add 0.6 times the thickness to the inside radius of the pipe rather than subtract 0.4 times the thickness from the outside radius. Thus, the inside diameter-based formula in the pressure vessel codes and Eqs. (7) and (8) of the piping Code are consistent.

A comparison of hoop stress calculated using the Lame equation versus the Boardman Eq. (7) is provided in Fig. 4.2. Remarkably, the deviation of the Boardman equation from the Lame equation is less than 1% for D/t ratios greater than 5:1. Thus, the Boardman equation can be directly substituted for the more complex Lame equation.

For thicker wall pipe, ASME B31.1 provides the following equation for the calculation of the y factor in the Note (b) of Table 104.1.2(A). Use of this equation to calculate y results in Eq. (7) matching the Lame equation for heavy wall pipe as well.

$$y = \frac{d}{D_O + d}$$

The factor y depends on temperature. At elevated temperatures, when creep effects become significant, creep leads to a more even distribution of stress across the pipe wall thickness. Thus, the factor y increases, leading to a decrease in the calculated required wall thickness (for a constant allowable stress).

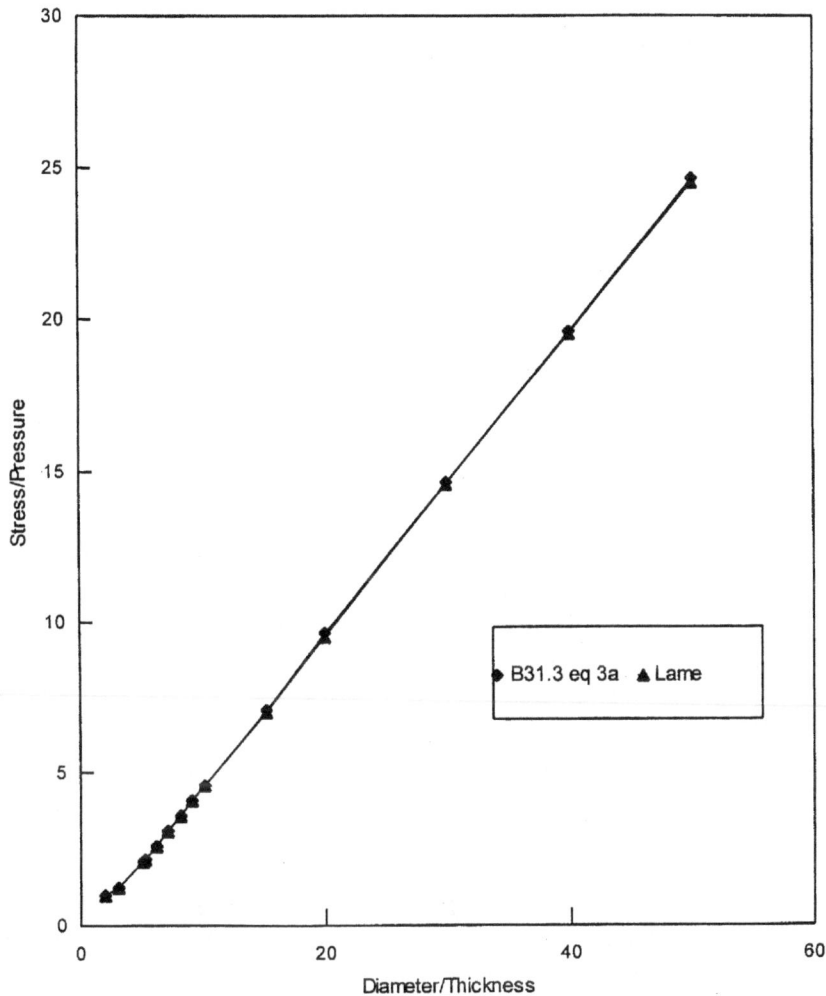

FIG. 4.2
COMPARISON OF LAME AND BOARDMAN EQUATIONS

The following additional equation is in ASME B31.1.

$$t_\mathrm{m} = \frac{Pd = 2SEA + 2y\mathrm{PA}}{2\left[\mathrm{SE} + Py - P\right]} \tag{8}$$

where:

d = inside diameter

Equation (8) is the same as Eq. (7) but with $(d+2t)$ substituted for D and the equation rearranged to keep thickness on the left side. This equation can provide a different thickness than Eq. (7) because Eq. (7) implicitly assumes that the additional thickness, A, is on the inside, whereas Eq. (8) implicitly assumes it is on the outside. If it were assumed to be on the inside, there would be an additional $P2A$ added to the numerator of Eq. (8). Alternatively, d could be taken as the inside diameter in the corroded condition.

The thickness of gray and ductile iron pipe in other than steam service may, as an alternate to Eq. (7), be determined from relevant standards. See para. 104.1.2(B). The thickness in steam service must be determined using Eq. (7).

Additional minimum thickness requirements are specified to provide added mechanical strength, beyond what is required to satisfy burst requirements, in para. 104.1.2(C).

Insert 4.1   Sample Wall Thickness Calculation

What is the required thickness of NPS 2 threaded A53 Grade B seamless pipe for the following conditions?:

- Design pressure = 150 psi
- Design temperature = 500°F
- Corrosion allowance (CA) = 1/16 in.
- Thread depth = 0.07 in.
- A = CA + thread depth = 1/16 + 0.07
- A = 0.133 in.
- SE = 17,100 psi
- W = 1.0
- y = 0.4
- D_o = 2.375
- t_m = 150(2.375)/[2(17,100 + 0.4×150)] + 0.13
- t_m = 0.010 + 0.133 = 0.143 in.

The minimum required nominal pipe wall thickness, considering mill tolerance, is

$$t_n = \frac{t_m}{0.875} = 0.16 \text{ in.}$$

Schedule 80, XS pipe, with a nominal wall thickness of 0.218 in. is acceptable.

Also, per Table 114.2.1 (Table 4.1 herein), 150 psi is less than the maximum permissible pressure for threaded NPS 2 pipe which is 600 psi.

Insert 4.2   Basic Stress Calculations for Cylinders Under Pressure

The average (through-thickness) circumferential and longitudinal (axial) stresses in a cylinder due to internal pressure can be calculated from equilibrium considerations. The circumferential stresses can be calculated from a longitudinal section, as shown in Fig. 4.3. The forces acting on that section must equilibrate, or, per Newton's law, the parts on either side of the section will start accelerating away from each other.

The pressure force acting on the section is $2rP$, where r is the inside radius and P is the internal pressure. The circumferential force in the pipe wall resisting the pressure force is $2t\sigma_c$, where t is the thickness (times two, because there are two sides), and σ_c is the average circumferential stress. These must be equal, and solving for σ_c, one arrives at the following equation:

$$\sigma_c = \frac{P_r}{t}$$

The longitudinal stress can be determined by making a girth cut (as a guillotine cut) on the pipe, as shown in Fig. 4.4. The pressure force acting on the section is $\pi r^2 P$ and the longitudinal force in the pipe wall resisting this pressure force is $2\pi r t \sigma_l$. Note that, to be more precise, the mean radius of the pipe should be used to calculate the area of the pipe wall, but using the inside radius is generally close enough and conservative. The longitudinal stress can be calculated by equating these two forces and solving for σ_l:

$$\sigma_l = \frac{P_r}{2t}$$

Thus, the longitudinal stress in a cylinder due to internal pressure is about one-half of the circumferential stress. This is quite convenient in the design of piping, because the wall thickness is determined based on pressure design. This leaves at least one-half of the strength in the longitudinal direction available for supporting the pipe weight.

A common example of the fact that the stress in the circumferential direction is twice that in the longitudinal direction can be found when cooking a hot dog. A hot dog has a pressure-containing skin. When the internal temperature reaches the point where the fluids contained inside begin to vaporize, the hot dog skin is pressurized. When the skin is overpressurized and fails, the split is always longitudinal, transverse to the direction of highest stress, the circumferential direction. Hot dogs, at least in the experience of this author, never experience guillotine failures during cooking (Fig. 4.4).

4.3   PRESSURE DESIGN FOR STRAIGHT PIPE UNDER EXTERNAL PRESSURE

For straight pipe under external pressure, there is a membrane stress check in accordance with Eq. (7) or (8) (or Eqs. (11) or (12) in the creep regime) of ASME B31.1 (the equation for internal pressure) as well as a buckling check in accordance with the external pressure design rules of the ASME BPVC, Section VIII, Division 1 (paras. UG-28, UG-29, and UG-30).

Flanges, heads, and stiffeners that comply with ASME BPVC, Section VIII, Division 1, para. UG-29 are considered stiffeners. The length between stiffeners is the length between such components. The buckling pressure is a function of geometry parameters and material properties.

Buckling pressure calculations in ASME BPVC, Section VIII, Division 1 require first calculation of a parameter A, which is a function of geometry, and then a parameter B, which depends on parameter A and a material property curve. The charts that provide the parameter B account for plasticity that occurs between the proportional limit of the stress-strain curve and the 0.2% offset yield stress. The chart for determination of parameter A is provided in Fig. 4.5. A typical chart for B is provided in Fig. 4.6.

Two equations are provided for calculating the maximum permissible external pressure. The first uses parameter B, as follows:

$$p = \frac{4B}{3D/t}$$

where:

B = parameter from material curves in ASME BPVC, Section II, Part D, Subpart 3
D = inside diameter (note that the B&PV Code takes dimensions as in the corroded condition)
p = allowable external pressure
t = pressure design thickness

The second equation is for elastic buckling and is necessary to use when the value of parameter A falls to the left of the material property curves that provide parameter B. This equation is as follows:

$$p = \frac{4AE}{3}$$

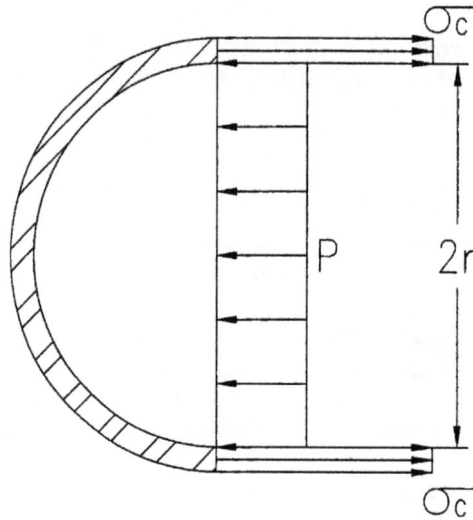

FIG. 4.3
EQUILIBRIUM AT A CIRCUMFERENTIAL CUT

where:

A = parameter from geometry curves in ASME BPVC, Section II, Part D, Subpart 3, Fig. G (included herein as Fig. 4.5)

E = elastic modulus from material curves in ASME BPVC, Section II, Part D, Subpart 3 (e.g. Fig. 4.6).

The second equation is based on elastic buckling, so the elastic modulus is used. Note that a chart of parameter B could be used, with the linear elastic portion of the curve extended to lower values of B, but this would unnecessarily enlarge the charts.

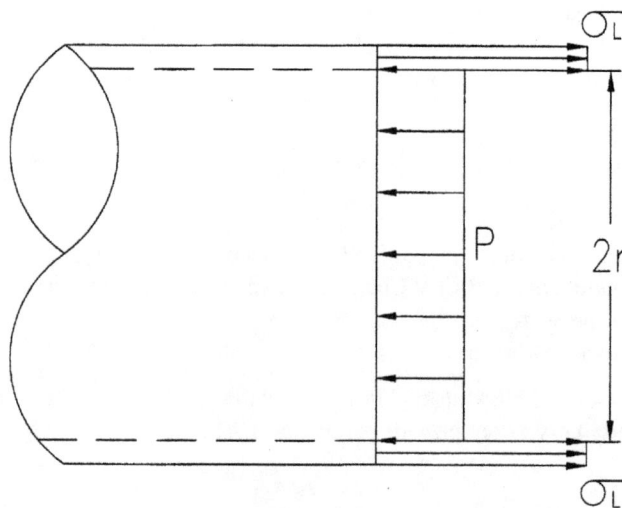

FIG. 4.4
EQUILIBRIUM AT A LONGITUDINAL CUT

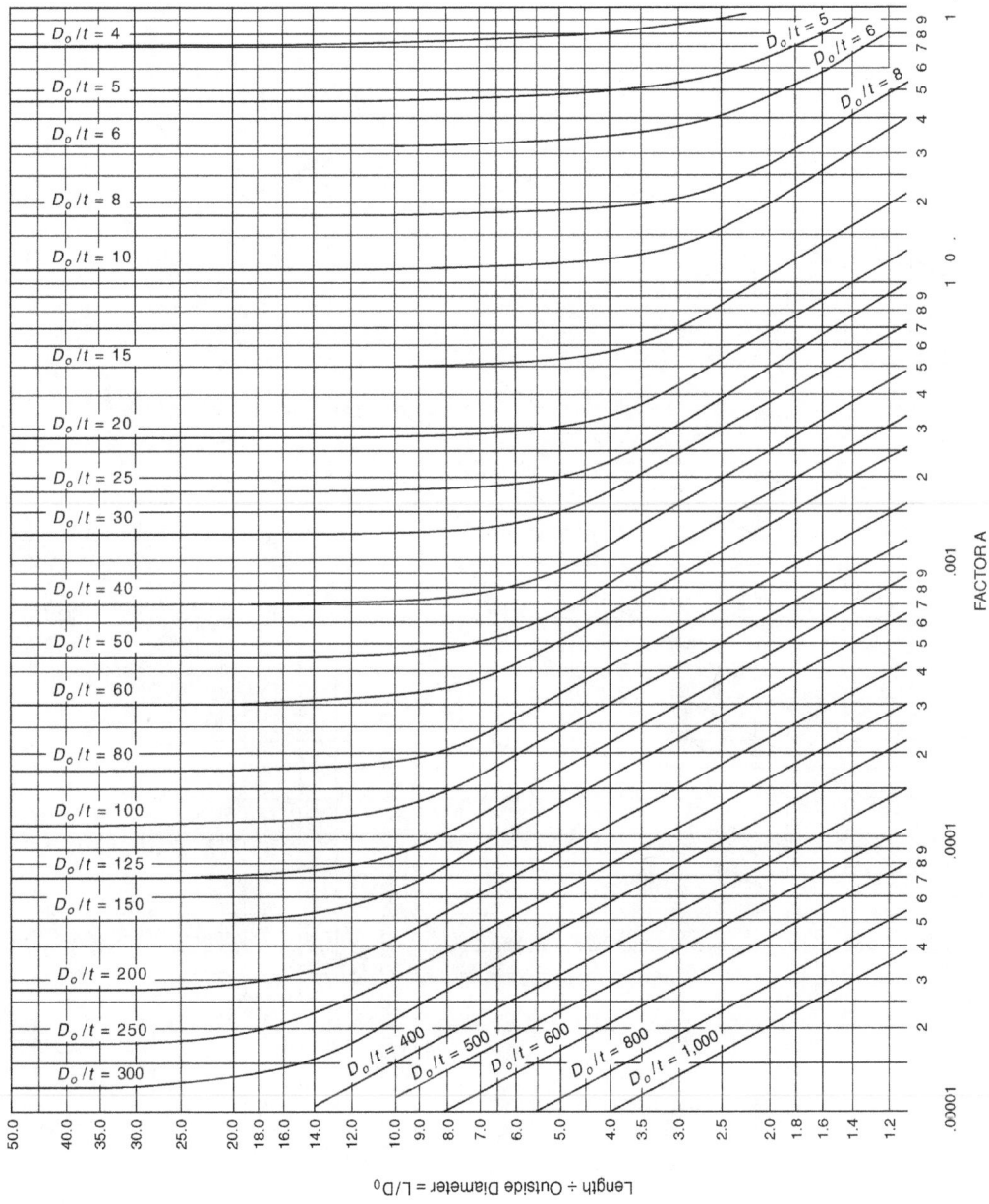

FACTOR A

FIG. 4.5

CHART FOR DETERMINING A (ASME BPVC, SECTION II, PART D, SUBPART 3, FIG. G). TABLE G CITED IN THE FIGURE IS GIVEN IN ASME BPVC, SECTION II

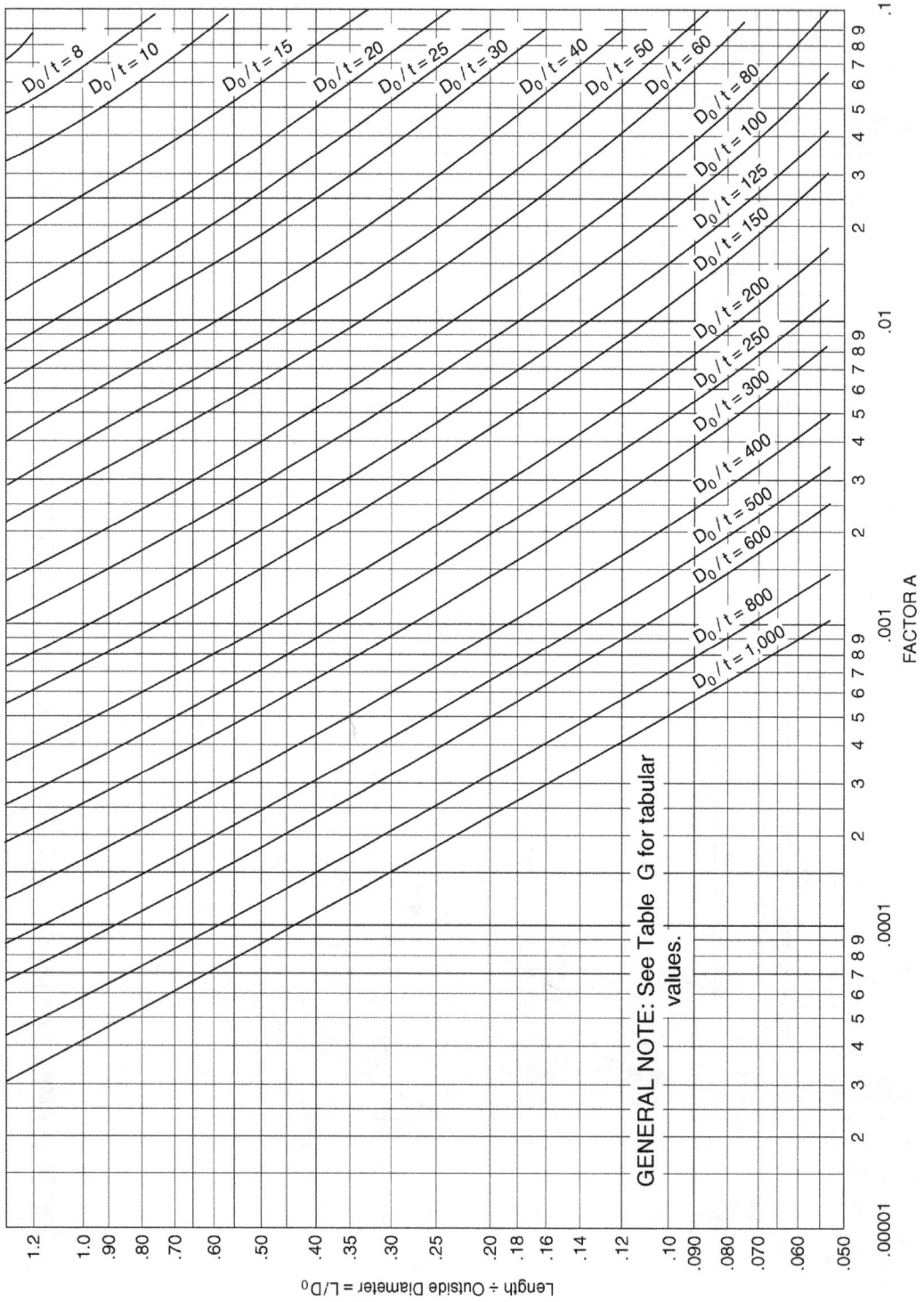

FIG. 4.5
CONTINUED

FACTOR A

Length ÷ Outside Diameter = L/D_0

$D_0/t = 8$
$D_0/t = 10$
$D_0/t = 15$
$D_0/t = 20$
$D_0/t = 25$
$D_0/t = 30$
$D_0/t = 40$
$D_0/t = 50$
$D_0/t = 60$
$D_0/t = 80$
$D_0/t = 100$
$D_0/t = 125$
$D_0/t = 150$
$D_0/t = 200$
$D_0/t = 250$
$D_0/t = 300$
$D_0/t = 400$
$D_0/t = 500$
$D_0/t = 600$
$D_0/t = 800$
$D_0/t = 1,000$

GENERAL NOTE: See Table G for tabular values.

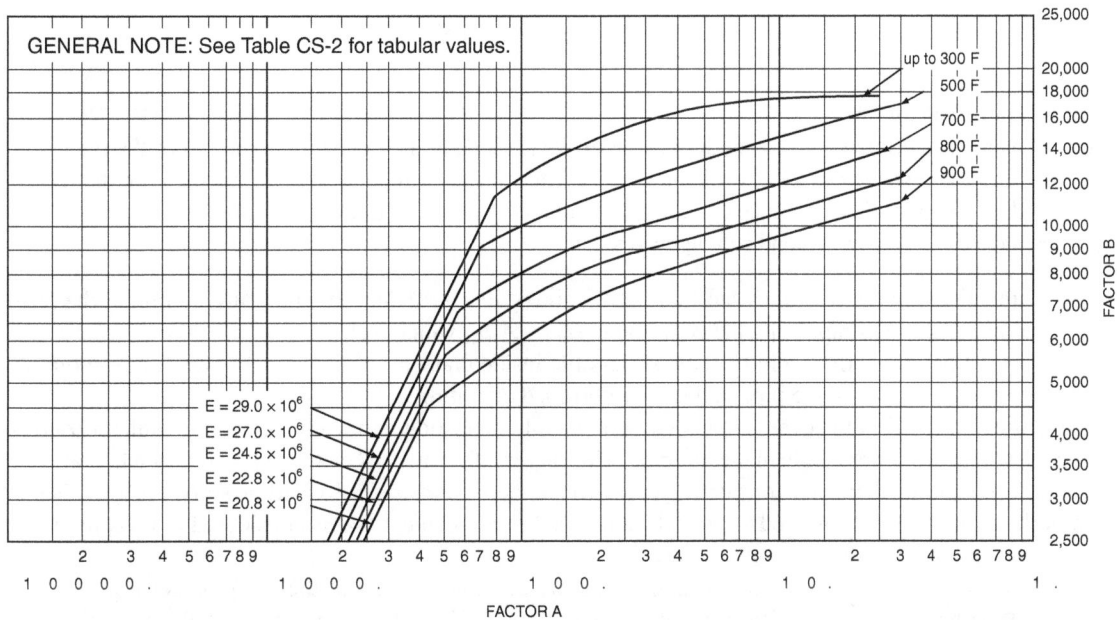

FIG. 4.6
TYPICAL CHART FOR DETERMINING B (ASME BPVC, SECTION II, PART D, SUBPART 3, FIG. CS-2). TABLE CS-2 CITED IN THE FIGURE IS GIVEN IN ASME BPVC, SECTION II

The ASME BPVC, Section VIII procedures include consideration of the allowable out-of-roundness in pressure vessels, and uses a design margin of 3. While pipe is not generally required to comply with the same out-of-roundness tolerance as is required for pressure vessels, this has historically been ignored and has not led to any apparent problems.

The basis for the ASME BPVC, Section VIII approach is provided in Bergman (1960), Holt (1960), Saunders and Windenburg (1960), Windenburg and Trilling (1960), and Windenburg (1960). A newer buckling evaluation procedure, provided in ASME BPVC Code Case 2286, is more relevant to piping as it permits consideration of combined loads, including external pressure, axial load, and gross bending moment. It is not presently explicitly recognized in ASME B31.1, but could be considered as permitted by the Introduction.

4.4 PRESSURE DESIGN OF WELDED BRANCH CONNECTIONS

The pressure design of branch connections is based on a rather simple approach, although the resulting design calculations are the most complex of the design-by-formula approaches provided in the Code. A branch connection cuts a hole in the run pipe. The metal removed is no longer available to carry the forces due to internal pressure. An area replacement concept is used for those branch connections that do not either comply with listed standards or with certain designs (see Section 4.7 herein). The area of metal removed by cutting the hole, to the extent that it was required for internal pressure, must be replaced by extra metal in a region around the branch connection. This region is within the limits of reinforcement, defined later.

The simplified design approach is limited to branches where the angle α (angle between branch and run pipe axes) is at least 45 deg.

Where the above limitations are not satisfied, the designer is directed to para. 104.7 (see Section 4.15 herein). Alternatives in that paragraph include proof testing and finite element analysis.

The area A_7 is the required reinforcement area and is defined as follows:

$$A_7 = (t_{mh} - A)d_1(2 - \sin\alpha)$$

where:

A = additional thickness

d_1 = inside centerline longitudinal dimension of the finished branch opening in the run of the pipe

t_{mh} = required minimum thickness of run pipe as determined from equations (7) or (8). Note that B31.1 does not presently reference equations (11) and (12) in the definition; but should.

α = angle between branch and run pipe axes

A_7 is the metal removed, A_6, times the factor $(2 - \sin\alpha)$. For a 90 deg., or right angle branch, the term $(2 - \sin\alpha)$ is equal to 1, simplifying the equation to $(t_{mh} - A)d_1$, which is A_6.

In this equation, d_1 is effectively the largest possible inside diameter of the branch pipe. It is appropriate to use the inside diameter of the pipe in the fully corroded condition.

The angle α is used in the evaluation because a lateral connection, a branch connection with an α other than 90 deg., creates a larger hole in the run pipe. This larger hole must be considered in d_1. For a lateral, d_1 is the branch pipe inside diameter, considering corrosion–erosion allowance, divided by $\sin\alpha$. The $(2 - 2\sin\alpha)$ term in the equation for $A7$ is used to provide additional reinforcement that is considered to be appropriate because of the geometry of the branch connection.

The required minimum thickness, t_{mh}, is the pressure design thickness of the run pipe per Eqs. (7), (8), (11) or (12) with one exception. If the run pipe is welded and the branch does not intersect the weld, the weld quality factor E should not be used in calculating the wall thickness. The weld quality factor only reduces the allowable stress at the location of the weld. The present Code text does not mention Eqs. (11) or (12) which are used when the material is in the creep regime; however, if applicable, that equation should presumably be used to determine wall thickness unless the branch does not intersect the weld, in which case, the pipe would be treated as seamless in the thickness calculation.

Only the pressure design thickness is used in calculating the required area since only the pressure design thickness was required to resist internal pressure. Corrosion allowance and mill tolerance at the hole are obviously of no consequence.

The area removed, A_7, must be replaced by available area around the opening. This area is available from excess wall thickness that may be available in the branch and run pipes as well as added reinforcement, and the fillet welds that attach the added reinforcement. This metal must be relatively close to the opening of the run pipe to reinforce it. Thus, there are limits, within which any metal area must be to be considered to reinforce the opening. The areas and nomenclatures are illustrated in Fig. 4.7.

The limit of reinforcement along the run pipe, taken as a dimension from the centerline of the branch pipe where it intersects the run pipe wall is d_2, defined as follows:

$$d_2 = \text{greater of} \left[d_1, (T_b - A) + (T_h - A) + d_1/2\right]$$

However, d_2 is not permitted to exceed D_{oh}.

where:

A = allowance (mechanical, corrosion, erosion)

D_{oh} = outside diameter of header pipe

T_b = measured or minimum thickness of branch permissible under purchase specification

T_h = measured or minimum thickness of header permissible under purchase specification

d_2 = half-width or reinforcing zone

The limit of reinforcement along the branch pipe measured from the outside surface of the run pipe is L_4. L_4 is the lesser of $2.5(T_h - A)$ and $2.5(T_b - A) + t_r$, where:

t_r = thickness of attached reinforcing pad (when the reinforcement is not of uniform thickness, it is the height of the largest 60 deg. right triangle supported by the run and branch outside diameter projected surfaces and lying completely within the area of integral reinforcement; see Fig. 4.7, Example B).

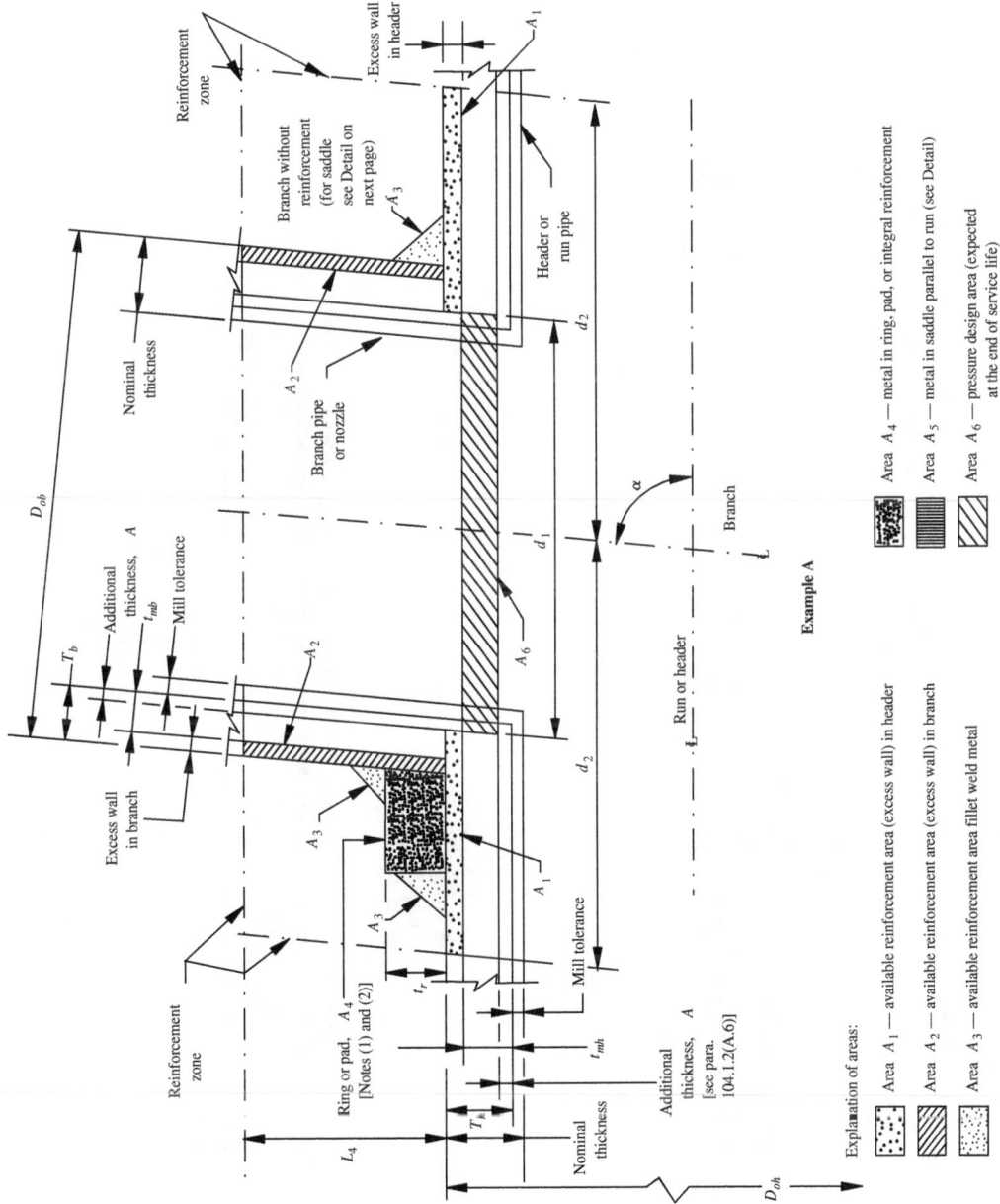

Explanation of areas:

Area A_1 — available reinforcement area (excess wall) in header
Area A_2 — available reinforcement area (excess wall) in branch
Area A_3 — available reinforcement area fillet weld metal

Area A_4 — metal in ring, pad, or integral reinforcement
Area A_5 — metal in saddle parallel to run (see Detail)
Area A_6 — pressure design area (expected at the end of service life)

FIG. 4.7
REINFORCEMENT OF BRANCH CONNECTIONS (ASME B31.1, FIG. 104.3.1(D))

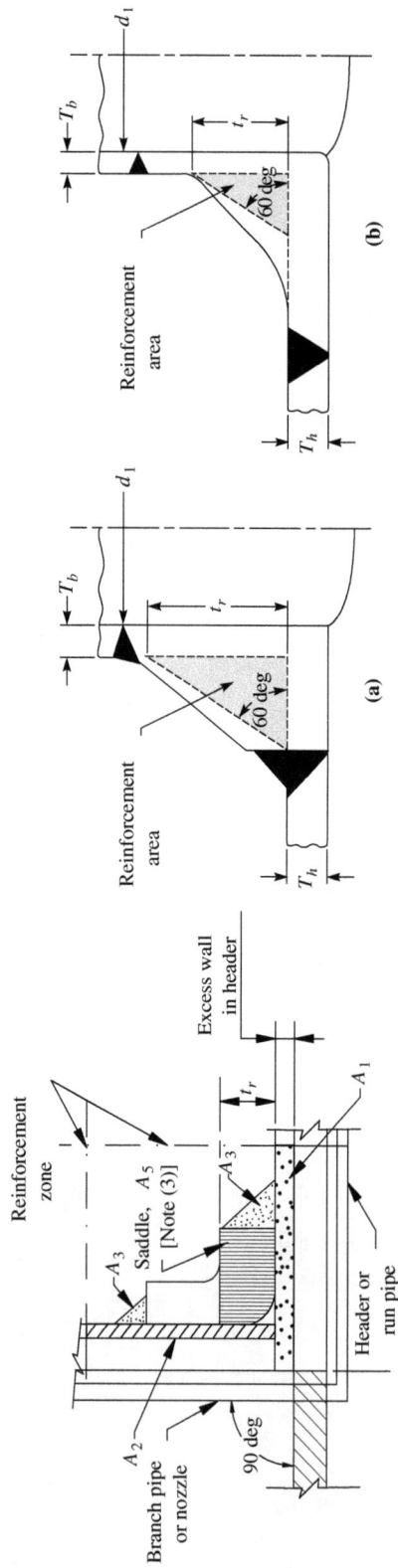

**Detail
for Example A**

Example B

**FIG. 4.7
CONTINUED**

GENERAL NOTES:

(a) This Figure illustrates the nomenclature of para. 104.3.1(D).

(b) Required reinforcement area $\geq A_7 \geq A_6 (2 - \sin\alpha)$ $\geq (t_{mh} - A)d_1 (2 - \sin\alpha)$.

(c) Available reinforcement areas $\geq A_1 + A_2 + A_3 + A_4 + A_5$ (as applicable).

(d) Available reinforcement areas $\geq$ required reinforcement area.

NOTES:

(1) When a ring or pad is added as reinforcement (Example A), the value of reinforcement area may be taken in the same manner in which excess header metal is considered, provided the weld completely fuses the branch pipe, header pipe, and ring or pad. Typical acceptable methods of welding which meet the above requirement are shown in Fig. 127.4.8(D), sketches (c) and (d).

(2) Width to height of rings and pads shall be reasonably proportioned, preferably on a ratio as close to 4:1 as the available horizontal space within the limits of the reinforcing zone along the run and the outside diameter of the branch will permit, but in no case may the ratio be less than 1:1.

(3) Reinforcement saddles are limited to use on 90 deg branches (Example A Detail).

The reinforcement within this zone is required to exceed A_7. This reinforcement consists of excess thickness available in the run pipe (A_1); excess thickness available in the branch pipe (A_2); additional area in the fillet weld metal, (A_3); metal area in ring, pad, or integral reinforcement (A_4); and metal in a reinforcing saddle along the branch (A_5) (see Fig. 4.7, Example A.). These can be calculated as follows:

$$A_1 = (2d_2 - d_1)(T_h - t_{mh})$$

$$A_2 = 2L_4(T_b - t_{mb})/\sin \alpha$$

A_3 is the area provided by deposited weld metal beyond the outside diameter of the run and branch and for fillet weld attachments of rings, pads, and saddles within the limits of reinforcement.

A_4 is the area provided by a reinforcing ring, pad, or integral reinforcement.

A_5 is the area provided by a saddle on 90 deg. branch connections. See Fig. 4.7, Example A.

The area A_4 is the area of properly attached reinforcement and the welds that are within the limits of reinforcement. For the reinforcement to be considered effective, it must be welded to the branch and run pipes. Minimum acceptable weld details are provided in Fig. 127.4.8(D). The ASME B31.1 Code does not require the designer to specify branch connection weld size because generally acceptable minimum sizes are specified by the Code. Furthermore, the ASME B31.1 Code differs from the B&PV Code in that strength calculations for load paths through the weld joints are not required.

If metal with a lower allowable stress than the run pipe is used for reinforcement, the contributing area of this reinforcement must be reduced proportionately. No additional area credit is provided for reinforcement materials with a higher allowable stress.

Note that branch connections of small bore pipe by creating a socket or threaded opening in the run pipe wall are permitted with certain limitations, as stated in para 104.3.1(B.4).

4.5   PRESSURE DESIGN OF EXTRUDED OUTLET HEADER

An extruded outlet header is a branch connection formed by extrusion, using a die or dies to control the radii of the extrusion. Paragraph 104.3.1(G) provides area-replacement rules for such connections; they are applicable for 90 deg. branch connections where the branch pipe centerline intercepts the run pipe centerline, and where there is no additional reinforcement. Figure 4.8 [ASME B31.1, Fig. 104.3.1(G)] shows the geometry of an extruded outlet header. Extruded outlet headers are subject to minimum and maximum external contour radius requirements, depending on the diameter of the branch connection.

A similar area-replacement calculation as described in Section 4.4 for fabricated branch connections is provided, except that the required replacement area is reduced for smaller branch-to-run diameter ratios. The replacement area is from additional metal in the branch pipe, additional metal in the run pipe, and additional metal in the extruded outlet lip.

4.6   ADDITIONAL CONSIDERATIONS FOR BRANCH CONNECTIONS
UNDER EXTERNAL PRESSURE

Branch connections under external pressure are covered in para. 104.3.1. The same rules described in Sections 4.4 and 4.5 above are used. However, only one-half of the area described in Sections 4.4, covering welded branch connections, requires replacement. In other words, only one-half of the area A_7 requires replacement. Also, the thicknesses used in the calculation are the required thicknesses for the external pressure condition.

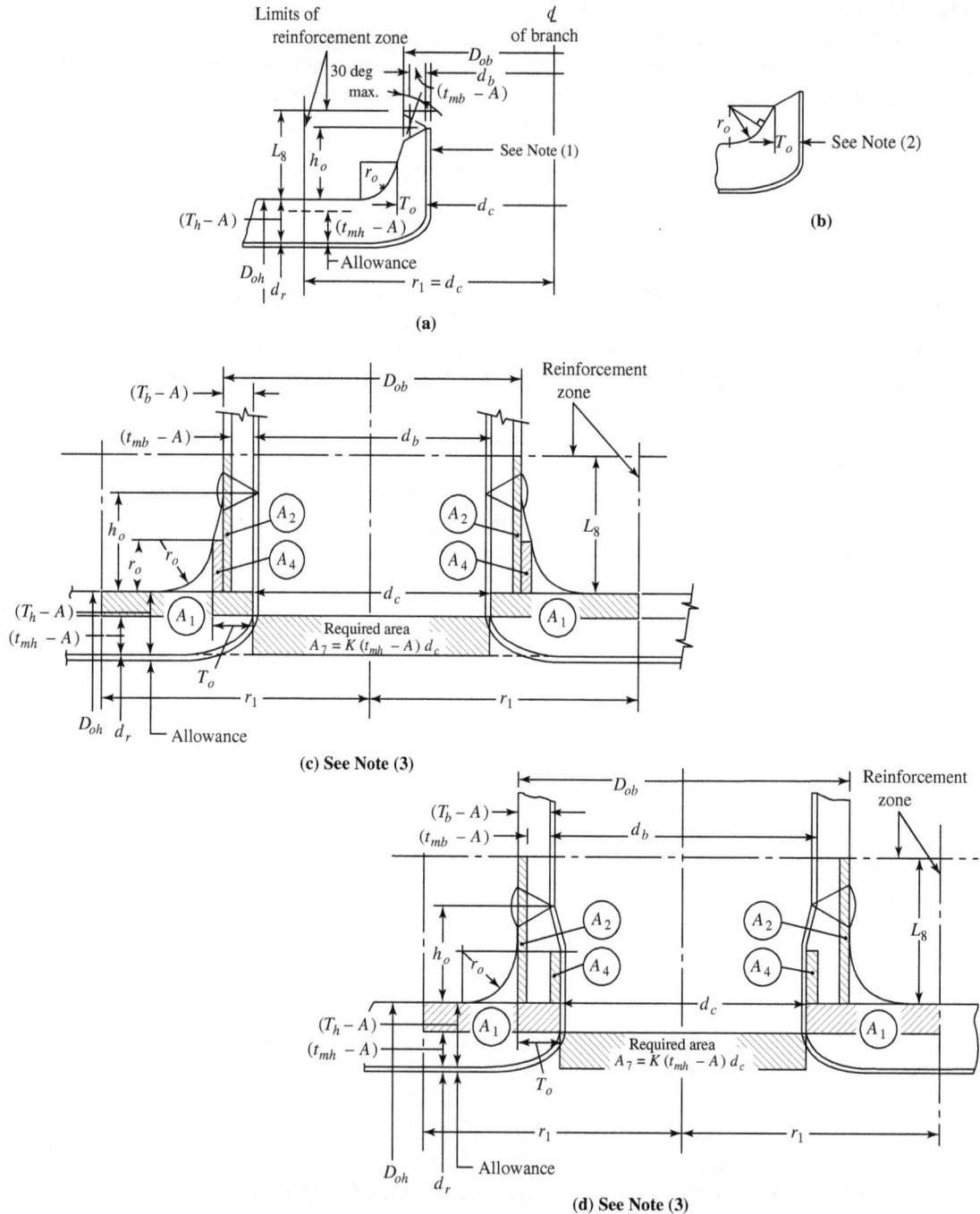

FIG. 4.8
REINFORCED EXTRUDED OUTLETS (ASME B31.1, FIG. 104.3.1(G))

NOTES:
(1) Taper bore inside diameter (if required) to match branch pipe 1:3 maximum taper.
(2) Sketch to show method of establishing T_o when the taper encroaches on the crotch radius.
(3) Sketch is drawn for condition where k p 1.00.

4.7   BRANCH CONNECTIONS THAT ARE PRESUMED TO BE ACCEPTABLE

Some specific types of branch connections are presumed to be acceptable. This includes fittings listed in Table 126.1 (e.g., ASME B16.9 tees, MSS SP-97 branch outlet fittings) and the following [para. 104.3.1(C)]:

(1) For branch connections NPS 2 or less that do not exceed one-fourth of the nominal diameter of the run pipe, threaded or socket welding couplings or half couplings (Class 3000 or greater) are presumed to provide sufficient reinforcement as long as the minimum thickness of the coupling within the reinforcement zone is at least as thick as the unthreaded branch pipe.

(2) Small branch connections, NPS 2 or smaller as shown in ASME B31.1 Fig. 127.4.8(G) (these are partial penetration weld branch connections for NPS 2 and smaller branch fittings), provided the thickness of the weld joint (not including the cover fillet) is at least equal to the thickness of schedule 160 pipe of the branch size, are acceptable.

Integrally reinforced branch outlet fittings and integrally reinforced extruded outlets that satisfy the area replacement requirements or are qualified by burst or proof tests or calculations substantiated by successful service of similar design [para. 104.3.1(D.2.7)] are also acceptable.

4.8   PRESSURE DESIGN OF BENDS AND ELBOWS

Bends are required to have, after bending, a wall thickness that satisfies Eqs. (3) or (4). These equations are based on the Lorenz equation (ca. 1910) which provides a means of calculating the required wall thickness.

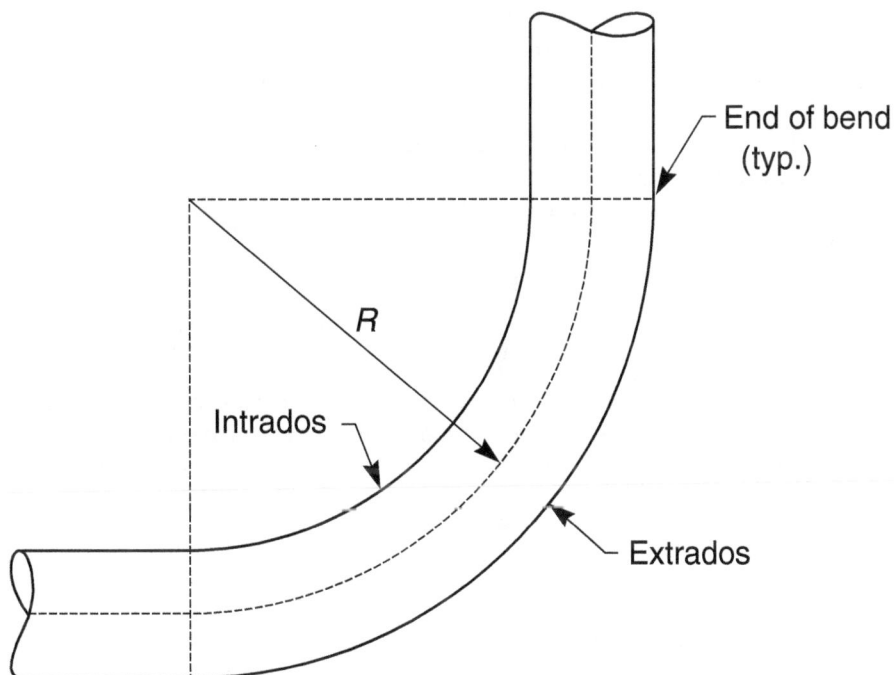

FIG. 4.9
NOMENCLATURE FOR PIPE BENDS (ASME B31.1, FIG. 102.4.5)

The Lorenz equation is basically the equation for a toroid. If the intrados and extrados (see Fig. 4.9 for an illustration showing the location of the intrados and extrados) had the same wall thickness, the inside would be subjected to higher hoop stress than straight pipe, and the outside would be subjected to lower hoop stress than straight pipe. A simple way to envision this is that the inside has less metal over the curve, and the outside has more metal over the curve. The Lorenz equation for an elbow or bend is given by:

$$t_m = \frac{P_{D_O}}{2(SE\,I\!I + Py)} + A \tag{3}$$

where the terms are as defined in Section 4.2 for Eq. (7), except for I, which is a stress index that accounts for the difference in hoop stress due to internal pressure in bends versus straight pipe. Equation (4) is provided similarly as for straight pipe, based on inside diameter rather than outside diameter.

On the inside curve of the bend, the intrados, we have:

$$I = \frac{4(R/D_o)-1}{4(R/D_o)-2} \tag{5}$$

On the outside of the bend, or the extrados, we have:

$$I = \frac{4(R/D_o)+1}{4(R/D_o)+2} \tag{6}$$

On the side of the elbow, or the crown, $I = 1.0$ (i.e., the hoop stress is the same as in straight pipe). The thickness variation from the intrados to the extrados is required to be gradual, and the requirements are stated to apply at the midspan of the bend. The thickness at the ends is required to satisfy the required thickness for straight pipe per para. 104.1.2.

Because of the bending process, the thickness tends to increase in the intrados, or inside curve of the elbow, and decrease on the extrados, or outside curve of the elbow. ASME B31.1 provides minimum recommended thickness of the pipe, prior to bending, in Table 102.4.5, which, based on experience, results in a pipe thickness after bending that is at least equal to the required wall thickness of straight pipe. Part of the reason for providing these new rules is due to the practice of fabricating elbows by forming two "clamshells" out of plate and welding them together. This produces a bend of uniform thickness, and the thickness on the intrados would be too thin if it simply satisfied the required thickness for straight pipe.

Elbows in accordance with standards listed in Table 126.1 (e.g., B16.9 elbows) are acceptable for their rated pressure-temperature.

4.9   PRESSURE DESIGN OF MITERS

Miter joints and miter bends are covered by para. 104.3.3. Miters in a miter bend are either widely spaced or closely spaced. The criteria for closely spaced versus widely spaced are contained in B31.1 Table D-1. If the following equation is satisfied, the miter is closely spaced; otherwise, it is widely spaced.

$$s < r(1 + \tan\theta)$$

where:

r = mean radius of pipe
s = chord length between miter joints, taken along pipe centerline
θ = one-half angle between adjacent miter axes (see Fig 4.10; the axes are the extension of the line of miter cuts to where they intercept)

If the miters are widely spaced and the half-angle satisfies the following equation (with θ in degrees), no further consideration is required. The miter cut is simply considered to be equivalent to a girth butt-welded joint.

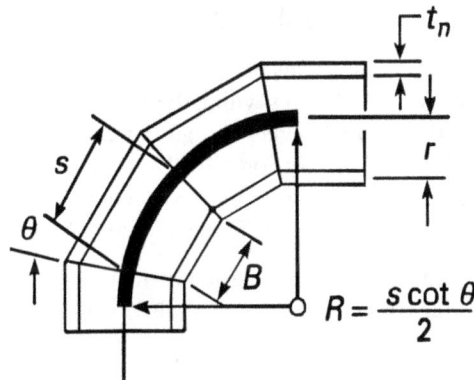

FIG. 4.10
ILLUSTRATION OF MITER BAND SHOWING NOMENCLATURE
(ASME B31.1, TABLE D-1)

$$\theta < 9\sqrt{\frac{t_n}{r}}$$

where:

t_n = nominal wall thickness of the pipe

The required wall thickness of other miters depends on whether they are closely spaced or widely spaced. For closely spaced miter bends, the required pressure design wall thickness is per the following equation:

$$t_s = t_m \frac{2 - r/R}{2(1 - r/R)}$$

where:

R = bend radius of miter bend
t_m = minimum required thickness for straight pipe
t_s = required thickness of miter segment

For widely spaced miters, the following equation provides the required pressure design wall thickness:

$$t_s = t_m(1 + 0.64\sqrt{r/t_s}\,\tan\theta)$$

This equation must be solved iteratively since the required thickness is on both sides of the equation.

There are additional pressure limitations for miters. These are 10 psi (70 kPa) and less, above 10 psi (70 kPa) but not exceeding 100 psi (700 kPa), and above 100 psi (700 kPa). The above equations can be used for design of the miter bends up to 100 psi under the following conditions: the thickness is not less than required for straight pipe; the contained fluid is nonflammable, nontoxic, and incompressible, except for gaseous vents to atmosphere; the number of full pressure cycles is less than 7000 during the expected lifetime of the piping system; and full penetration welds are used in joining miter segments.

For above 100 psi, or when the above conditions are not satisfied, the design is required to be qualified per para. 104.7, with additional qualifications to the para. 104.7 requirements stated in para. 104.3.3(C).

For use up to 100 psi, the following requirements must be satisfied:

(1) Angle θ must not exceed 22.5 deg. (e.g., 2 cut miter for 90 deg. bend, minimum).

(2) The minimum length of the miter segment at the crotch (the shortest length in a miter segment), B, must be at least $6t_n$ where t_n is the pipe nominal wall thickness.

The above two conditions need not be satisfied if the pressure is limited to 10 psi (70 kPa).

4.10 PRESSURE DESIGN OF CLOSURES

Closures are covered in para. 104.4.1. Components in accordance with standards listed in Table 126.1, such as ASME B16.9 pipe caps, can be used for closures within their specified pressure-temperature ratings. The other options provided in ASME B31.1 are to either design the closure in accordance with either ASME BPVC, Section I, PG-31, or to ASME BPVC, Section VIII, Division 1, UG-34 and UW-13, or to qualify it as an unlisted component in accordance with para. 104.7 (see Section 4.15 herein).

Openings in closures are covered in para. 104.4.2. These requirements are summarized as follows:

(1) If the opening is greater than one-half of the inside diameter of the closure, it is required to be designed as a reducer per para. 104.6. While not an ASME B31.1 requirement, the ASME B31.3 requirement that if the opening is in a flat closure, it be designed as a flange, is appropriate and should be considered.

(2) Small openings and connections using branch connection fittings that comply with para. 104.3.1(C) (by the reference to para. 104.3.1) are considered to be inherently adequately reinforced.

(3) The required area of reinforcement is the inside diameter of the finished opening times the required thickness of the closure. The ASME BPVC, Section VIII, Division 1 rules that only require one-half of that area for flat heads are not applicable.

(4) The available area of reinforcement should be calculated per the rules in ASME B31.1 contained in para. 104.3.1.

(5) Rules for multiple openings follow para. 104.3.1(D.2.5) rules for multiple openings (by the reference to para. 104.3.1).

4.11 PRESSURE DESIGN OF FLANGES

Most flanges are in accordance with standards listed in Table 126.1, such as ASME B16.5 and, for larger flanges, ASME B16.47. When a custom flange is required, design by analysis is permitted by para. 104.5.1. ASME B31.1 refers to the rules for flange design contained in ASME BPVC, Section VIII, Division 1, Appendix 2, but uses the allowable stresses and temperature limits of ASME B31.1. In addition, the fabrication, assembly, inspection, and testing requirements of ASME B31.1 are governing.

4.12 PRESSURE DESIGN OF BLIND FLANGES

Most blind flanges are in accordance with standards listed in Table 126.1, such as ASME B16.5. When designing a blind flange, the rules of ASME BPVC, Section I for bolted flat cover plates are applicable (these are contained in PG-31). Additionally, the ASME B31.1 design pressure and allowable stresses are to be used.

4.13 PRESSURE DESIGN OF BLANKS

Blanks are flat plates that get sandwiched between flanges to block flow. A design equation for permanent blanks is provided in para. 104.5.3, as follows:

$$t = d_6 \sqrt{\frac{3P}{16SE}} \tag{14}$$

where:

d_6 = inside diameter of gasket for raised or flat-face flanges, or the gasket pitch diameter for retained, gasketed flanges

Other terms are as defined in Section 4.2 herein.

Mechanical and corrosion–erosion allowances must be added to the pressure design thickness calculated from Eq. (14).

Blanks used for test purposes are required to be designed per the foregoing equation, except that the test pressure is used and SE may be taken, if the test fluid is incompressible (e.g., not a pneumatic test), at 95% of the specified minimum yield strength of the blank material.

4.14 PRESSURE DESIGN OF REDUCERS

Most reducers in piping systems are in accordance with the standards listed in Table 126.1. This is the only provision for reducers in para. 104.6 (which is not helpful when one is referred from para. 104.4.2 to this paragraph for large-diameter openings in closures). However, pressure design per para. 104.7 is also an option.

4.15 SPECIALLY DESIGNED COMPONENTS

If a component is not in accordance with a standard listed in Table 126.1, and the design rules provided elsewhere in para. 104 are not applicable, para. 104.7.2 is applicable. This paragraph requires that some calculations be done in accordance with the design criteria provided by the Code and be substantiated by one of several methods. The most important element of this paragraph is considered to be the substantiation; the aforementioned calculations are not generally given much consideration. The methods to verify the pressure design include the following:

(1) Extensive, successful service experience under comparable conditions with similarly proportioned components of the same or like material.

(2) Experimental stress analysis, such as described in the B&PV Code ASME BPVC, Section VIII, Division 2, Annex 5-F.

(3) A proof test conducted in accordance with ASME B16.9, MSS SP-97, or ASME BPVC, Section I, A-22. The option for witnessing by the Authorized Inspector was removed in the 1999 addenda. This prior provision was not necessarily practical and could create difficulty for the manufacturer, since proof tests may be conducted to qualify a line of components well before being sold for any specific piping system.

(4) Detailed stress analysis (e.g., finite element method) with results evaluated in accordance with ASME BPVC, Section VIII, Division 2, Part 5, except the basic allowable stress from Appendix A is required to be used in place of S_m. These are the design-by-analysis rules in the B&PV Code.

Of the above, the methods normally used to qualify new unlisted components are proof testing and detailed stress analysis.

It should be noted that the Code permits interpolation between sizes, wall thicknesses, and pressure classes, and also permits analogies among related materials. Extrapolation is not permitted.

The issue of how to determine that the above has been done in a satisfactory manner is addressed in the 1999 addenda. Earlier editions of the Code only provided for witnessing of the proof test for boiler external piping. However, this is not practical when the manufacturer performs proof tests to qualify a line of piping components. Obviously, all the potential future Authorized Inspectors could not be gathered for this event. Furthermore, the other methods are of at least equal concern, and their review may be more appropriately done by an engineer rather than an Inspector. As a result of these concerns, the requirement was added that documentation showing compliance with the above means of pressure design verification must be available for the owner's approval and, for boiler external piping, available for the Authorized Inspector's review. The owner's review could be done by an inspector or some other qualified individual.

While MSS SP-97 and ASME B16.9 provide a clear approach for determining that the rating of a component is equivalent or better to matching straight pipe, they do not provide defined procedures for determining a rating for a component that may have a unique rating, which may differ from matching straight pipe. The procedure generally used here is to establish a pressure–temperature rating by multiplying the proof pressure by the ratio of the allowable stress for the test specimen to the actual tensile strength of the test specimen. In the proposed ASME B31H Standard, this would be reduced by a testing factor depending on the number of tests. An example of this approach is provided in Biersteker et al (1991).

The proposed standard ASME B31H, *Standard Method to Establish Maximum Allowable Design Pressure for Piping Components*, is under development by the ASME and may eventually add to or replace the existing proof test alternatives in para. 104.7.2. This draft standard provides procedures to either determine if a component has a pressure capacity at least as great as a matching straight pipe or to determine a pressure-temperature rating for a component.

LIMITATIONS ON COMPONENTS AND JOINTS

5.1 OVERVIEW

ASME B31.1 includes limitations on components and joints in the design chapter, Chapter II. These are contained in Part 3, Selection and Limitations of Piping Components; and in Part 4, Selection and Limitations of Piping Joints. This chapter (5) combines the limitations with pressure design and other considerations, on a component-by-component basis.

5.2 VALVES

Most valves in ASME B31.1 piping systems are in accordance with standards listed in Table 126.1. These standards include the following:

(1) ASME B16.10, Face-to-Face and End-to-End Dimensions of Valves
(2) ASME B16.34, Valves-Flanged, Threaded, and Welding End
(3) AWWA C500, Metal-Seated Gate Valves for Water Supply Service
(4) AWWA C504, Rubber-Seated Butterfly Valves (with limitation regarding stem retention)
(5) AWWA 509 Resilient Seated Gate Valves for Water Supply Service
(6) MSS SP-42, Corrosion-Resistant Gate, Globe, Angle, and Check Valves With Flanged and Butt-Welded Ends (Classes 150, 300 & 600) (with limitation regarding stem retention)
(7) MSS SP-67, Butterfly Valves (with limitation regarding stem retention)
(8) MSS-SP-68 High Pressure Butterfly Valves with Offset Design
(9) MSS SP-80, Bronze Gate, Globe, Angle and Check Valves
(10) MSS SP-88, Diaphragm Valves
(11) MSS-SP-105 Instrument Valves for Code Applications

Listed valves are accepted for their specified pressure ratings. Valves that are not in accordance with one of the listed standards can be accepted as unlisted components in accordance with para. 102.2.2. The pressure–temperature rating for such valves should be established in accordance with para. 104.7.2. The manufacturer's recommended rating is not permitted to be exceeded.

Additional requirements are provided in para. 107. These include the following:

(1) requirements for marking (para. 107.2);
(2) requirement for use of outside screw threads for valves NPS 3 (DN 75) and larger for pressure above 600 psi (4150 kPa) (para. 107.3);

(3) prohibition of threaded bonnet joints where the seal depends on the thread tightness for steam service at pressure above 250 psi (1,750 kPa) (para. 107.5); and

(4) requirements for bypasses (para. 107.6).

Additional requirements for valves in boiler external piping (steam-stop valves, feedwater valves, blowoff valves, and safety valves) are provided in para. 122.1.7.

Requirements for safety-relief valves for ASME B31.1 piping are also covered in para. 107.8. Safety-relief valves on boiler external piping are required to be in a accordance with ASME BPVC, Section I (by reference to para. 122.1.7(D.1)). Safety-relief valves for non-boiler external piping are required to be in accordance with ASME BPVC, Section VIII, Division 1, paras. UG-126 through UG-133. An exception for valves with set pressures 15 psig (100 kPa (gage)) and lower is that ASME Code Stamp and capacity certification are not required. Safety-relief valves for non-boiler external reheat piping are required to be in accordance with ASME BPVC, Section I, PG-67 through PG-73.

Appendix II provides non-mandatory rules for the design of safety-valve installations.

For piping containing toxic fluids [para. 122.8.2(D)], steel valves are required, and bonnet joints with tapered threads are prohibited. Also, special consideration should be given to valve design to prevent stem leakage. Permitted bonnet joints include union, flanged with at least four bolts; proprietary, attached by bolts, lugs, or other substantial means, and having a design that increases gasket compression as fluid pressure increases; or threaded with straight threads of sufficient strength, with metal-to-metal seats and a seal weld.

5.3 FLANGES

Most flanges in ASME B31.1 piping systems are in accordance with listed standards. These listed standards include the following:

(1) ASME B16.1, Cast Iron Pipe Flanges and Flanged Fittings—25, 125, 250, and 800 Class

(2) ASME B16.5, Pipe Flanges and Flanged Fittings

(3) ASME B16.24, Cast Copper Alloy Pipe Flanges and Flanged Fittings Class 150, 300, 400, 600, 900, 1500, and 2500

(4) ASME B16.42, Ductile Iron Pipe Flanges and Flanged Fittings, Classes 150 and 300

(5) ASME B16.47, Large Diameter Steel Flanges, NPS 26 through NPS 60

(6) AWWA C115, Flanged Ductile-Iron Pipe with Threaded Flanges

(7) AWWA C207, Steel Pipe Flanges for Water Works Service, Sizes 4 in. through 144-in. (100 mm through 3600 mm)

(8) MSS SP-51, Class 150LW Corrosion-Resistant Cast Flanges and Flanged Fittings

(9) MSS SP-106, Cast Copper Alloy Flanges and Flanged Fittings, Class 125, 150 and 300

Flanges that are listed in Table 126.1 are accepted for their specified pressure ratings. Flanges that are not in accordance with one of the listed standards can be designed using the rules of ASME BPVC, Section VIII, Division 1, Appendix 2, with appropriate allowable stress and design pressure (see para. 104.5.1), or qualified using para. 104.7.2. ASME B31.1 states that the ASME BPVC, Section VIII rules are not applicable when the gasket extends beyond the bolt circle, which is also a limitation stated in ASME BPVC, Section VIII.

Paragraphs 104.5, 108, 112, and 122.1.1 provide additional requirements for flanges, including the following:

(1) ASME B16.5 slip-on flanges are not permitted for higher than Class 300 flanges [para. 104.5.1(A)].

(2) When bolting Class 150 steel flanges to matching cast iron flanges, the steel flange is required to be flat face to prevent overloading the cast iron flange (para. 108.3). Use of full-face gaskets with flat-face flanges helps the flange resist rotation from the bolt load.

(3) Class 250 cast iron are permitted to be used with raised-face Class 300 steel flanges (para. 108.3).

(4) Table 112 provides detailed requirements for flange bolting, facing, and gaskets. These depend on flange class and material.

(5) Slip-on flanges for boiler external piping are not permitted to exceed NPS 4 (DN 100) and are required to be double-welded [para. 122.1.1(F)].

(6) Hub-type flanges for boiler external piping are not permitted to be cut from plate material [para. 122.1.1(H)].

(7) Socket-weld flanges are limited for boiler external piping to NPS 3 (DN 75) for Class 600 and lower and NPS 2 for Class 1500 [para. 122.1.1(H)].

A double-welded slip-on flange has a weld between the pipe and the flange hub and between the pipe and the bore of the flange. A single-welded slip-on flange only has the weld to the flange hub.

5.4   FITTINGS, BENDS, MITERS, AND BRANCH CONNECTIONS

Most fittings in ASME B31.1 piping systems are in accordance with standards listed in Table 126.1. These listed standards include the following:

(1) ASME B16.3, Malleable Iron Threaded Fittings

(2) ASME B16.4, Gray Iron Threaded Fittings

(3) ASME B16.9, Factory-Made Wrought Steel Butt-Welding Fittings

(4) ASME B16.11, Forged Fittings, Socket-Welding and Threaded

(5) ASME B16.14, Ferrous Pipe Plugs, Bushings, and Lock-nuts With Pipe Threads

(6) ASME B16.15, Cast Bronze Threaded Fittings, Classes 125 and 250

(7) ASME B16.18 Cast Copper AlloySolder-Joint Fittings

(8) ASME B16.22, Wrought Copper and Copper Alloy Solder Joint Pressure Fittings

(9) ASME B16.50, Wrought Copper and Copper Alloy Braze-Joint Pressure Fittings

(10) AWWA C110, Ductile-Iron and Gray-Iron Fittings, 3 in. through 48 in. (75 mm through 1,200 mm), for Water and Other Liquids

(11) AWWA C208, Dimensions for Fabricated Steel Water Pipe Fittings

(12) MSS SP-43, Wrought and Fabricated Butt Welding Fittings for Low Pressure, Corrosion Resistant Applications

(13) MSS SP-75, Specifications for High Test Wrought Butt-Welding Fittings

(14) MSS SP-79, Socket-Welding Reducer Inserts

(15) MSS SP-83, Class 3000 Steel Pipe Unions, Socket-Welding and Threaded

(16) MSS SP-95, Swaged Nipples and Bull Plugs

(17) MSS SP-97, Integrally Reinforced Forged Branch Outlet Fittings-Socket Welding, Threaded, and Butt-Welding Ends

(18) PFI ES-24, Pipe Bending Methods, Tolerances, Process and Material Requirements

Listed fittings are accepted for their specified pressure ratings. Note that some fittings are simply specified to have equivalent pressure ratings to matching straight seamless pipe. Per para. 102.2.2, these fittings are rated for the same allowable pressure as seamless pipe of the same nominal thickness with material having the same allowable stress. Fittings that are not in accordance with one of the listed standards can be qualified in accordance with para. 104.7.

Branch connections are required to be designed per para. 104.3. These rules are described in Sections 4.4 through 4.7 herein. They permit fabricated branch connections designed using area replacement rules, branch connections per standards listed in Table 126.1, certain designs that do not require reinforcement (such as couplings of limited size), and branch connections qualified per para. 104.7.2. Fabricated branch connections are not permitted when the pipe contains toxic fluids (para. 122.8.2) unless other listed branch connections are not available.

5.5 BOLTING

Requirements for bolting are provided in para. 108.5 and Table 112. Bolts, bolt studs, nuts, and washers are required to comply with applicable standards and specifications listed in Table 126.1 and Table 112. The listed standards include the following:

    (1) ANSI B18.2.1, Square and Hex Bolts and Screws—Inch Series
    (2) ANSI B18.2.4.6M, Metric Heavy Hex Nuts
    (3) ANSI B18.22M, Metric Plain Washers
    (4) ANSI B18.22.1, Plain Washers
    (5) ASME B1.1, Unified Inch Screw Threads
    (6) ASME B1.13M, Metric Screw Threads—M Profile
    (7) ASME B18.2.2, Square and Hex Nuts (Inch Series)
    (8) ASME B18.2.3.5M, Metric Hex Bolts
    (9) ASME B18.2.3.6M, Metric Heavy Hex Bolts
    (10) ASME B18.21.1, Lock Washers (Inch Series)

Requirements for flanged joints relative to bolt strength are provided in Table 112. The table essentially requires the use of low-strength rather than high-strength bolting under conditions where high-strength bolts may overload the flange. This includes lower pressure class cast iron flanges with ring-type (but not full-face-type) gaskets. Low-strength bolting has a specified minimum yield strength of 30 ksi (207 MPa) or less.

Miscellaneous additional bolting requirements are provided in para. 108.5.

5.6 WELDED JOINTS

Welded joints are covered by para. 111 and are required to follow the ASME B31.1 rules for fabrication and examination. In addition, the following specific rules are also provided.

Weld backing rings are generally permitted to be left in after the weld is completed. However, if their presence will result in severe corrosion or erosion, para. 111.2.2 requires removal of the ring and grinding the internal joint face smooth. Where this is not practical, consideration should be given to not using a backing ring or to using a consumable insert ring.

Socket-welded joints are generally permitted, with the following exceptions. ASME B31.1 states that special considerations should be given to further restricting the use of socket-welded piping joint where temperature or pressure cycling or severe vibration is expected, or where it may accelerate crevice corrosion. In addition, local erosion at the socket joint can occur if the service is erosive. Socket welds are also prohibited in sizes larger than NPS 2-1/2 (DN 65) for piping containing toxic fluids [para. 122.8.2(C)].

The dimensions of the socket joint are required to conform to either ASME B16.5 for flanges or ASME B16.11 for other socket-welding components. Weld dimensions are required to comply with the fabrication rules of ASME B31.1 [Figs. 127.4.4(B) and 127.4.4(C)]. Socket joints are generally part of listed components, so their pressure design is satisfied by the component standard and compliance with the fillet weld size requirements of the fabrication rules of ASME B31.1.

Seal welds are permitted to be used to prevent leakage of threaded joints. However, the seal weld is not permitted to be considered as contributing to the strength of the joint. The fabrication requirements for seal welds are contained in para. 127.4.5 and require that the weld be made by a qualified welder and that the weld completely cover the threads.

5.7 THREADED JOINTS

Most threaded joints in ASME B31.1 piping systems are made with taper threads in accordance with ASME B1.20.1. Figure 5.1 shows a taper thread. Other threads may be used if the tightness of the joint is achieved by seal welding, or a seating surface other than the threads, and where experience of test has demonstrated that such threads are suitable.

Threaded joints are prohibited from use at temperatures greater than 925°F (496°C), and where severe erosion, crevice corrosion, shock, or vibration is expected to occur. The maximum permitted pressure as a function of pipe size is provided for steam and hot-water service [above 220°F (105°C)] in Table 5.1 herein. These restrictions do not apply to threaded access holes with plugs that are used for openings for radiographic examination. Furthermore, not all the thread limitations apply to threaded connections for insertion-type instrumentation (see para. 114.2.3).

Threaded joints that are to be seal-welded should not use thread-sealing compounds.

The minimum thickness for threaded pipe is standard weight per ASME B36.10M. Additional thickness limits, schedule 80 minimum, are provided for certain steam and water services, as described in Section 4.2 herein (para. 114.3).

Use of threaded joints for piping containing flammable or combustible liquids, flammable gas, and toxic fluids (liquid or gas) is discouraged. When their use is unavoidable, specific limitations and requirements are provided in para. 122.7.3(A), 122.8.1(B.1), and 122.8.2(C) for these three services, respectively. A common requirement of these paragraphs is that the pipe thickness be at least extra strong and that it be assembled with extreme care to ensure leak-tightness. Threaded joints are prohibited for flammable gases, toxic gases or liquids, and nonflammable nontoxic gases in para. 114.2.1(B) for sizes in excess of those specified in Table 5.1 herein, but with exceptions cited in paras. 122.8(B) and 122.8.2(C.2).

5.8 TUBING JOINTS

Tubing joints are covered by para. 115, including flared, flare-less, and compression-type tube fittings. While compliance with a standard listed in Table 126.1 is one option, no standards for this type of fitting are presently listed. While some standards exist, such as SAE standards, many tubing joints that are used are proprietary fittings that are qualified as unlisted components.

Unlisted tube fittings must be qualified in accordance with para. 104.7.2. Also, para. 115 requires performance testing of a suitable quantity of the type, size, and material of the fittings. This is to test the joint under simulated service conditions, including vibration, fatigue, pressure cycles, low temperature, thermal expansion, and hydraulic shock.

Fittings are not permitted to be used at pressure–temperature conditions exceeding the recommendations of the manufacturer.

5.9 MISCELLANEOUS JOINTS

Caulked or leaded joints (para. 116) are only permitted for cold-water service. Soldered joints (para. 117) are only permitted for nonflammable and nontoxic fluids, and in systems that are not subject to shock or vibration. Brazed joints (para. 117) are prohibited from flammable or toxic fluids in areas where there is a

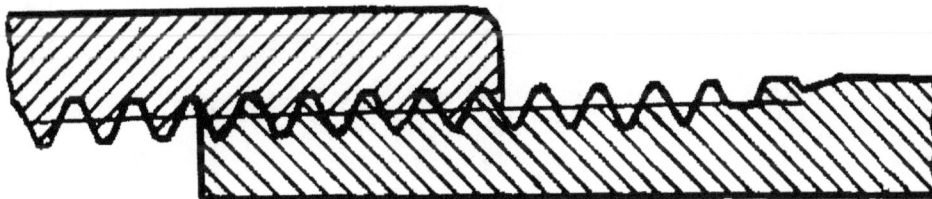

FIG. 5.1
TAPER THREAD

TABLE 5.1
THREADED JOINT LIMITATIONS (ASME B31.1, TABLE 114.2.1)

Maximum Nominal Size, in.	Maximum Pressure	
	psi	kPa
3	400	2 750
2	600	4 150
1	1,200	8 300
¾ and smaller	1,500	10 350

GENERAL NOTE: For instrument, control, and sampling lines, refer to para. 122.3.6(A.5).

fire hazard. Provisions are required to prevent disengagement of caulked joints and resist the effects of longitudinal forces from internal pressure.

For other joints, such as coupling-type, mechanical-gland-type, bell-type, and packed joints, the separation of the joint must be prevented by a means that has sufficient strength to withstand the anticipated conditions of service. Pressure tends to pull these joints apart. Pipe joints that require friction characteristics or resiliency of combustible materials for mechanical or leak-tightness are prohibited from use in piping containing flammable or combustible liquids inside buildings. Unions are permitted up to size NPS 3 (DN 80) (para. 115) per MSS SP-83 (see Section 5.4).

DESIGN REQUIREMENTS FOR SPECIFIC SYSTEMS

6.1 OVERVIEW

ASME B31.1 provides requirements for piping systems in Part 6 of Chapter II. Requirements for boiler external piping that were transferred from ASME BPVC, Section I in 1972, as well as requirements for systems with specific types of fluids or equipment, can be found here.

Boiler external piping is covered in para. 122.1 and includes requirements for the following systems:

    (1) steam piping;
    (2) feedwater piping;
    (3) blowoff and blowdown piping; and
    (4) drains.

Very specific design requirements, including design pressure–temperature, valving, valve design, and materials are provided. Background on these requirements is provided in Bernstein and Yoder (1998) and Mackay and Pillow (2011).

Requirements are also provided for the following specific types of piping systems:

    (1) blowoff and blowdown piping in non-boiler external piping (para. 122.2);
    (2) instrument, control, and sampling piping (para. 122.3);
    (3) spray-type desuperheater piping for use on steam generators, main steam, and reheat piping (para. 122.4);
    (4) pressure-relief piping (para. 122.6);
    (5) temporary piping systems (para. 122.10);
    (6) steam trap piping (para. 122.11); and
    (7) district heating and steam distribution systems (para. 122.14).

Requirements are provided for systems involving the following components or equipment:

    (1) pressure-reducing valves (para. 122.5); and
    (2) pump discharge piping (para. 122.13).

Requirements are provided for systems handling the following types of fluids:

(1) flammable or combustible liquids (para. 122.7),
(2) flammable gases (para. 122.8.1),
(3) toxic fluids (gas or liquid) (para. 122.8.2),
(4) non-flammable non-toxic gas (e.g., air, oxygen, carbon dioxide, and nitrogen) (para. 122.8.3), and
(5) corrosive liquids and gases (para. 122.9).

6.2 BOILER EXTERNAL PIPING

Rules governing boiler external piping are provided in 122.1. Note that additional requirements can be found the ASME BPVC Section I. This piping includes steam piping, feedwater piping, blowoff piping, blowdown piping and drains. "Blowoff piping are operated intermittently to remove accumulated sediment from equipment and/or piping, or to lower boiler water level in a rapid manner. Blowdown systems are primarily operated continuously to control the concentrations of dissolved solids in the boiler water." (ASME B31.1) Blowoff systems are subject to flashing, two phase flow, and consequential vibration, and as a result, have more stringent design requirements than blowdown piping.

The following is intended to provide a general overview of requirements, organized in a different fashion than as presented in the Code to show commonality of requirements between different systems. It is not intended to cover all the rules in 122.1, and it does not cover the various exceptions provided for miniature boilers.

Most of the time, the piping is to be designed for the maximum pressure and temperature conditions with the pressure taken as the boiler maximum allowable working pressure (MAWP) or the lowest set pressure of any relief valve on the steam drum. The design temperature, other than for superheated steam, is generally taken as the saturation temperature at the corresponding pressure. The expected temperature is required to include the manufacturer's maximum temperature tolerance. The following describe exceptions:

1. For all boiler external piping, the minimum design pressure is 100 psi (690 kPa [gage]).
2. For steam piping from the superheater outlet, the design pressure shall not be less than the lowest of
   a. the lowest set pressure of any safety valve on the superheater and
   b. 85% of the lowest set pressure of any safety valve on the steam drum.

3. For single boilers feeding single prime movers, when there is automatic combustion control equipment that is responsive to steam pressure, the design pressure may be set at the greater of
   a. the throttle inlet pressure plus 5% and
   b. 85% of the lowest set pressure of any safety valve on the steam drum and
   c. the expected maximum sustained operating pressure.

4. When two steam stop valves are required, the design pressure for the piping between them may be set as low as the greater of
   a. the expected maximum operating pressure and
   b. 85% of the lowest safety valve set pressure on the steam drum.

5. Feedwater (except for a forced flow steam generators) and blowoff piping are considered to be in shock service. The feedwater piping can be subject to pressure transients associated with pump stop and start and valve operation. The blowoff piping is subject to flashing of the condensate being let down to a lower pressure. In consideration of shock service, in both systems, the design pressure is required to exceed the boiler MAWP by the lesser of 25% or 225 psi (1550 kPa).

6. Forced flow steam generators with no fixed steam and water line may have the design conditions for the boiler external piping based on the expected maximum sustained operating conditions to which the pressure part will be subjected. These conditions may vary through the system. For the steam piping, the design throttle pressure plus 5% is used, if greater than the maximum expected operating condition.

7. The minimum design temperature for drain piping is 220°F (105°C).

The use of non-ferrous materials for boiler external piping is limited to NPS 3 and smaller. Blowoff and blowdown piping are required to be steel, minimum Schedule 80, unless the design pressure is 100 psig (690 kPa [gage]) or lower, in which case non-ferrous pipe may be used, and the fittings may be bronze, cast iron, malleable iron, ductile iron, or steel. Galvanized piping is prohibited. The use of expansion joints is prohibited (para. 122.1.1(I)).

The feedwater piping and blowdown piping size are required to be at least the same size as the connection to the boiler.

Some of the valve requirements are described below.

- All boiler steam discharge outlets (except to safety valves) are required to have a stop valve, which is the limit of the boiler external piping. An exception is a single boiler discharging to a prime mover installation. Requirements for steam stop valves are provided in para. 122.1.7(A). In some circumstances with two or more boilers connected to a common header, two stop valves are required, in a double block and bleed arrangement set up to permit safe access to a boiler that is taken off line, while other boilers connected to the header remain in operation.

- The feedwater boiler external piping generally includes the feedwater stop valve and check valve. If a feedwater heater is between an upstream check valve and the stop valve, the boiler external piping ends at the stop valve. For feedwater piping for forced flow steam generators with no fixed steam and water line, the boiler external piping may end at the stop valve provided that check valve(s) with pressure ratings at least equal to the boiler inlet design pressure are included in the non-boiler external piping downstream of the feed pumps.

- If the feedwater is supplying more than one boiler, regulating valves are required. If there is no stop valve between the regulating valve and the check valve, the regulating valve is included in boiler external piping. If there is a stop valve between the regulating valve and the check valve, the boiler external piping ends at the intervening stop valve, and the regulating valve is included in non-boiler external piping.

- Blowoff piping is required to have two valves (with some exceptions), at least one of which is generally required to be slow opening (12.1.7(C)). Additional requirements, such as a prohibition against valve types that can accumulate sediment, also apply.

- Blowdown piping may have a single valve

- Drain piping may use a single valve, with certain conditions as described in 122.1.5(D), if it is intended only to be used when the boiler is not under pressure.

- Valves for feedwater and blowoff piping (shock service) shall be at least Class 125 cast iron or bronze, or Class 150 steel or bronze. Blowoff cast iron valves are limited to NPS 2-1/2 and are required to be Class 250. Also, for blowoff valves, if the design pressure is higher than 250 psig [1725 kPa (gage)], steel valves, minimum Class 300, are required.

6.3 OTHER SYSTEM REQUIREMENTS

Specific requirements on layout, size and design conditions for blowoff and blowdown piping in non-boiler external piping (beyond the limit of the boiler external piping) are provided in para. 122.2.

Para. 122.3 specifies what is considered instrument, control, and sampling piping that is within the scope of, and to be designed in accordance with ASME B31.1. Essentially, all the valves, fittings, tubing, and piping are included in the scope. Instruments themselves such as pressure gages are not within the scope. Specific requirements for this piping are detailed in para. 122.3.

Requirements for piping for flammable gases, toxic fluids, and non-flammable non-toxic gasses are provided in para. 122.8. They provide limitations on materials of construction, joints, and components. The use of explosive concentrations of flammable gases is prohibited unless the piping is designed to withstand an internal explosion. Further, vent lines are required to be routed in a way to avoid explosive concentrations during venting and are to be subjected to a hazard analysis (see para. 122.8.1). In addition, as per para. 122.8.2(H), piping for toxic fluids is required to be pneumatic leak tested, or given a sensitive leak test and a hydrostatic leak test.

DESIGN FOR SUSTAINED AND OCCASIONAL LOADS

7.1 PRIMARY LONGITUDINAL STRESSES

The wall thickness of pipe is nearly always selected based on the thickness required for internal pressure and allowances. The piping is then supported sufficiently such that the longitudinal stress (this is the stress in the axial direction of the pipe) is within Code limits and deflection is within acceptable limits.

Deflection limits are not Code requirements but are generally accepted practice. Table 121.5, Suggested Support Spacings, is based on a deflection of 0.1 in. (2.5 mm). Less stringent deflection limits may be acceptable, and more stringent limits may be required for lines that must avoid pockets caused by sagging of the line. At elevated temperatures, significant sagging the line can occur over time. Insert 7.1 describes a method for evaluating long-term deflection of piping due to creep.

It is fortunate that the longitudinal pressure stress is half of the hoop stress in a cylinder. What this means is that if the pipe is designed for pressure, at least half of the strength in the longitudinal direction remains available for weight and other sustained loads.

The allowable stress from Appendix A must be divided by the weld joint efficiency, E, to determine a basic allowable stress for evaluation of longitudinal loads. The weld joint efficiency and casting quality factor only apply to pressure design.

Insert 7.1 Span Limits for Elevated Temperature Piping

When a piping system is at a temperature sufficiently high for the material to creep, it is possible that significant sagging (deflection due to creep) can occur over time. Methods are readily available for calculation of elastic deflection of the pipe, and there are commonly used pipe span tables. However, such methods are not available for calculating the long-term deflection due to creep, which can be many times the initial elastic deflection for pipe operating in the creep regime of the material. Span tables, piping stress analysis programs, and generally available design methods only consider the elastic deflection of the pipe.

Becht and Chen (2000) developed closed form equations for calculating creep deflection for simple spans, in order to develop span tables for 980°C (1800°F) pipe for the Marble Hill Nuclear Reactor Pressure Vessel Annealing Demonstration Project. Closed form integrals to predict creep deflection of simply supported, cantilever and fixed end beams were developed. These equations follow.

The Norton creep equation is assumed.

$$\epsilon_c = B\,\sigma^n t$$

I_c and K, in the following equations, are defined as follows.

$$I_c = I\,\frac{8}{\left(3+1\frac{1}{n}\right)\cdot\sqrt{\pi}}\cdot\frac{1-\left(\frac{r_i}{r_o}\right)^{3+\frac{1}{n}}}{1-\left(\frac{r_i}{r_o}\right)^4}\cdot\frac{\Gamma\left(1+\frac{1}{2\cdot n}\right)}{\Gamma\left(1.5+\frac{1}{2\cdot n}\right)}$$

$$K = \frac{2\cdot B\cdot t}{h}\cdot\left(\frac{h}{2\cdot I_c}\right)^n\cdot\left(\frac{w}{2}\right)^n$$

Slope and deflection for simply supported beam

$$y_p = K\cdot\int_0^{\frac{1}{2}}|(L\cdot x - x^2)^n|dx$$

$$y = \int_0^{\frac{1}{2}} y_p + K\cdot\int_0^x |(x\cdot L - x^2)^n|dx\,dx$$

Slope and deflection for cantilever beam

$$y_p = K\cdot\int_0^{\frac{1}{2}}(x^2)^n\,dx$$

closed form solution

$$y = \frac{2\cdot B\cdot t}{h}\cdot\left(\frac{h}{2\cdot I_c}\right)^n\cdot\left[\frac{w\cdot(L)^2}{2}\right]^n\cdot\frac{L^2}{2\cdot(n+1)}$$

Slope and deflection for fixed ended beam
  Slope at inflection point

$$y_{ps} = K\cdot\int_0^s -\|[6\cdot L\cdot x - (L)^2 - 6\cdot x^2]^n\|dx$$

$$s = 0.2113L$$

deflection at inflection point

$$y_s = \int_O^s K \cdot [|\,16 \cdot L \cdot x - (L)^2 - 6x^2\,|]^n \cdot (x - s)\, dx$$

deflection at mid span

$$y = y_s - \int_s^{\frac{1}{2}} K \cdot [|\,16 \cdot Lx - (L)^2 - 6 \cdot x^2\,|]^n \cdot (x - s)\, dx$$

where

B	constant in creep equation
h	outside diameter of pipe
I	moment of inertia of pipe
I_c	fictitious moment of inertia used to calculate outer fiber stress for a creeping beam
L	beam length
n	stress exponent in creep equation
r_o	outside radius of pipe
r_i	inside radius of pipe
s	distance from end of fixed-fixed beam to inflection point
t	time
w	weight of pipe and contents per unit length
x	dimension along beam length
y	beam deflection
y_1	extreme fiber distance from the neutral axis
y_p	slope of deflected beam
y_{ps}	slope of fixed-fixed beam at inflection point
$\in_c$	creep strain
Γ	gamma function
σ	stress

Evaluating creep deflection provided insights into the problem of establishing allowable spans for high-temperature piping. The allowable span was found to be relatively insensitive to the allowable deflection and the constant B in the creep equation, within reasonable limits. It is the sensitivity of creep rate to stress, and, in turn, the sensitivity of the stress to span length, that dominates the creep deflection problem. Considering that stress is proportional to the span length to the second power, and assuming that creep strain rate is proportional to stress to, say, the sixth power, knowing that deflection is proportional to strain, we find that creep deflection rate is highly sensitive to span length.

Creep deflection goes up exponentially with span length, so that beyond a given length, at least at relatively high temperatures, deflection is unacceptable, and below that length, deflection is not particularly significant. Figure 7.1 shows the creep deflection of a simply supported Schedule 10S 304L stainless steel line, assuming 44 kg/m (30 lb/ft) steel and insulation weight with vapor as the contents, as a function of length at 815°C (1500°F) and for a duration of 1000 hours. One can see that,

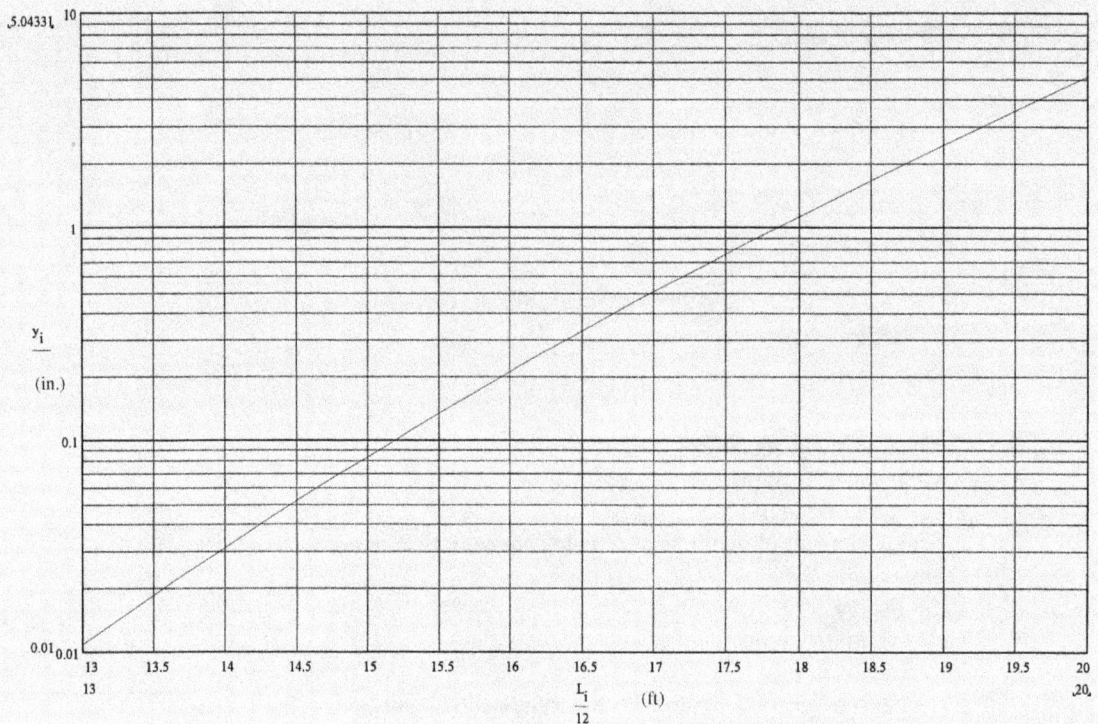

FIG. 7.1
CREEP DEFLECTION OF SIMPLY SUPPORTED BEAM AT 1000 HR VERSUS SPAN,
815°C (1500°F)

between span lengths of 5.2 and 5.5 m (17 and 18 ft), the deflection increases from 13 mm (½ in.) to over 25 mm (1 in.). At 4.9 m (16 ft), the deflection is about 5 mm (0.2 in.).

Figure 7.2 shows deflection as a function of span length, temperature and end restraint condition. These curves are based on 304 L material. The duration for the chart is 1000 hours and the temperature is 870°C (1600°F). They illustrate the effect of span length and support condition on deflection. The project for which this work was done used dual stamped material, so that the L-grade material properties were appropriate to evaluate the elevated temperature behavior of the material.

Figure 7.3 shows a chart for 304 L Schedule 20S pipe with 29 kg/m (20 lb/ft) insulation and a duration of 100,000 hours, about 10 years, for elastic plus creep deflection as well as for elastic deflection only. Included on the chart are results based on simply supported and fixed supports. Note that the allowable span length based on allowable stress consideration only, per ASME B31.3, is about 4.9 m (16 ft) for simply supported and 7 m (23 ft) for fixed supports. The allowable span length, based on 13 mm (0.5 in.) permissible elastic deflection and a simply supported condition would be 9.4 m (31 ft). It is obvious that these span lengths based on these conventional criteria would result in excessive long-term deflection for this elevated temperature pipe.

Creep deflection should be considered in determining allowable span lengths for elevated temperature piping. The equation provided herein can be used to develop simplified span tables for specific applications.

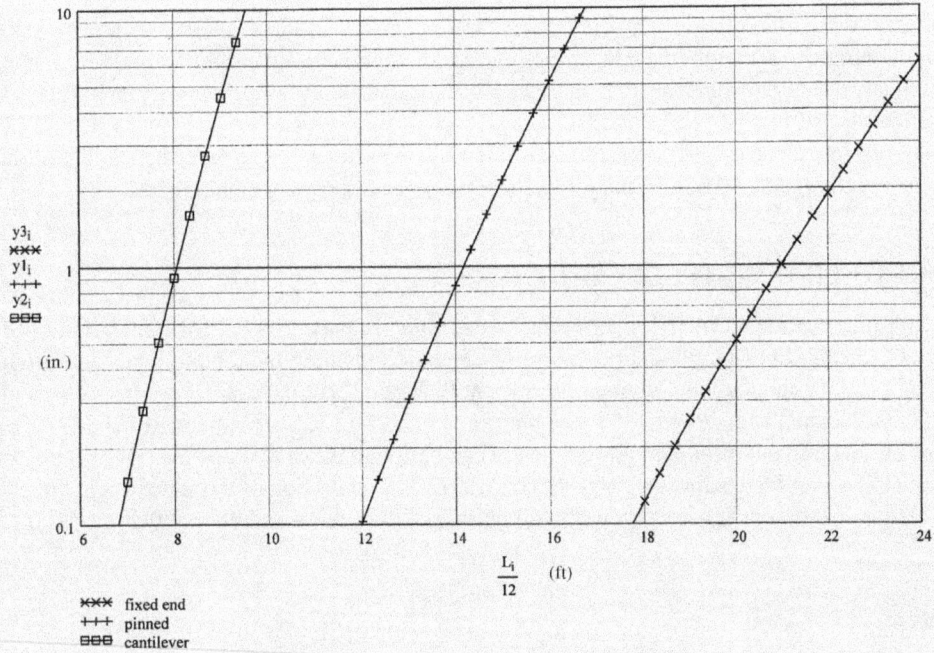

FIG. 7.2
CREEP DEFLECTION VERSUS SPAN LENGTH AT 1000 HR FOR DIFFERENT RESTRAINT CONDITIONS, 870°C (1600°F)

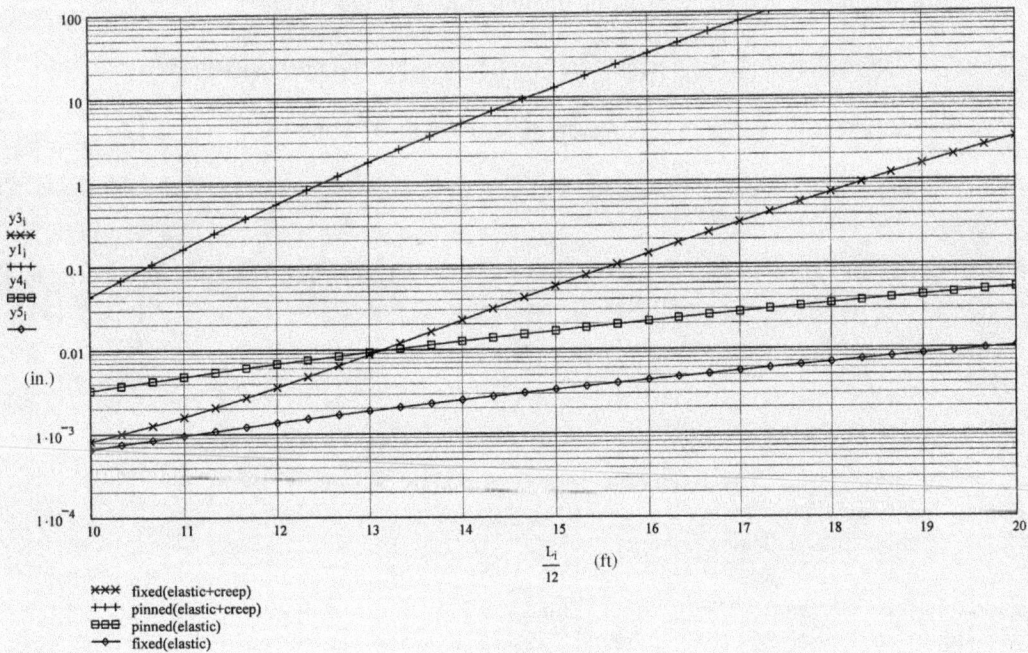

FIG. 7.3
COMPARISON OF CREEP AND ELASTIC DEFLECTION OF BEAMS AT 100,000 HR VERSUS SPAN LENGTH FOR PINNED AND FIXED RESTRAINT, 815°C (1500°F)

These sustained and the occasional load stresses should encompass all of the load-controlled, primary-type stresses to which the pipe is subjected. They will either fall in the sustained or occasional category depending on duration. While the weight of the pipe and contents are sustained, forces and moments from wind, earthquakes, and phenomena, such as dynamic loads caused by water hammer, are considered to be occasional loads.

The equations for calculating stresses from sustained and occasional loads, described in the following sections, can be found in para. 104.8 of Chapter II, Part 2, Analysis of Piping Components.

7.2   SUSTAINED LONGITUDINAL STRESS

ASME B31.1 provides Eq. (11A) [(Eq. 11B) for metric units] in para. 104.8.1 for calculation of SL, which is defined as the sum of the longitudinal stresses. The equation includes the longitudinal force from internal pressure and bending and torsional moment. It does not include shear or axial forces from loads other than pressure. The calculated stress is limited to the basic allowable stress, Sh, per para. 102.3.2(D). The discussion regarding dividing out the weld joint efficiency can be found in para. 102.3.2(A.3).

Equation (15) is provided below, in consistent units. The metric version of this equation in the Code is not provided below; it is based on inconsistent metric units and, thus, has an additional factor of 1000 in one of the terms.

$$S_L = \frac{PD_o}{4t_n} + \frac{0.75iM_A}{Z} \leq 1.0S_h \tag{15}$$

where:

D_o =   outside diameter of pipe

M_A =   resultant moment loading on cross section from weight and other sustained loads; this is a vector summation of both bending and torsional moments

P =   internal design gage pressure

S_L =   sum of the longitudinal stresses from pressure, weight, and other sustained loads

Z =   section modulus

i =   stress intensification factor; however, the product of 0.75 and i shall not be taken as less than 1.0

t_n =   nominal wall-thickness of pipe

An alternate to $PD_o/4t_n$ is provided in para. 102.3.2(A.3). The alternate equation for determining the longitudinal pressure stress is provided below.

$$S_{lp} = \frac{pd_n^2}{D_n^2 - d_n^2}$$

where:

D_n =   nominal outside diameter of pipe

S_{lp} =   longitudinal pressure stress

d_n =   nominal inside diameter of pipe

As noted in the Code (para. 104.8), Eq. (15) (and Eq. [16] in the next section) is not correct for some systems with expansion joints. The longitudinal pressure stress term in the equation may be incorrect. An example would be a straight pipe with elbows on either end, anchors at each elbow, and a bellows expansion joint (without pressure thrust-restraining hardware such as tie rods) in the center of the run, with a bellows mean diameter larger than the pipe inside diameter. The longitudinal pressure thrust forces are carried by the anchors, rather than through the pipe. The force acting on the pipe will be reversed from tension to compression and will be the pressure times the difference in areas between circles defined by the mean diameter of the bellows (bellows pressure thrust area) and the inside diameter of the pipe. See Becht (1989) and the Standards of the Expansion Joint Manufacturers Association for more detailed discussions.

It is assumed that the appropriate intensification (actually stress index) for sustained loads is 0.75 times the stress intensification factor for thermal loads. The stress intensification factor for thermal loads is based on fatigue. The stress index for sustained loads should be based on collapse loads for piping components. It turns out that the 0.75 factor is appropriate for elbows. The relationship between fatigue stress intensification and collapse loads may not be the same for other components; however, the 0.75 factor has been deemed appropriate for design purposes. This is an area of continuing work and debate for the ASME B31 Mechanical Design Committee.

Note that ASME B31.1 calculates longitudinal stress from sustained (and occasional) loads using nominal dimensions of the pipe. Mill tolerance, allowances for erosion and corrosion, and mechanical allowances (e.g., thread or groove depth) are not subtracted from the pipe wall-thickness, either in calculation of the loads or in calculation of the stress.

The section modulus used in the calculation of stress for reduced outlet branch connections is a special case. An effective section modulus is used in conjunction with the branch connection stress intensification factor. See Section 9.5 herein for the effective section modulus.

7.3   LIMITS OF CALCULATED STRESSES FROM OCCASIONAL LOADS

ASME B31.1 provides Eq. (16) in para. 104.8.2 for evaluation of longitudinal stresses from occasional loads such as wind and earthquake in combination with sustained loads. The same as Eq. (15) for sustained loads, the longitudinal force from internal pressure and bending and torsional moment are evaluated. It does not include shear or axial forces from loads other than pressure. The calculated stress is limited to the basic allowable stress, Sh, per para. 102.3.2(D) times a factor k. The discussion regarding dividing out the weld joint efficiency can be found in para. 102.3.2(B).

Equation (16) is provided below, in consistent units. The metric version is not provided below; it is based on inconsistent metric units and, thus, has an additional factor of 1000 in the moment terms.

$$\frac{PD_o}{4t_n} + \frac{0.75iM_A}{Z} + \frac{0.75iM_B}{Z} \leq kS_h \qquad (16)$$

where:

M_B = resultant moment loading on cross section from occasional loads

k = 1.15 for occasional loads acting for no more than 8 hours at any one time and no more than 800 hours/year; 1.2 for occasional loads acting for no more than 1 hr. at any one time and no more than 80 hr/yr.

The factor k provides an increase in allowable stress for occasional loads. It used to be 1.2, which was consistent with ASME BPVC, Section VIII, Division 1, but was changed to the above, which is consistent with the allowances for pressure-temperature variations from normal operation.

The moment, M_B, is the moment loads from occasional loads such as thrusts from relief/safety-valve installations, flow transients, wind, and earthquake.

The treatment of earthquake forces as an occasional load is generally considered to be extremely conservative, and a new set of rules, ASME B31E, has been prepared but is not yet referenced by ASME B31.1.

The same considerations as are discussed in Section 7.2 herein for effective section modulus, use of 0.75i, thickness used in the calculation, alternate calculation for longitudinal stress from pressure, and expansion joints also apply to Eq. (16).

Design Criteria for Thermal Expansion

8.1 ALLOWABLE STRESS FOR THERMAL EXPANSION

The allowable stress for thermal expansion and other deformation-induced stresses is substantially higher than for sustained loads. This is because of the difference between load-controlled conditions, such as weight and pressure, and deformation-controlled conditions, such as thermal expansion or end displacements (e.g., from thermal expansion of attached equipment).

When a load-controlled stress is calculated, it is an actual stress value; it is governed by equilibrium. For example, the stress in a bar when a tensile force is applied to it is the force divided by the area of the bar. In the case of thermal stresses, however, it is the value of strain that is known. The elastically calculated stress is simply the strain value times the elastic modulus. This makes essentially no difference until the stress exceeds the yield strength of the material. In that case, the location on the stress–strain curve for the material is determined based on the calculated stress for load-controlled or sustained loads. The location on the stress–strain curve for the material is determined based on the calculated strain (or elastically calculated stress divided by elastic modulus) for deformation-controlled (e.g., thermal expansion) loads. This is illustrated in Fig. 8.1. Because the stress analyses are based on the assumption of elastic behavior, it is necessary to discern between deformation-controlled and load-controlled conditions to properly understand the post-yield behavior.

It is considered desirable for the piping system to behave in a substantially elastic manner so that the elastic stress analysis is valid. Having plastic deformation at every cycle carries with it uncertainties with respect to strain concentration and can be potentially far more damaging than it is calculated to be in the elastic analysis. One way to address this would be to limit the total stress range to yield stress. However, this would be overly conservative and result in unnecessary expansion loops and joints. Rather, the concept of shakedown to elastic behavior is used in the Code.

The allowable thermal expansion stress in the Code is designed to result in shakedown to elastic behavior after a few operating cycles. The basic allowable thermal expansion stress range is provided in para. 102.3.2(B), as follows:

$$S_A = f(1.25S_c + 0.25S_h) \tag{1A}$$

where:

S_A = allowable displacement stress range

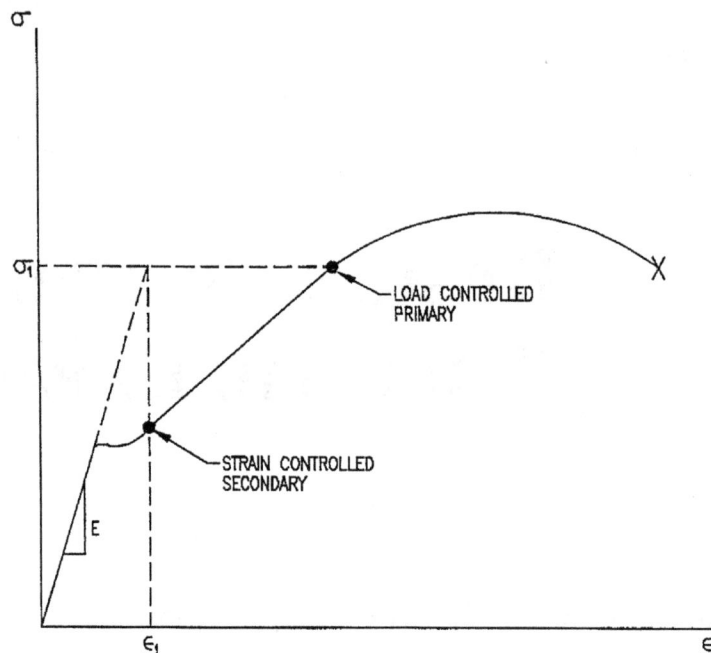

FIG. 8.1
LOAD-CONTROLLED VERSUS DEFORMATION-CONTROLLED BEHAVIOR. σ = STRESS
ε = STRAIN, E = ELASTIC MODULUS

S_c = basic material allowable stress at the minimum (cold) temperature from the Allowable Stress Tables (note that weld efficiency and casting quality factors should be backed out of this allowable stress value; see para. 102.3.2[B]).

S_h = basic material allowable stress at the maximum (hot) temperature from the Allowable Stress Tables (note that weld efficiency and casting quality factors should be backed out of this allowable stress value; see para. 102.3.2[B]).

f = cyclic stress range factor for the total number of equivalent reference stress range cycles, N, determined from the following equation.

$$f = 6/N^{0.2} \leq 1.0 \tag{1C}$$

N = total number of equivalent reference displacement stress range cycles expected during the service life of the piping. A minimum value for f is 0.15, which results in an allowable displacement stress range for a total number of equivalent reference displacement stress range cycles greater than 10^8 cycles.

For the purposes of Eq. (1A), S_c and S_h are limited to 20 ksi (140 MPa) for materials with a minimum tensile strength (at room temperature) of over 70 ksi (480 MPa), unless otherwise justified. This is because of a concern that the f factor may be unconservative for high strength steels.

Note that S_c and S_h should be taken at the maximum and minimum metal temperatures for the cycle under consideration.

Equation (1A) assumes that the sustained stress consumes the entire allowable sustained stress. Paragraph 102.3.2(B) provides that when the full allowable sustained stress, S_h, is not used, the difference between the longitudinal sustained stress, S_L, and the allowable, S_h, may be added to S_A, as follows.

$$S_A = f\left[1.25(S_c + S_h) - S_L\right] \tag{1B}$$

where:

S_L = longitudinal stress from sustained loadings

The allowable total stress range is $1.25f (S_h + S_c)$, from which any stress used for sustained loads, S_L, is subtracted, with the remainder permitted for thermal expansion stress range.

The allowable thermal expansion stress range can exceed the yield strength for the material, since both S_c and S_h may be as high as two-thirds of the yield strength (or 90% of yield strength for austenitic stainless steel and similar alloys). However, it is anticipated that the piping system will shake down to elastic behavior if the stress range is within this limit.

This behavior is illustrated in Fig. 8.2, which is based on the assumption of elastic-perfectly plastic material behavior. Consider, for example, a case where the elastically calculated thermal expansion stress range is two times the yield strength of the material. Remember, since it is a deformation-controlled condition, one must actually move along the strain axis to a value of stress divided by elastic modulus. In the material, assuming elastic-perfectly plastic behavior, the initial start-up cycle goes from point A to point B (yield) to point C (strain value of twice yield). When the system returns to ambient temperature, the system returns to zero strain and the piping system will unload elastically until it reaches yield stress in the reverse direction. If the stress range is less than the twice yield, there is no yielding on the return to ambient temperature. On returning to the operating condition, the system returns from point D to point C, elastically. Thus, the cycling will be between points D and C, which is elastic. The system has essentially self-sprung and is under stress from displacement conditions in both the ambient and operating conditions.

If twice yield is exceeded, shakedown to elastic cycling does not occur. An example would be if the elastically calculated stress range was three times the yield strength of the material. In this case, again referring to Fig. 8.2, the start-up goes from point A to point B (yield) to point E. Shutdown results in yielding in the reverse direction, going from point E to point F to point D. Returning to the operating condition again results in yielding, from point D to point C to point E. Thus, each operating cycle results in plastic deformation and the system has not shaken down to elastic behavior.

This twice-yield condition was the original consideration. Since the yield strength in the operating and ambient conditions are different, the criteria become that stress range must be less than the hot yield strength

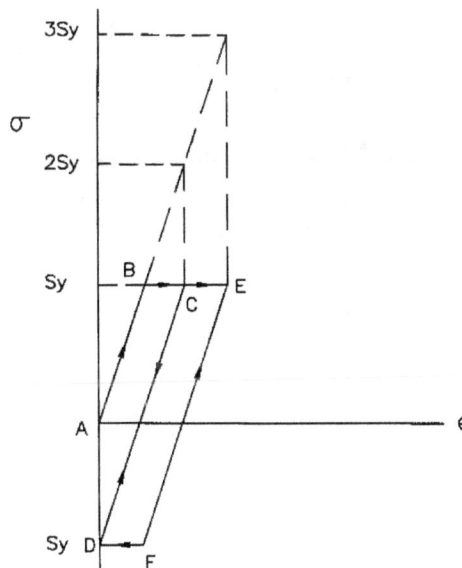

FIG. 8.2
STRESS–STRAIN BEHAVIOR ILLUSTRATING SHAKEDOWN

plus the cold yield strength, which, because of the allowable stress criteria, must be less than 1.5 times the sum of S_c and S_h (note that the original ASME B31.1 criteria limited the allowable stress to 62.5% of yield, so the original factor that was considered was 1.6). This 1.5 (1.6 originally) factor was reduced, for conservatism, to 1.25. Furthermore, this total permissible stress range is reduced by the magnitude of sustained longitudinal stress to calculate the permissible thermal expansion stress range. This is represented in equation (1B). Equation (1A) simply assumes that $S_L = S_h$, the maximum permitted value, and assigns the remainder of the allowable stress range to thermal expansion.

The same stress limit as reflected in Eq. (1B) also works in the creep regime. Deformation controlled stresses relax to a stress value sufficiently low that no further creep occurs. This stress value is the hot relaxation strength, S_H. Stress–strain behavior under the condition of creep, is illustrated in Fig. 8.3. The initial start-up cycle, which can include some yielding, goes from point A to point B. During operation, the stresses relax to the hot relaxation strength, S_H, at which point no further relaxation occurs, point C. When the system returns to ambient temperature, the system returns to zero strain and the piping system will unload elastically until it reaches yield stress in the reverse direction. If the stress range is less than S_H plus to cold yield strength, there is no yielding on the return to ambient temperature. This is illustrated by going from point C to point D. On returning to the operating condition, the system returns from point D to point C elastically. Thus, if the stress range is less than the cold yield strength plus the hot relaxation strength, shakedown to elastic behavior also occurs at elevated temperature.

If S_H is considered to be 1.25 S_h, then elevated temperature shakedown also is achieved with the Code allowable for displacement stresses, S_A. The anticipated behavior over time, with multiple shut downs, and a gradual relaxation process, is illustrated in Fig. 8.4.

Figure 8.3 also shows the behavior when the allowable stresses are exceeded at elevated temperatures. In this case, the startup goes from A to E. Stresses relax to point F. When the system returns to ambient tempera-

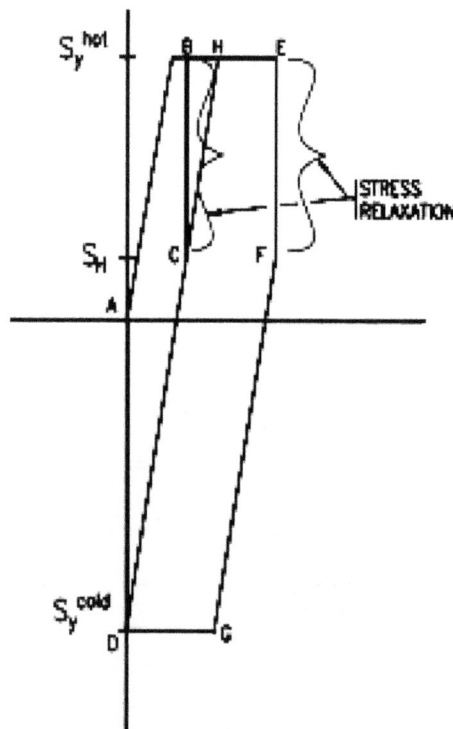

FIG. 8.3
STRESS–STRAIN BEHAVIOR ILLUSTRATING ELEVATED TEMPERATURE SHAKEDOWN

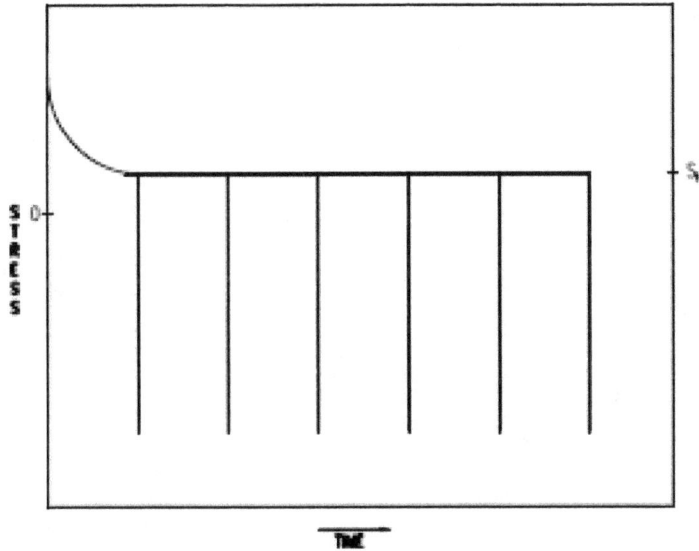

FIG. 8.4
CYCLIC STRESS HISTORY WITH SHAKEDOWN

ture, yielding in the reverse direction occurs, going from point F to G to D. Returning to operating condition again results in yielding, from point D to H to E. Since high stresses are re-established, another relaxation cycle then must occur. The behavior of this system over time is illustrated in Fig. 8.5.

Even though the stress range is limited to result in shakedown to elastic behavior, there remains the potential for fatigue failure if there is a sufficient number of cycles. Therefore, the f factor is used to reduce the allowable stress range when the number of cycles exceeds 7000. This is about once per day for 20 years.

Figure 8.6 provides the basic fatigue curve for butt-welded pipe developed by A. R. C. Markl (1960a) for carbon steel pipe. A safety factor of two on stress was applied to this curve, giving a design fatigue curve. It

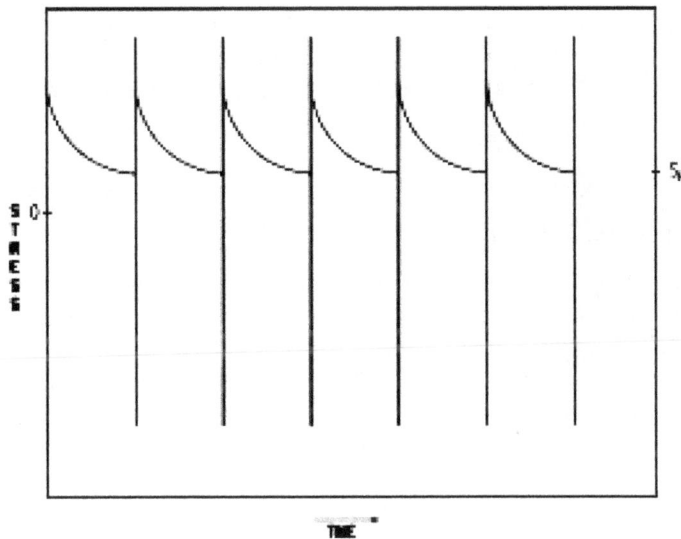

FIG. 8.5
CYCLIC STRESS HISTORY WITHOUT SHAKEDOWN

For A106, Gr. B Pipe at Room Temperature
Sh=Sc=20,000 psi

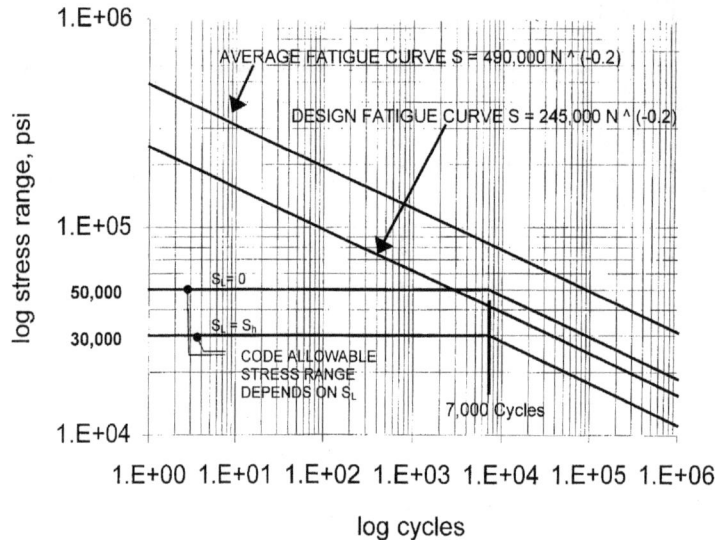

FIG. 8.6
MARKL FATIGUE CURVE FOR BUTT-WELDED STEEL PIPE

can be observed that the allowable thermal expansion stress range for A106 Grade B, carbon steel pipe, prior to application of an f factor, intercepts the design fatigue curve at about 7000 cycles. For higher numbers of cycles, the allowable stress is reduced by the f factor to follow the fatigue curve, as per Eq. (1C).

As of the 2012 edition of ASME B31.1, the factor f is not permitted to exceed 1.0. At the present, increasing the maximum permissible f factor to 1.2 is being considered. An f factor of 1.2 corresponds to 3125 cycles. The rationale for allowing a factor as high as 1.2 is that stresses are permitted to be as high as two times yield when $f = 1.2$. Thus, the desired shakedown behavior is maintained. It backs out a little of the conservatism introduced when the original criteria were developed.

Papers written by Markl (1960a, 1960b, 1960c, 1960d) describe the development of this methodology.

Insert 8.1   What About Vibration?

ASME B31.1 is a new design code, and typically, piping systems are not analyzed for vibrating conditions during design. While the Code does require that piping be designed to eliminate excessive and harmful effects of vibration (para. 101.5.4), this is typically done by attempting to design systems to not vibrate, rather than by performing detailed vibration analysis. However, there are cases, such as certain reciprocating compressor piping systems, for which detailed vibration assessments are performed, but these are exceptions to the general rule. As such, evaluation of vibration tends to be a post-construction exercise.

Excellent guidance for the evaluation of vibrating piping systems can be found in Part 3 of ASME Standard OM-S/G, ASME OM-3, *Requirements for Preoperational and Initial Start-Up Vibration*

Testing of Nuclear Power Plant Piping Systems. Vibration can be evaluated via the procedures contained therein based on peak measured velocity of the vibrating pipe or calculated stress. A screening velocity criterion that is generally very conservative, 0.5 in./second peak velocity, is provided. Endurance limit stress ranges are also provided.

Based on the endurance limit stress ranges, an *f* factor for an unlimited number of cycles may be derived. The "endurance limit"[1] stress ranges are provided for carbon steel and stainless steel. Including all the factors provided in the document gives the "endurance limit"[2] stress range of 106 MPa (15.4 ksi) for carbon steel. Assuming a typical S_A of 345 MPa (50 ksi) (1.25 S_c + 1.25 S_h), and considering that the Code flexibility analysis equations calculate about one-half of the actual peak stress, we find an "endurance limit" *f* factor of 0.15 (at about 10^8 cycles). This "endurance limit" *f* factor was derived using ASME B31.3 allowable stresses. To the extent that those in ASME B31.1 are lower, the factor is more conservative. The "endurance limit" *f* factor for stainless steel would be higher, if calculated by the same procedure. However, the same "endurance limit" *f* factor is applied to all materials, consistent with existing rules. This endurance limit was added in the 2007 addendum of ASME B31.1.

Caution should be exercised in the fatigue design when the number of cycles is between about 2,000,000 cycles and the "endurance limit" *f* factor. It has been shown (Hinnant and Paulin, 2008) that the slope of the fatigue curve developed by Markl, from which the *f* factors were developed, is significantly flatter than has been observed in more recent fatigue tests of welded structures. While the Markl curve has a slope of 5 on cycles to 1 on stress on a log-log graph, more recent data indicates the slope should be 3. The newer data includes much more data in the higher cycle regime, and the difference is apparent. At higher cycles, fatigue design using the Markl equation becomes less conservative, and potentially unconservative. This data is being considered by the committees.

8.2   HOW TO COMBINE DIFFERENT DISPLACEMENT CYCLE CONDITIONS

A designer may have more than one thermal expansion or another displacement cycle condition to be considered. Cycles at lower stress ranges are substantially less damaging than cycles at higher stress ranges. ASME B31.1 uses the cycle with the highest displacement stress range (full displacement cycle), S_E, to compare to the allowable stress range, S_A. However, cycles with lower displacement stress ranges are converted into equivalent numbers of full displacement cycles to determine the *f* factor. This procedure, described in para. 102.3.2(B), uses the following Eq. (2) in the Code:

$$N = N_E + \sum\left[q_i^5 N_i\right] \quad \text{for } i = 1, 2, ..., n \tag{2}$$

where:

N_E = number of cycles of reference displacement stress range, S_E

N_i = number of cycles associated with displacement stress range, S_i

n = total number of displacement stress conditions to be considered

q_i = S_i/S_E

S_i = any computed stress range other than the reference displacement stress range

In the past, the above equation was based on temperature range; however, it was generalized by changing from temperature range to stress range.

---

[2]The term "endurance limit" stress range is used for convenience although endurance limit is typically stress amplitude.

Following this procedure, the maximum stress range is limited to S_A, which satisfies the shakedown limit. Lesser cycles are converted into equivalent (with respect to fatigue damage) numbers of cycles at S_E to determine if an f factor less than one is required to protect against fatigue failure.

Table 8.1 provides an example of combining several displacement cycle conditions with different stress ranges and numbers of cycles. Note that because of the sensitivity of fatigue damage to stress (the fifth power in the equation), displacement cycles at significantly lower stress ranges than S_E produce very little damage or a significantly reduced number of equivalent cycles.

TABLE 8.1

COMBINATION OF DIFFERENT DISPLACEMENT CYCLES[1]

Load Case	Design Cycles	Calc. Stress Range	S_i/S_E	N_{equiv}
1	1,000	30,000	1	1000
2	7,000	10,000	0.33	27
3	20,000	6,000	0.20	6

NOTE:

(1) Total equivalent number of cycles with stress range of 30,000 is 1033 < 7000. The stress range reduction factor $f = 1$.

FLEXIBILITY ANALYSIS

9.1 FLEXIBILITY ANALYSIS

In flexibility analysis, the response of the system to loads is calculated. Flexibility analysis is used to determine the response to thermal loads.

Flexibility analysis is essentially a beam analysis model on pipe centerlines. Some fundamental principles are used in piping flexibility analysis that simplifies the analysis procedures. These include the following:

(1) The analysis is based on nominal wall thickness of the pipe.

(2) The effect of components such as elbows and tees on piping flexibility and stress are considered by inclusion of flexibility factors and stress intensification factors.

(3) For thermal stresses, only moment and torsion are typically included. Stresses from shear and axial loads are generally not significant. However, the designer should consider axial forces when they are significant. The following para. 319.2.3(c), from ASME B31.3, describes this issue:

> "Average axial stresses (over the pipe cross section) due to longitudinal forces caused by displacement strains are not normally considered in the determination of displacement stress range, since this stress is not significant in typical piping layouts. In special cases, however, consideration of average axial displacement stress is necessary. Examples include buried lines containing hot fluids, double wall pipes, and parallel lines with different operating temperatures, connected together at more than one point."

(4) The modulus of elasticity at 70°F (21°C) is normally used in the analysis. Paragraph 119.6.4 states that the calculation of S_E shall be based on the modulus of elasticity, E_C, at room temperature. For a more detailed discussion, see Section. 9.8 herein.

Flexibility factors for typical components are included in Appendix D of ASME B31.1. The flexibility factor is the length of straight pipe having the same flexibility as the component divided by the centerline length of the component. They can be used in hand calculations of piping flexibility and are included in all modern piping stress analysis programs.

Additional flexibility is introduced in the system by elbows. Elbows derive their flexibility from the fact that the cross section ovalizes when the elbow is bent. This ovalization reduces the moment of inertia of the pipe cross section, reducing its stiffness and increasing flexibility. Note that the presence of a flange at the end of an elbow will reduce the ability of the elbow to ovalize, and thus Appendix D provides reduced flexibilities for elbows with flanges welded to one or both ends.

9.2   WHEN FORMAL FLEXIBILITY ANALYSIS IS REQUIRED

ASME B31.1 requires that the designer perform an analysis unless certain exemption criteria are met, as provided in para. 119.7.1. An analysis does not necessarily mean a computer stress analysis, but can be by simplified, approximate, or comprehensive methods. These vary from simple charts and methods such as Kellogg's guided cantilever method to detailed computer stress analysis of the piping system.

There are three exemptions from analysis. These are the following:

(1) systems that duplicate a successfully operating installation system or replaces a system with a satis-factory service record;

(2) a system that can be adjudged adequate by comparison with previously analyzed systems; and

(3) a system that is of uniform size, has no more than two anchors and no intermediate restraints, is designed for essentially non-cylic service (less than 7000 total cycles), and satisfies the approximate equation that follows.

The first method is basically a method of grandfathering a successful design. The difficulty comes when trying to determine how long a system must operate successfully to demonstrate that the design is accept-able. Considering that some systems may cycle less than one time per year, and the design criteria is based on fatigue considerations, the fact that a piping system has not had a fatigue failure for some period of time provides little indication that it actually complies with the Code or that it will not fail sooner or later. This is left as a judgment call, but should not be blindly used to accept a design without careful consideration.

The second method relies on the judgment of an engineer or designer who, based on his or her experience, can determine that a system has adequate flexibility.

The third method uses a simplified equation that has limited applicability. The requirements for use of the equation are that the system is of uniform size, has no more than two points of fixation, has no intermediate restraints, and satisfies the following equation:

$$\frac{DY}{(L-U)^2} \leq 30\,(208,300)\,\frac{S_A}{E_C}$$

(On the right-hand side of the equation, 30 is used with the units in U.S. Customary units and 208,000 is used with the units in SI units.)
where:

D = nominal pipe size, in. (mm)
E_C = modulus of elasticity at room temperature, psi (kPa)
L = developed length of piping between anchors, ft. (m)
S_A = allowable displacement stress range, psi (kPa)
U = anchor distance, straight line between anchors, ft. (m)
Y = resultant movements to be absorbed by pipe lines, in. (mm); displacements of equipment to which the pipe is attached (end displacements of the piping system) should be included

This equation tends to be very conservative. However, there are a number of warning statements within the Code regarding limits to applicability. One such warning is for nearly straight sawtooth runs. In addition, the equation provides no indication of the end reactions, which would need to be considered in any case for load-sensitive equipment. With the wide availability of PC-based pipe stress programs, the use of this equa-tion, which has significant limitations, is much less frequent than it may have once been.

9.3   WHEN COMPUTER STRESS ANALYSIS IS TYPICALLY USED

The Code does not indicate when computer stress analysis is required. It is difficult to generalize when a particular piping flexibility problem should be analyzed by computer methods since this depends on the type of service, actual piping layout and size, and severity of temperature. However, there a quite a few guidelines in use by various organizations that indicate which types of lines should be evaluated by computer in a project. These tend to be the lines that are at higher combinations of line size and temperature, or for larger lines that are attached to load-sensitive equipment. One set of recommended criteria is provided in the following list:

(a) In the case of general piping systems, according to the following line size/flexibility temperature criteria:

   (1) all 2 in. (DN 50) and larger lines with a design differential temperature over 500°F (260°C);

   (2) all 4 in. (DN 100) and larger lines with a design differential temperature exceeding 400°F (205°C);

   (3) all 8 in. (DN 200) and larger lines with a design differential temperature exceeding 300°F (150°C);

   (4) all 12 in. (DN 300) and larger lines with a design differential temperature exceeding 200°F (90°C); and

   (5) all 20 in. (DN 500) and larger lines at any temperature.

(b) All 3 in. (DN 75) and larger lines connected to rotating equipment.
(c) All 4 in. (DN 100) and larger lines connected to air-fin heat exchangers.
(d) All 6 in. (DN 150) and larger lines connected to tankage.
(e) Double-wall piping with a design temperature differential between the inner and outer pipe greater than 40°F (20°C).

Again, it is emphasized that the intent of the listed criteria is to identify in principle only typical lines that should be considered at least initially for detailed stress analysis. Obviously, the final decision for whether or not a computer analysis should be performed depends on the complexity of the specific piping layout under investigation and the sensitivity of equipment to piping loads.

Just because a line may pass some exemption from computer stress analysis does not mean that it is exempt from other forms of analysis nor that it will always meet the Code criteria if analyzed in detail. What it is intended to be is a screen that separates the more trouble-free types of systems from those that are more subject to overload or overstress. The lines exempted from computer stress analysis are considered more likely to be properly laid out with sufficient flexibility by an experienced designer.

9.4   STRESS INTENSIFICATION FACTORS

Stress intensification factors are used to relate the stress in a component to the stress in nominal thickness straight pipe. As discussed in the prior section, the analysis is based on nominal pipe dimensions, so the calculated stress would be the stress in straight pipe unless some adjustment is made. The stress can be higher in components such as branch connections. Stress intensification factors that relate the stress in components to that in butt-welded pipe have been developed from "Markl" fatigue testing of piping components. These generally follow the procedures developed by A. R. C. Markl (1960a).

As mentioned previously, Markl developed a fatigue curve for butt-welded pipe. This was based on displacement-controlled fatigue testing, bending the pipe in a cantilever-bending mode. Figure 9.1 shows a Markl-type fatigue test machine. Using a butt-welded pipe fatigue curve had several practical advantages. One is that the methodology was being developed for butt-welded pipe, and the stress analyst typically does not know where the welds will be in the as-constructed system. Using a butt-welded pipe fatigue curve as the baseline fatigue curve provides that butt welds could be anywhere in the system. Furthermore, from a testing standpoint, appropriate fatigue curves could not readily be developed for straight pipe without welds in a cantilever-bending mode, since the failure will occur at the point of fixity, where effects of the method of anchoring the pipe could significantly affect results.

FIG. 9.1
MARKL-TYPE FATIGUE TESTING MACHINE WITH VARIOUS CONFIGURATIONS
(COURTESY OF PAULIN RESEARCH GROUP)

The stress intensification factors were developed from component fatigue testing. The stress intensification factor is the nominal stress from the butt-welded pipe fatigue curve at the number of cycles to failure in the component test, divided by the nominal stress in the component. The nominal stress in the component is the range of bending moment at the point of failure divided by the section modulus of matching pipe with nominal wall-thickness. In a flexibility analysis, it is precisely this nominal stress that is calculated. When the nominal stress is multiplied by the stress intensification factor, and then compared to the fatigue curve for butt-welded pipe, one can determine the appropriate number of cycles to failure of the component.

While the stress intensification factors for some components depend on the direction of loading, ASME B31.1 uses the highest stress intensification factor for all loading directions. While ASME B31.1 originally provided different stress intensification factors of in-plane and out-plane bending moments (see Figure 9.2 for in-plane versus out-plane directions), it was changed to simply use the highest value, the out-plane stress intensification factor, to be consistent with the nuclear Code.

One of the commonly unknown aspects of piping flexibility analysis per the ASME B31 Codes is that in piping stress analysis, the calculated stress range from bending loads is about one-half of the peak stress range. This is because the stress concentration factor for typical as-welded pipe butt welds is two. Since the stresses are compared to a butt-welded pipe fatigue curve, one-half of the actual peak stresses are calculated. Thus, the theoretical stress, for example, in an elbow caused by bending loads is two times what is calculated in a piping flexibility analysis following Code procedures. This is not significant when performing standard design calculations, since the Code procedures are self-consistent. However, it can be very significant when trying to do a more detailed analysis-for example, in a fitness-for-service assessment.

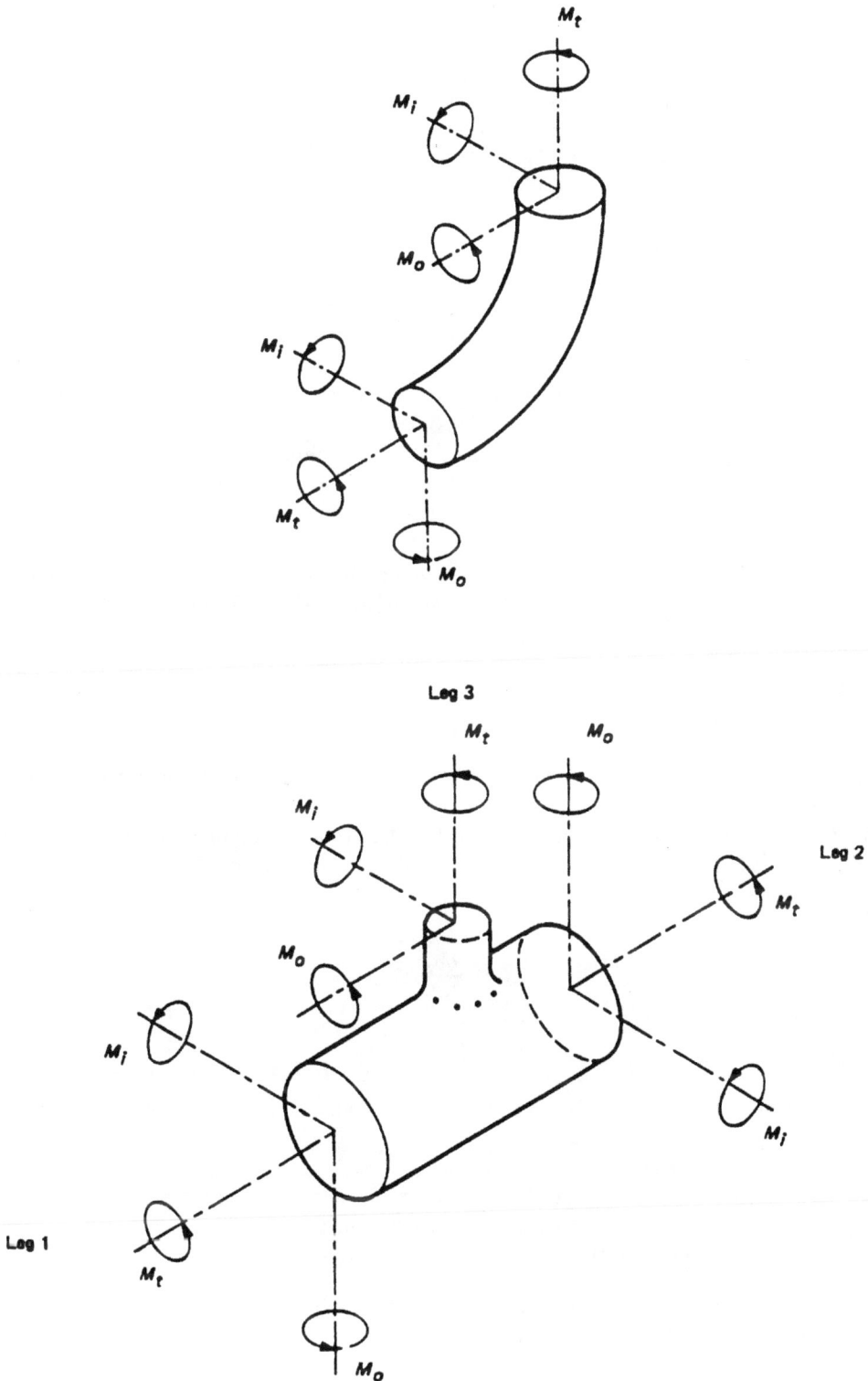

FIG. 9.2
IN-PLANE, OUT-PLANE AND TORSIONAL BENDING MOMENTS IN BENDS AND BRANCH CONNECTIONS (ASME B31.3, FIGS. 319.4.4A AND 319.4.4B)

One example occurs when attempting to perform a creep damage (remaining life) assessment. When calculating stresses from piping thermal expansion via a flexibility analysis, these calculated stresses must be multiplied by a factor of two to arrive at the actual stress condition. This should not be confused with the differences between stress range and stress amplitude, which is an additional consideration.

Another commonly misunderstood item is where the peak stresses are in an elbow, as this is counterintuitive. With in-plane bending (see Fig. 9.2 for in-plane versus out-plane directions) of an elbow, the highest stresses are not at the intrados or extrados but in the elbow side walls (crown). They are through-wall bending stresses from the ovalization of the elbow. Again, this is not significant in design, but can be so in failure analysis or fitness-for-service evaluations.

Stress intensification factors were developed for a number of common components by Markl around 1950. More recently, as a result of some findings of non-conservatism and in the development of newer products, additional fatigue tests have been performed. Stress intensification and flexibility factors are provided in Appendix D of ASME B31.1, which are for use in the absence of more directly applicable data (see para. 119.7.3). This means that a designer could use different factors if based on more applicable data. The stress intensification factors in Appendix D are based on committee judgment based on available fatigue test data. The B31 Mechanical Design Committee is performing continued evaluations of available and new data to improve these stress intensification factors. A comprehensive study (Paulin, 2012), including testing, has been completed and work has been initiated to document these factors in ASME B31J.

A new standard detailing the procedures for performing fatigue testing to develop stress intensification factors has been published. This is ASME B31J, *Standard Method for Determining Stress Intensification Factors (i-Factors) for Metallic Piping Components*.

9.5 FLEXIBILITY ANALYSIS EQUATIONS

ASME B31.1 provides the following set of equations for calculating the stress from thermal expansion loads in piping systems.

The reference displacement stress range, S_E, is calculated from the combination of stresses from bending and shear stresses caused by torsion. The stress, S_E, is calculated per Eq. (17) of ASME B31.1, as follows:

$$S_E = \frac{iM_C}{Z} \leq S_A \tag{17}$$

where:

M_C = resultant moment loading range on the cross section due to reference displacement load range
S_E = reference displacement stress range
S_A = allowable stress range for expansion stress
Z = section modulus of pipe (note that this is the section modulus of nominal matching pipe), or equivalent section modulus for reducing branch connections
i = stress intensification factor

A metric equation is also provided with inconsistent units so it includes an additional factor of 1000.

The moment is determined as the square root of the sum of the squares of the two moments and torsion, as follows:

$$M_C = \sqrt{M_x^2 + M_y^2 + M_x^2}$$

Insert 9.1   How to Increase Piping Flexibility

There are a number of means of increasing piping flexibility. The intent of this insert is to describe some general concepts in order to better enable the designer to develop increased flexibility in piping Systems

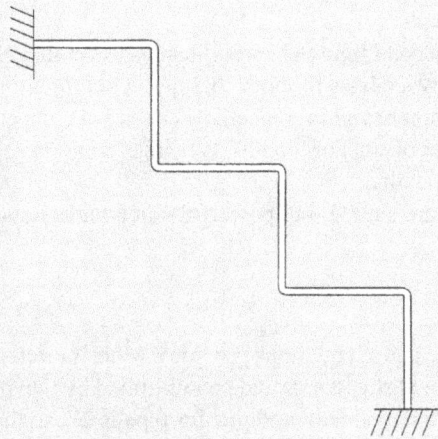

FIG. 9.3
PIPING LAYOUT 1

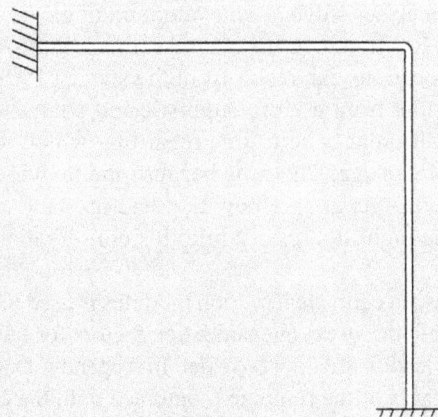

FIG. 9.4
PIPING LAYOUT 2

One alternative is to add expansion joints. However, this is usually undesirable. Although it may make the job easier for the analyst, it typically reduces the overall reliability of the piping system. Expansion joints should be the solution of last resort.

Here are a few fundamental concepts in flexibility:

(a) Flexibility of a pipe in bending is roughly proportional to the cube of the length of a straight run of pipe. Therefore, longer straight runs are much more flexible than a series of shorter runs chopped up into short lengths and separated by elbows.

(b) Flexibility can be provided by a length of piping perpendicular to the direction of thermal expansion that must be accommodated.

(c) In general, the further the center of gravity of the piping system is from a straight line between piping system end points (for a two-anchor system), the lower are the thermal expansion stresses.

These concepts are illustrated in the classic, near-sawtooth piping configuration. The question is whether the system in Fig. 9.3 or that in Fig. 9.4 is more flexible. The reader should look and decide before reading the next paragraph.

The thermal stresses in the second figure are significantly lower than those in the first figure. To resolve any doubt, run a flexibility analysis for both cases. The first figure is considered a near-sawtooth pipe configuration. The fact that it has additional elbows does not compensate for the fact that the pipe has been chopped into a number of smaller lengths. By either measure (a) or (c) above, Fig. 9.4 can be visually discerned to be more flexible.

The concepts related above are simple, but powerful when understood, in application to flexibility problems.

Note that, while not obvious, the above equations provide a Mohr's circle-type combination of normal and torsional-shear stresses. The torsional stresses should be multiplied by two prior to combination with normal stresses by root mean square, but the torsional modulus for pipe is $2Z$, so the factors of two cancel out.

There is no distinction between in-plane and out-plane moments and stress intensification factors. While the ASME B31 Code originally provided separate stress intensification factors for the two moment directions, it was revised in ASME B31.1 to use the same value, the higher of the two, in both directions, and also for torsion. This is more conservative. However, the user should be cautioned when trying to do more sophisticated analyses, such as an evaluation of existing piping systems. When trying to locate, by analysis, the critical location in terms of stress (this is done to determine locations for creep damage inspection), the wrong locations can result from the standard ASME B31.1 flexibility analysis. For example, comparison of results from a more sophisticated analysis and a standard ASME B31.1 flexibility analysis led to quite different conclusions regarding which was the critical elbow in terms of maximum stress and potential damage. This was because the higher out-plane stress intensification factors were used for in-plane moments at an elbow that had in-plane moments. That elbow was then incorrectly calculated to have the highest stress. While this conservatism does not cause a problem in design, it can in life assessments.

For reducing branch connections, an equivalent section modulus is used in the calculation of stresses in the branch connection. These tie in with the stress intensification factors for reducing branch connections. Two formulations for the stress intensification are now provided in Appendix D. One simply requires knowledge of the nominal diameter and thickness of the run pipe (combined with the equivalent section modulus) and the other requires much more detailed information, with these dimensions illustrated in Fig. D-1 of Appendix D. Each of these two formulations use a different equivalent section modulus. The equivalent section modulus used with the simpler formulation of the equation for the stress intensification factor is per the following equation:

$$Z = \pi r_b^2 t_e$$

where:

r_b = branch mean cross-sectional radius

t_e = effective branch wall thickness, lesser of t_{nh} or it_{nb} in Eq. (17); when used in the sustained and occasional load Eqs. (15) and (16), the lesser of t_{nh} and $0.75it_{nb}$, with $0.75i$ not permitted to be less than 1.0, is used

t_{nh} = nominal thickness of header or run pipe

t_{nb} = nominal thickness of branch pipe

The equivalent section modulus used with the more complex formulation of the equation for stress intensification factor is as per the following equation:

$$Z = \pi r_m'^2 T_b$$

where:

T_b = effective thickness of branch reinforcement (see Fig. D-1)

r_m' = mean radius of branch pipe (see Fig. D-1 and para. 104.8.4)

9.6 COLD SPRING

Cold spring is the deliberate introduction of a cut short in the system to offset future thermal expansion. It is used to reduce loads on equipment; however, it does not affect the strain range. As such, the cold spring does not affect the calculation of the displacement stress range, S_E, or the allowable stress range, S_A.

The range of loads on equipment from thermal expansion is also unchanged by cold spring. However, the magnitude of the load at any given operating condition can be changed. Cold spring is typically used to reduce the load in the operating condition. It does this by shifting the load to the non-operating, ambient condition.

The effectiveness of cold spring is generally considered to be questionable, but it is occasionally the only reasonable means to satisfy equipment load limits. It should not be used indiscriminately. While it may provide an easy way for an analyst to solve an equipment load problem, there are a number of considerations. Its implementation in the field is generally difficult to achieve accurately. After the plant has been operated, deliberately installed cold spring can be misunderstood as being piping misalignment and "corrected." Furthermore, when evaluating an existing piping system that is designed to include cold spring, it is highly questionable what the actual condition is.

Because of difficulties in accomplishing the desired cold spring, ASME B31.1 only permits credit for two-thirds of the cold spring that is designed into the system.

Equations are provided in para. 119.10 for calculation of the maximum reaction force or moment, including cold spring, at the hot and cold conditions. The equations are only applicable when an equal amount of cold spring is specified in all directions. Currently, any piping requiring cold spring would most likely be evaluated using computer flexibility analysis. In such an analysis, the load in the operating condition is required to be calculated using two-thirds of the design cold spring and the load in the ambient condition should (this is not specified in the Code) be calculated using the full cold spring.

9.7 ELASTIC FOLLOW-UP/STRAIN CONCENTRATION

The analysis procedures in the Code essentially assume that the strain range in the system can be determined from an elastic analysis. That is, strains are proportional to elastically calculated stresses. The stress range is limited to less than two times the yield stress, in part to achieve this. However, in some systems, strain concentration or elastic follow-up occurs. A classic reference for elastic follow-up due to creep conditions is Robinson (1960).

As an example, consider a cantilevered pipe with a portion adjacent to the fixed end constructed with a reduced diameter or thickness pipe or lower-yield strength material that has the free end laterally displaced. The elastic analysis assumes that strains will be distributed in the system in accordance with the elastic stiffnesses. However, consider what happens when the locally weak section yields. As the material yields, a greater proportion of additional strain due to displacement occurs in the local region because its effective stiffness has been reduced by yielding of the material. Thus, there is plastic strain concentration in the local region. In typical systems, this strain concentration is generally not considered to be significant. However, it can be highly significant under specific conditions, such as unbalanced systems. Paragraph 119.3 provides warnings regarding these conditions.

An example of a two-bar system under axial compression is provided to illustrate elastic follow-up, although the concern in piping is generally bending. However, the axial compression case illustrates the problem.

The problem is illustrated in Fig. 9.5. The elastic distribution of the total displacement, Δ, between bar A and bar B is as follows.

$$\Delta_A = \frac{K_B}{K_A + K_B} \Delta$$

$$\Delta_B = \frac{K_B}{K_A + K_B} \Delta$$

where

Δ_A	=	displacement absorbed by bar A
Δ_B	=	displacement absorbed by bar B
K_A	=	$A_A E_A / L_A$, elastic stiffness of bar A
K_B	=	$A_B E_B / L_B$, elastic stiffness of bar B
A_A	=	area of bar A
A_B	=	area of bar B
E_A	=	elastic modulus of bar A
E_B	=	elastic modulus of bar B
L_A	=	length of bar A
L_B	=	length of bar B
Δ	=	total axial displacement imposed on two bars

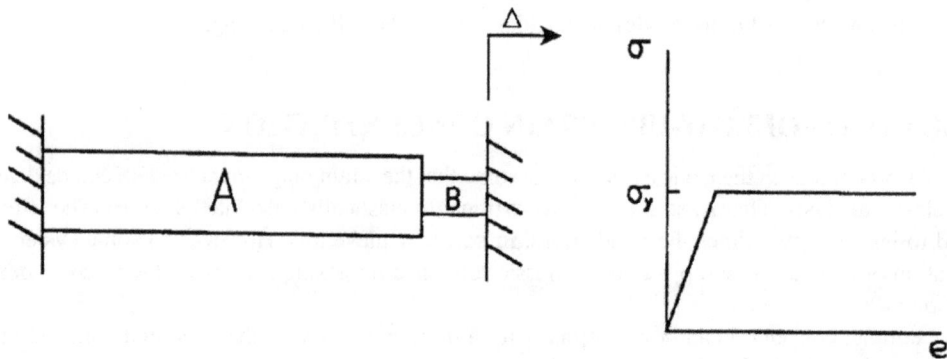

FIG. 9.5
STRAIN CONCENTRATION TWO-BAR MODEL

For elastic, perfectly plastic behavior of the material with a yield stress σ_y, the stress in bar B cannot exceed yield, so the load in bar A cannot exceed $\sigma_y A_B$. Therefore, after bar B starts to yield, the displacement in each bar is given by

$$\Delta_A^{e-p} = \frac{\sigma y A_A}{K_A}$$

$$\Delta_B^{e-p} = \Delta - \Delta_A^{e-p}$$

where

Δ_A^{e-p} = actual displacement absorbed by A, elastic plastic case
Δ_B^{e-p} = actual displacement absorbed by B, elastic plastic case

The strain concentration is the actual strain in bar B, considering elastic plastic behavior, divided by the elastically calculated strain in bar B. Since strain is displacement divided by length, the strain concentration is given as

$$\text{strain concentration} = \frac{\Delta_B^{e-p}}{\Delta_B} = \frac{\left[\Delta - (\sigma_y A_B / K_A)\right]/L_B}{\left[(K_A/(K_A + K_B))\Delta\right]/L_B}$$

This can be rearranged as

$$\text{strain concentration} = \left(1 + \frac{K_B}{K_A}\right)\left(1 - \frac{\sigma_y A_B}{\Delta K_A}\right)$$

If we consider a general region under lower stress or in a stronger condition coupled with a local region under higher stress or with weaker material (e.g., lower σ_y), then the more flexible is the general region (stiffer is the local region), the more severe is the elastic follow-up. This can be seen by considering the above equation for strain concentration. As the stiffness of bar B increases relative to bar A, the plastic strain concentration becomes more severe.

For a system subject to plastic strain concentration, the simplest solution in design is typically to limit the thermal expansion stress range to less than the yield strength of the material. This avoids plastic strain concentration by keeping the component elastic. Although the stresses due to sustained loads such as weight and pressure usually do not need to be added to those due to thermal expansion when satisfying this limit, proper consideration of this requires a very detailed understanding of the phenomenon. Thus, it is generally preferable to conservatively add the stresses due to sustained loads to the thermal expansion stresses in this type of evaluation.

Under creep conditions, elastic follow-up can have very severe effects. This is due to the implicit assumption in the code allowable stress basis that thermal expansion stresses will relax. Near-yield-level stresses cannot be sustained very long at the very high temperatures permitted in ASME B31.3 without rapid creep rupture failures. With elastic follow-up, creep strain in the local region does not result in a corresponding reduction in thermal expansion load/stress. In severe cases of elastic follow-up, rapid creep rupture failures can occur, even though the calculated stresses fall well within Code limits. The most straightforward solution to this condition is to design the system such that the level of thermal expansion stresses in the local region subject to elastic follow-up does not exceed S_h.

A well-known circumstance where elastic follow-up can occur in creep conditions is in refractory lined piping systems with local sections that are hot walled. For example, in fluid catalytic cracking units, it is common to use carbon steel pipe with an insulating refractory lining to carry high temperature fluid solids and flue gas, at temperatures to 760°C (1400°F) and higher. The metal temperature is much lower, as a result of the insulating lining. However, in some circumstances, hot wall sections such as hot wall valves are included in the system. The wall temperature for these components is generally the hot process temperature, so they will creep over time. In this circumstance, the thermal strain in the carbon steel portion of the system gradually transfers to the creeping hot walled section. Further, the carbon steel portion of the system acts as a spring, keeping the load on the hot walled section, preventing it from relaxing. The elastic analysis that was performed in design does not consider this behavior. Thus, piping systems that met the allowable thermal expansion stresses have cracked in service as a result of elastic follow-up. Becht (1988) describes an evaluation of such a system that had failed.

As an analogous situation, consider a cantilever beam. At the built in end, heat up the beam to creep temperatures. The elastic calculation predicts a distribution of strain over the entire beam. However, during service, the strain that was predicted to be in the low temperature portions of the beam will gradually shift to the creeping portion. Further, the low temperature portion of the beam will act as a spring, keeping a load on the high temperature portion, preventing it from relaxing.

9.8   EFFECT OF ELASTIC MODULUS VARIATIONS FROM TEMPERATURE

In typical flexibility analysis, the elastic modulus at ambient temperature [70°F (21°C)] is used in the analysis. Fatigue, in terms of strain range versus cycles to failure, is generally considered to be temperature independent at temperatures where creep effects are not significant. The Markl fatigue testing was done at ambient temperature, and, thus, an ambient temperature fatigue curve was developed. If one were to calculate the stress in a flexibility analysis using the elastic modulus at temperature, one would need to divide this by the elastic modulus at temperature and multiply this result by the ambient temperature elastic modulus to compare it to the Markl fatigue curve (dividing the stress by hot elastic modulus yields strain, which is then multiplied by ambient temperature elastic modulus to yield stress range at ambient temperature). This is the procedure used in fatigue analysis per the ASME BPVC, Section VIII, Division 2. Use of the ambient temperature elastic modulus in the calculation of stress directly results in the appropriate stress calculation for evaluation of fatigue.

SUPPORTS AND RESTRAINTS

10.1 OVERVIEW OF SUPPORTS

Requirements for supports and other devices to restrain the piping are provided in Chapter II of ASME B31.1, specifically in Sections 120, Loads on Pipe-Supporting Elements, and 121, Design of Pipe-Supporting Elements. The reference standard for pipe-support design and manufacture is MSS SP-58, *Pipe Hangers and Supports — Materials, Design, and Manufacture.* Fabrication requirements for pipe supports are provided in para. 130.

ASME B31.1 provides general requirements for piping supports as well as descriptions of conditions for which they must be designed. The support elements (e.g., springs and hanger rods) are within the scope of ASME B31.1, but the support structures to which they are attached are not within the scope of ASME B31.1. The supports must achieve the objectives in the design of the piping for sustained and occasional loads as well as thermal displacement.

10.2 MATERIALS AND ALLOWABLE STRESS

Pipe-support elements may be constructed from a variety of materials, including the materials listed in the allowable stress tables, materials listed in MSS SP-58, materials listed in Table 126.1, other metallic materials, and steel of unknown specification. The allowable stress can be taken from Appendix A or MSS SP-58. If it is not listed in either one, the allowable stress from ASME BPVC, Section II, Part D, Tables 1A and 1B (the allowable stress for ASME BPVC, Section I, Power Boilers) may be used. If the material is not listed in any of these tables, or if it is of unknown specification, para. 121.2 provides an allowable stress basis.

For steel of unknown specification, or of a specification not listed in Table 126.1 or MSS SP-58, the yield strength may be determined by test, and an allowable of 30% of the yield strength at room temperature used as the allowable stress up to 650°F (345°C). However, the allowable stress is not permitted to exceed 9500 psi (65.5 MPa).

The limits on shear and bearing are the same as for other components, which are 0.8 times the basic allowable stress in tension for shear and 1.6 times the basic allowable stress for bearing. Structural stability (i.e., buckling) must also be given due consideration for elements in compression. While there is no specific margin specified for structural elements in compression, a design margin of three would be consistent with other ASME B31.1 stability calculations.

Various increases and reductions in the allowable stress for supports are provided in para. 121.2, including the following:

(1) A reduction in allowable stress is applied to threaded hanger rods. The base material allowable stress is reduced 25% [para. 121.2(G)]. This is in addition to basing the stress calculation on the root area.

(2) The allowable stress in welds in support assemblies is reduced by 25%.

(3) The allowable stress may be increased by 20% for short-term overloading during operation (consistent with the occasional load criteria).

(4) The allowable stress may be increased to as high as 80% of the minimum yield strength at room temperature during hydrostatic testing [but not to exceed 16,000 psi (110.3 Mpa) for steel of unknown specifications or not listed in Table 126.1, Appendix A, or MSS SP-58].

Material that is welded to the pipe must be compatible for welding.

Although ASME B31.1 generally requires steel to be used for pipe supports (para. 123.3), cast iron and malleable iron castings are permitted for some specific types of supports. Cast iron per ASTM A48 may be used for bases, rollers, anchors, and parts of supports where the loading will be mainly compression (use in tension is not permitted because of the brittle nature of the material) [para. 121.7.2(C)]. Malleable iron per ASTM A47 may be used for pipe clamps, beam clamps, hanger flanges, clips, bases, swivel rings, and parts of pipe supports up to a maximum temperature of 450°F (230°C) [para. 121.7.2(D)]. Malleable iron is not recommended for services where impact loads (e.g., waterhammer, steam hammer, slug flow) are anticipated, because of the risk of brittle fracture.

10.3   DESIGN OF SUPPORTS

The discussion in section 120 of ASME B31.1 provides a discussion of the types of loads and conditions that must be considered. This includes loads caused by weight, pressure (e.g., with an unrestrained expansion joint), wind, and earthquake. It can also include dynamic loads such as waterhammer. Items such as hangers that are not intended to restrain, direct, or absorb piping movements from thermal expansion are required to permit free thermal expansion. Items that are intended to limit thermal expansion are required to be designed for the forces and moments resulting from it.

The pipe-support elements must be designed for all the loads that they can be subjected to, including surge, thermal expansion, and weight. In addition, while not specifically mentioned except for brackets, for supports that can slide the lateral load from friction must be considered.

The following list of objectives in the layout and design of piping and its supporting elements is provided in ASME B31.3, but is also applicable to ASME B31.1 piping systems. Many of the requirements are contained in paras. 120 and 121 and are provided below for information. The layout and design should be directed toward preventing the following:

(1) piping stresses in excess of those permitted;

(2) leakage at joint;

(3) excessive thrusts and moments on connected equipment (such as pumps and turbines);

(4) excessive stresses in the supporting (or restraining) elements;

(5) resonance with imposed or fluid-induced vibrations;

(6) excessive interference with thermal expansion and contraction in piping that is otherwise adequately flexible;

(7) unintentional disengagement of piping from its supports;

(8) excessive piping sag in piping requiring drainage slope;

(9) excessive distortion or sag of piping (e.g., thermoplastics) subject to creep under conditions of repeated thermal cycling; and

(10) excessive heat flow, exposing supporting elements to temperature extremes outside their design limits.

The spacing of supports is required to prevent excessive sag, bending, and shear stress in the piping. Suggested support spacings are provided in Table 121.5 (Table 10.1 herein); they may be used in lieu of performing calculations to determine the support spacing.

The allowable load for all threaded parts (e.g., rods and bolts) is required to be based on the root diameter [para. 121.7.2(A)]. Screw threads used for adjustment of hangers are required to be per ASME B1.1. Turnbuckles and adjusting nuts are required to have full thread engagement and all threaded adjustments are required to be provided with a means of locking (e.g., locknut) (para. 121.4).

Counterweight, hydraulic, and other constant effort supports are required to be provided with stops or other means to support the pipe in the event of failure.

Paragraphs 128.8.2, 102.4.4, and 104.3.4 require consideration of stresses in the pipe caused by pipe attachments. These include local stresses from lugs, trunnions, and supporting elements welded to the pipe or attached by other means. The stated criteria are performance based (e.g., shall not cause undue flattening of the pipe, excessive localized bending stress, or harmful thermal gradients in the pipe wall). Specific criteria are not provided. However, the design-by-analysis rules of ASME BPVC, Section VIII, Division 2, Part 5, would be considered an acceptable approach to evaluating these stresses. Approaches to evaluating stresses from loads on attachments are provided in Bednar (1986), WRC 107 (Wichman et al. 1979), and WRC 297 (Mershon et al. 1984).

TABLE 10.1
SUGGESTED PIPING SUPPORT SPACING (ASME B31.1, TABLE 121.5)

Nominal Pipe Size, NPS	Suggested Maximum Span			
	Water Service		Steam, Gas, or Air Service	
	ft	m	ft	m
1	7	2.1	9	2.7
2	10	3.0	13	4.0
3	12	3.7	15	4.6
4	14	4.3	17	5.2
6	17	5.2	21	6.4
8	19	5.8	24	7.3
12	23	7.0	30	9.1
16	27	8.2	35	10.7
20	30	9.1	39	11.9
24	32	9.8	42	12.8

GENERAL NOTES:

(a) Suggested maximum spacing between pipe supports for horizonta straight runs of standard and heavier pipe at maximum operating temperature of 750°F (400°C).

(b) Does not apply where span calculations are made or where there are concentrated loads between supports, such as flanges, valves, specialties, etc.

(c) The spacing is based on a fixed beam support with a bending stress not exceeding 2,300 psi (15.86 MPa) and insulated pipe filled with water or the equivalent weight of steel pipe for steam, service, and the pitch of the line is such gas, or air sag of 0.1 in. (2.5 mm) between supports is permissible.

Insert 10.1　Spring Design

Springs are used to provide continued support, while permitting the pipe to move vertically. There are variable-effort springs, which vary supporting force when the pipe moves up or down, and constant effort springs, which provide a nearly constant supporting force as the pipe moves up or down. Because it is normally desirable to limit the change in load in the spring caused by pipe movement (typically a 25% limit on load change), when the movement becomes large, constant-effort supports are considered.

The three primary considerations in selecting a variable-effort spring are for the spring to provide sufficient support (i.e., be strong enough), sufficient movement capability to accommodate the movement of the pipe, and sufficiently low stiffness to prevent excessive change in load as a result of the pipe movement.

Springs are typically selected from charts. An example is provided in Fig. 10.1. The first step is to find the design load and movement. The design load is used to make the initial selection of spring size. For example, if the load is 3600 lb, a size 13 or size 14 variable spring could be selected. The movement can then be considered by calculating the maximum permissible stiffness or by looking at the spring chart. The maximum stiffness can be determined from the equation

$$K_{max} = \frac{\text{force} \times \text{variation}}{\text{displacement}}$$

where

K_{max} = maximum spring stiffness
force = required supporting force from the spring
variation = allowable load variation, expressed as a fraction
displacement = vertical movement of piping

The allowable load variation is typically 25%, or a fraction of 0.25.

Continuing with the same example, for a movement of 1 in. and an allowable variation of 0.25, the maximum allowable spring stiffness would be 900 lb/in. Springs that are more flexible would be acceptable, because their load variation caused by the piping movement would be less than 25%. Looking at the bottom of the chart in Fig. 10.1, we see that a size 14, figure 268 spring would be acceptable. The stiffness of this spring is 800 lb/in., which is less than 900 lb/in. A size 13 spring is also acceptable from a stiffness standpoint.

The spring should be installed at a load such that the supporting force in the operating condition will be the desired supporting force, 3600 lb in this example. If the movement from ambient to operating condition is up, the spring supporting force will be reduced. Thus, the load in the installed, nonoperating or cold condition would need to be higher. The required load would be the desired force in the operating condition plus the spring stiffness times the displacement. In this example, with a size 14, figure 268 spring, the initial force in the cold condition would be 4400 lb [3600 (lb) + 1 (in.) × 800 (lb / in.)]. Noting that this force is within the operating range of a size 14 spring, 2800 lb to 4800 lb, we see that the spring is an acceptable choice. A size 13 spring would not be acceptable, because the required cold load, 4200 lb, exceeds the maximum design load of the spring, 3600 lb.

Another way to approach the problem is to look at the spring charts. Looking at a size 14, figure 268 spring, we can use the displacement scale at the left to determine what the cold setting of the spring

spring hangers

SIZE AND SERIES SELECTION

HOW TO USE HANGER SELECTION TABLE:

In order to choose a proper size hanger, it is necessary to know the actual load which the spring is to support and the amount and direction of the pipe line movement from the cold to the hot position.

Find the actual load of the pipe in the load table. As it is desirable to support the actual weight of the pipe when the line is hot, the actual load is the hot load.

To determine the cold load, read the spring scale, up or down, for the amount of expected movement. The chart must be read opposite from the direction of the pipe's movement. The load arrived at is the cold load.

If the cold load falls outside of the working load range of the hanger selected, relocate the actual or hot load in the adjacent column and find the cold load. When the hot and cold loads are both within the working range of a hanger, the size number of that hanger will be found at the top of the column.

Should it be impossible to select a hanger in a particular series such that both loads occur within the working range, consideration should be given to a variable spring hanger with a wider working range or a constant support hanger.

The cold load is calculated by adding (for up movement) or subtracting (for down movement) the product of spring rate times movement to or from the hot load. Cold load = hot load (±) movement x spring rate.

A key criteria in selecting the size and series of a variable spring is a factor known as variability. This is a measurement of the percentage change in supporting force between the hot and cold positions of a spring and is calculated from the formula:

$$\text{Variability} = \frac{\text{Movement x Spring Rate}}{\text{Hot Load}}$$

If an allowable variability is not specified, good practice would be to use 25% as specified by MSS-SP58.

load table in pounds: for selection of hanger size

		working range, in.			hanger size																									spring deflection, in.		
Quadruple	Triple	fig. 98	fig. 268	fig. 82	* 000	* 00	0	1	2	3	4	5	6	7	8	9	10	11	12	13	14	15	16	17	18	19	20	21	22	fig. 82	fig. 268	fig. 98
					7	19	43	63	81	105	141	189	252	336	450	600	780	1020	1350	1800	2400	3240	4500	6000	7990	10610	14100	18750	25005	0	0	0
					7	20	44	66	84	109	147	197	263	350	469	625	813	1063	1406	1875	2500	3375	4688	6250	8322	11053	14688	19531	26047			
					8	22	46	68	88	114	153	206	273	364	488	650	845	1105	1463	1950	2600	3510	4875	6500	8655	11495	15275	20313	27089			
					9	24	48	71	91	118	159	213	284	378	506	675	878	1148	1519	2025	2700	3645	5063	6750	8987	11938	15863	21094	28131			
0	0	0	0	0	10	26	50	74	95	123	165	221	294	392	525	700	910	1190	1575	2100	2800	3780	5250	7000	9320	12380	16450	21875	29173	¼	½	1
					11	28	52	76	98	127	170	228	305	406	544	725	943	1233	1631	2175	2900	3915	5438	7250	9652	12823	17038	22656	30215			
					12	30	54	79	101	131	176	236	315	420	563	750	975	1275	1688	2250	3000	4050	5625	7500	9985	13265	17625	23438	31256			
					12	31	56	81	105	136	182	244	326	434	581	775	1008	1318	1744	2325	3100	4185	5813	7750	10317	13706	18213	24219	32298			
2	1½	1	½	¼	14	34	58	84	108	140	188	252	336	448	600	800	1040	1360	1800	2400	3200	4320	6000	8000	10650	14150	18800	25000	33340	½	1	2
					14	35	59	87	111	144	194	260	347	462	619	825	1073	1403	1856	2475	3300	4455	6188	8250	10982	14592	19388	25781	34382			
					15	38	61	89	115	149	200	268	357	476	638	850	1105	1445	1913	2550	3400	4590	6375	8500	11315	15035	19975	26563	35424			
					16	40	63	92	118	153	206	276	368	490	656	875	1138	1488	1969	2625	3500	4725	6563	8750	11647	15477	20563	27344	36466			
4	3	2	1	½	17	41	65	95	122	158	212	284	378	504	675	900	1170	1530	2025	2700	3600	4860	6750	9000	11980	15920	21150	28125	37508	¾	1½	3
					18	43	67	97	125	162	217	291	389	518	694	925	1203	1573	2081	2775	3700	4995	6938	9250	12312	16362	21738	28906	38549			
					19	45	69	100	128	166	223	299	399	532	713	950	1235	1615	2138	2850	3800	5130	7125	9500	12645	16805	22325	29688	39591			
					20	47	71	102	132	171	229	307	410	546	731	975	1268	1658	2194	2925	3900	5265	7313	9750	12977	17247	22913	30469	40633			
6	4½	3	1½	¾	21	49	73	105	135	175	235	315	420	560	750	1000	1300	1700	2250	3000	4000	5400	7500	10000	13310	17690	23500	31250	41675	1	2	4
					21	50	74	108	138	179	241	323	431	574	769	1025	1333	1743	2306	3075	4100	5535	7688	10250	13642	18132	24088	32031	42717			
					22	53	76	110	142	184	247	331	441	588	788	1050	1365	1785	2363	3150	4200	5670	7875	10500	13975	18575	24675	32813	43759			
					23	55	78	113	145	188	253	339	452	602	806	1075	1398	1828	2419	3225	4300	5805	8063	10750	14307	19017	25263	33594	44801			
8	6	4	2	1	24	56	80	116	149	193	258	347	462	616	825	1100	1430	1870	2475	3300	4400	5940	8250	11000	14640	19460	25850	34375	45843	1¼	2½	5
					25	58	82	118	152	197	264	354	473	630	844	1125	1463	1913	2531	3375	4500	6075	8438	11250	14972	19902	26438	35156	46885			
					26	60	84	121	155	201	270	362	483	644	863	1150	1495	1955	2588	3450	4600	6210	8625	11500	15305	20345	27025	35938	47926			
					27	62	86	123	159	208	276	370	494	658	881	1175	1528	1998	2644	3525	4700	6345	8813	11750	15637	20787	27613	36719	48968			
10	7½	5	2½	1¼	28	64	88	126	162	210	282	378	504	672	900	1200	1560	2040	2700	3600	4800	6480	9000	12000	15970	21230	28200	37500	50010	1½	3	6
					28	66	89	129	165	214	288	386	515	686	919	1225	1593	2083	2756	3675	4900	6615	9188	12250	16302	21672	28768	38281	51052			
					29	68	91	131	169	219	284	394	525	700	938	1250	1625	2125	2813	3750	5000	6750	9375	12500	16635	22115	29375	39063	52094			
					30	70	93	134	172	223	300	402	536	714	956	1275	1658	2168	2869	3825	5100	6885	9563	12750	16967	22557	29963	39844	53136			
					31	72	95	137	176	228	306	410	546	728	975	1300	1690	2210	2925	3900	5200	7020	9750	13000	17300	23000	30550	40625	54178	1¾	3½	7
					spring rate — lb. per in.																											
				82			30	42	54	70	94	126	168	224	300	400	520	680	900	1200	1600	2160	3000	4000	5320	7080	9400	12500	16670	82		
			268		7	15	15	21	27	35	47	63	84	112	150	200	260	340	450	600	800	1080	1500	2000	2660	3540	4700	6250	8335	268		
		98					7	10	13	17	23	31	42	56	75	100	130	170	225	300	400	540	750	1000	1330	1770	2350	3125	4167	98		
	Triple						5	7	9	12	16	21	28	37	50	67	87	113	150	200	267	360	500	667	887	1180	1567	2083	2778			
Quadruple							4	5	7	9	12	16	21	28	38	50	65	85	113	150	200	270	375	500	665	885	1175	1563	2084			

*Available in fig. B268 & C-B268 only.

FIG. 10.1
VARIABLE-SPRING HANGER TABLE (COURTESY OF ANVIL INTERNATIONAL)

figs. 80-V and 81-H model R load travel table

constant supports

load in pounds for total travel in inches

hanger size no.	total travel* in inches: 1½	2	2½	3	3½	4	4½	5	5½	6	6½	7	7½	8	8½
1	144	108	86	72	62	54	48	43	39	36	33	31	29	27	
	173	130	104	87	74	65	58	52	47	43	40	37	35	33	
2	204	153	122	102	87	77	68	61	56	51	47	44	41	38	
3	233	175	140	117	100	88	78	70	64	58	54	50	47	44	
4	280	210	168	140	120	105	93	84	76	70	65	60	56	53	
5	327	245	196	163	140	123	109	98	89	82	75	70	65	61	
6	373	280	224	187	160	140	124	112	102	93	86	80	75	70	
7	451	338	270	225	193	169	150	135	123	113	104	97	90	85	
8	527	395	316	263	226	198	176	158	144	132	122	113	105	99	
9	600	450	360	300	257	225	200	180	164	150	138	129	120	113	
10	727	545	436	363	311	273	242	218	198	182	168	156	145	136	
11	851	638	510	425	365	319	284	255	232	213	196	182	170	160	
12	977	733	586	489	419	367	326	293	267	244	226	209	195	183	
13	1177	883	708	589	505	442	392	353	321	294	272	252	235	221	
14	1373	1030	824	687	589	515	458	412	375	343	317	294	275	258	
15	1573	1180	944	787	674	590	524	472	429	393	363	337	315	295	
16	1893	1420	1136	947	811	710	631	568	516	473	437	406	379	355	
17	2217	1663	1330	1109	950	832	739	665	605	554	512	475	443	416	
18	2540	1905	1524	1270	1089	953	847	762	693	635	586	544	508	476	
19		2025	1620	1350	1157	1013	900	810	736	675	623	579	540	506	448 476
20		2145	1716	1430	1226	1073	953	858	780	715	660	613	572	536	505
21		2335	1868	1557	1334	1168	1038	934	849	778	718	667	623	584	549
22		2525	2020	1683	1443	1263	1122	1010	918	842	777	721	673	631	594
23		2710	2168	1807	1549	1355	1204	1084	985	903	834	775	723	678	638
24		2910	2328	1940	1663	1455	1293	1164	1058	970	895	831	776	728	685
25		3110	2488	2073	1777	1555	1382	1244	1131	1037	957	889	829	778	732
26		3310	2648	2207	1891	1655	1471	1324	1204	1103	1018	946	883	828	779
27		3630	2904	2420	2074	1815	1613	1452	1320	1210	1117	1037	968	908	854
28		3950	3160	2633	2257	1975	1756	1580	1436	1317	1215	1129	1053	988	929
29		4270	3416	2847	2440	2135	1898	1708	1553	1423	1314	1220	1139	1068	1005
30		4535	3628	3023	2591	2268	2016	1814	1649	1512	1395	1296	1209	1134	1067
31		4795	3836	3197	2740	2398	2131	1918	1744	1598	1475	1370	1279	1199	1128
32		5060	4048	3373	2891	2530	2249	2024	1840	1687	1557	1446	1349	1265	1191
33		5295	4236	3530	3026	2648	2353	2118	1925	1765	1629	1513	1412	1324	1246
34		5625	4420	3683	3157	2763	2456	2210	2009	1842	1700	1579	1473	1381	1300
35			4696	3913	3354	2935	2609	2348	2135	1957	1806	1677	1565	1468	1381
36			4968	4140	3549	3105	2760	2484	2258	2070	1911	1774	1656	1553	1461
37			5240	4367	3743	3275	2911	2620	2382	2183	2015	1871	1747	1638	1541
38			5616	4680	4011	3510	3120	2808	2553	2340	2160	2006	1872	1755	1652
39			5988	4990	4277	3743	3327	2994	2722	2495	2303	2139	1996	1871	1761
40			6360	5300	4543	3975	3533	3180	2891	2650	2446	2271	2120	1988	1871
41			6976	5813	4983	4360	3876	3488	3171	2907	2683	2491	2325	2180	2052
42			7588	6323	5420	4743	4216	3794	3449	3162	2919	2710	2529	2371	2232
43			8200	6833	5857	5125	4556	4100	3727	3417	3154	2929	2733	2563	2412
44			8724	7270	6231	5453	4847	4362	3965	3635	3355	3116	2908	2726	2566
45			9284	7737	6631	5803	5158	4642	4220	3868	3571	3316	3095	2901	2731
46			9760	8133	6971	6100	5422	4880	4436	4067	3754	3486	3253	3050	2871
47			10376	8647	7411	6485	5764	5188	4716	4323	3991	3706	3459	3243	3052
48			10988	9157	7848	6868	6104	5494	4995	4578	4226	3924	3663	3434	3232
49			11600	9667	8286	7250	6444	5800	5273	4833	4462	4143	3867	3625	3412
50				10367	8886	7775	6911	6220	5655	5183	4785	4443	4147	3888	3659
51				11067	9486	8300	7378	6640	6036	5533	5108	4743	4427	4150	3906
52				11847	10154	8885	7898	7108	6462	5923	5468	5077	4739	4443	4181
53				12623	10820	9468	8415	7574	6886	6311	5826	5410	5049	4734	4455
54				13400	11486	10050	8933	8040	7309	6700	6185	5743	5360	5025	4730
55				14713	12611	11035	9809	8828	8026	7356	6791	6306	5885	5518	5193
56				16023	13734	12018	10682	9614	8740	8011	7396	6867	6409	6009	5655
57				17333	14857	13000	11556	10400	9455	8666	8000	7429	6933	6500	6118
58				18423	15791	13818	12282	11054	10049	9211	8503	7896	7369	6809	6503
59				19510	16723	14633	13007	11706	10642	9755	9006	8362	7804	7316	6886
60				20600	17667	15450	13733	12360	11236	10300	9508	8829	8240	7725	7271
61				21890	18763	16418	14593	13134	11940	10945	10103	9382	8756	8209	7726
62				23176	19865	17383	15451	13906	12642	11588	10697	9833	9270	8691	8180
63				24463	20968	18348	16309	14678	13344	12231	11291	10484	9785	9174	8634
"B" average inches	1⅜	1⅞	2¼	2¾	3¼	3¾	4⅛	4⅝	5⅛	5½	6	6½	6⅞	7⅜	7⅞

table continued

table continued

*NOTE: Total Travel equals Actual Travel plus 1 inch or 20% (whichever is greater), rounded up to the nearest ½ inch as applicable. Constant supports are readily available for travel and loads not listed in this table. Dimensions and lug locations may vary from those shown on the following pages.

ph-120

FIG. 10.2
CONSTANT EFFORT-SPRING HANGER TABLE (COURTESY OF ANVIL INTERNATIONAL)

constant supports

load travel table *(continued from opposite page)* *hanger size Nos. 64 to 110 on next page*

load in pounds for total travel in inches

hanger size no.	total travel* in inches: 9	9½	10	10½	11	11½	12	12½	13	13½	14	14½	15	15½	16
1															
2															
3															
4															
5															
6															
7															
8															
9															
10															
11															
12															
13															
14															
15															
16															
17															
18															
19	423 450	401 426	381 405												
20	477	452	429												
21	519	492	467												
22	561	532	505												
23	602	571	542												
24	647	613	582												
25	691	655	622												
26	736	697	662												
27	807	764	726												
28	878	832	790												
29	949	899	854												
30	1008	956	907												
31	1066	1009	959												
32	1124	1065	1012												
33	1177	1115	1059												
34	1228	1163	1105												
35	1304	1236	1174	1053 1118	1005 1067	962 1021	922 978	885 939	851 903	819 870	790 838				
36	1380	1307	1242	1183	1129	1080	1035	994	955	920	887				
37	1456	1379	1310	1248	1191	1139	1092	1048	1008	970	936				
38	1560	1478	1404	1337	1276	1221	1170	1123	1080	1040	1003				
39	1663	1576	1497	1426	1361	1302	1247	1198	1151	1109	1069				
40	1767	1674	1590	1514	1445	1383	1325	1272	1223	1178	1136				
41	1938	1836	1744	1661	1585	1516	1453	1395	1341	1292	1246				
42	2108	1997	1897	1807	1724	1649	1581	1518	1459	1405	1355				
43	2278	2158	2050	1952	1863	1782	1708	1640	1577	1518	1464				
44	2423	2296	2181	2077	1983	1896	1817	1745	1678	1615	1558				
45	2579	2443	2321	2210	2110	2018	1934	1857	1785	1719	1658				
46	2711	2568	2440	2324	2218	2122	2033	1952	1877	1807	1743				
47	2882	2730	2594	2470	2358	2255	2162	2075	1995	1921	1853				
48	3052	2891	2747	2616	2497	2389	2289	2198	2113	2035	1962				
49	3222	3053	2900	2762	2636	2522	2417	2320	2231	2148	2071				
50	3456	3274	3110	2962	2827	2704	2592	2488	2392	2304	2221	2001 2145	1934 2073	1871 2006	1813 1944
51	3689	3495	3320	3162	3018	2887	2767	2656	2554	2459	2371	2289	2213	2142	2075
52	3949	3741	3554	3384	3231	3090	2962	2843	2734	2632	2538	2451	2369	2293	2221
53	4208	3986	3787	3606	3442	3293	3156	3030	2913	2805	2705	2612	2524	2443	2367
54	4467	4231	4020	3828	3654	3495	3350	3216	3092	2978	2871	2772	2680	2593	2513
55	4904	4646	4414	4203	4012	3838	3678	3531	3395	3269	3152	3044	2942	2847	2759
56	5341	5060	4807	4518	4370	4180	4006	3846	3698	3561	3433	3315	3204	3101	3004
57	5778	5474	5200	4952	4727	4521	4333	4160	4000	3852	3714	3586	3466	3355	3250
58	6141	5818	5527	5263	5024	4806	4606	4422	4251	4094	3947	3811	3684	3565	3454
59	6503	6161	5853	5574	5320	6089	4877	4682	4502	4335	4180	4036	3902	3776	3658
60	6867	6505	6180	5885	5618	5374	5150	4944	4754	4578	4414	4262	4120	3987	3863
61	7297	6912	6567	6254	5969	5710	5472	5254	5051	4864	4690	4529	4378	4236	4104
62	7725	7319	6953	6621	6320	6046	5794	5562	5348	5150	4965	4795	4635	4485	4346
63	8154	7725	7339	6989	6671	6361	6116	5871	5645	5436	5242	5061	4892	4734	4587
"B" average inches	8¼	8¾	9⅛	9⅝	10⅛	10⅝	11	11½	12	12⅜	12⅞	13⅜	13⅞	14¼	14¾

*NOTE: Total Travel equals Actual Travel plus 1 inch or 20% (whichever is greater), rounded up to the nearest ½ inch as applicable.

Constant supports are readily available for travel and loads not listed in this table. Dimensions and lug locations may vary from those shown on the following pages.

FIG. 10.2 CONTINUED

fig. 80-V and 81-H model R

constant supports

See pages ph-120, 121, for sizes 1 to 63

load travel table

load in pounds for total travel in inches

hanger size no.	total travel* in inches																
	4	4½	5	5½	6	6½	7	7½	8	8½	9	9½	10	10½	11	11½	12
64	19225	17089	15380	13982	12818	11831	10986	10253	9613	9047	8544	8094	7690	7323	6990	6686	6408
65	20100	17866	16080	14618	13400	12370	11486	10720	10050	9459	8933	8463	8040	7657	7308	6991	6700
66	22068	19615	17654	16049	14711	13580	12610	11769	11034	10385	9808	9291	8827	8406	8024	7675	7356
67	24033	21362	19226	17478	16021	14790	13733	12817	12016	11310	10681	10119	9613	9154	8738	8359	8011
68	26000	23111	20800	18909	17333	16000	14857	13866	13000	12236	11555	10947	10400	9904	9454	9043	8666
69	27635	24564	22108	20098	18423	17007	15792	14738	13818	13005	12282	11635	11054	10527	10048	9611	9211
70	29268	26015	23414	21266	19511	18011	16725	15609	14634	13773	13008	12323	11707	11149	10642	10179	9755
71	30900	27466	24720	22473	20599	19016	17657	16480	15450	14542	13733	13010	12360	11770	11235	10747	10300
72	32835	29186	26268	23880	21889	20207	18763	17512	16418	15452	14593	13825	13134	12508	11939	11420	10945
73	34768	30904	27814	25286	23177	21396	19868	18542	17384	16362	15452	14639	13907	13244	12641	12092	11589
74	36700	32622	29360	26691	24466	22585	20972	19573	18350	17271	16311	15452	14680	13980	13344	12764	12233
75	38800	34489	31040	28218	25866	23878	22172	20693	19400	18259	17244	16336	15520	14780	14108	13495	12933
76	40900	36355	32720	29746	27266	25170	23372	21813	20450	19248	18178	17221	16360	15580	14871	14225	13633
77	43000	38222	34400	31273	28666	26462	24572	22933	21500	20236	19111	18105	17200	16380	15635	14955	14333
78	45335	40297	36268	32971	30222	27899	25906	24178	22668	21335	20149	19088	18134	17269	16484	15768	15111
79	47668	42371	38134	34668	31779	29335	27239	25422	23834	22432	21185	20070	19067	18158	17332	16579	15889
80	50000	44444	40000	36364	33332	30770	28572	26666	25000	23530	22222	21052	20000	19046	18180	17390	16666
81	52500	46666	42000	38182	35000	32309	30000	27999	26250	24707	23333	22105	21000	19998	19089	18260	17500
82	55000	48888	44000	40000	36665	33847	31429	29333	27500	25883	24444	23157	22000	20951	20000	19129	18333
83	57500	51111	46000	41819	38332	35386	32858	30666	28750	27060	25555	24210	23000	21903	20907	20000	19166
84			49200	44728	40998	37847	35144	32799	30750	28942	27333	25894	24600	23427	22361	21390	20500
85			52400	47637	43665	40309	37429	34932	32750	30824	29111	27578	26200	24950	23816	22781	21832
86			55400	50364	46165	42616	39572	36932	34625	32589	30777	29157	27700	26379	25179	24085	23082
87			58400	53091	48665	44924	41715	38932	36500	34354	32444	30736	29200	27807	26543	25389	24332
88			61400	55819	51165	47232	43858	40932	38375	36119	34111	32315	30700	29236	27906	26694	25582
89			66000	60000	54998	50771	47144	43999	41250	38825	36666	34736	33000	31426	29997	28694	27500
90					61331	56617	52572	49065	46000	43295	40888	38736	36800	35045	33451	31998	30665
91					67164	62002	57573	53732	50375	47413	44777	42420	40300	38378	36633	35041	33582
92					73500	67848	63001	58799	55125	51884	49000	46420	44100	41996	40087	38345	36749
93					80830	74617	69287	64665	60625	57060	53888	51051	48500	46187	44087	42171	40415
94					87500	81540	75716	70665	66250	62355	58888	55788	53000	50472	48177	46084	44165
95							78930	73665	69063	65002	61388	58156	55250	52615	50222	48040	46040
96							82145	76665	71875	67649	63888	60525	57500	54757	52268	50000	47915
97							85360	79665	74688	70296	66388	62893	59750	56900	54313	51953	49790
98							87500	82665	77500	72943	68888	65261	62000	59043	56358	53909	51665
99								85998	80625	75884	71666	67893	64500	61423	58631	56083	53748
100								87500	83750	78826	74444	70524	67000	63804	60903	58257	55831
101									86875	81767	77221	73156	69500	66185	63176	60430	57914
102									87500	84708	80000	75787	72000	68566	65448	62604	60000
103										87500	83610	79210	75250	71661	68402	65430	62706
104											87221	82629	78500	74756	71357	68256	65414
105											87500	86050	81750	77851	74311	71082	68122
106												87500	85000	80946	77265	73908	70831
107													87500	84469	80628	77125	73914
108														87500	83992	80342	77000
109															87446	83646	80163
110															87500	86950	83330
"B" dim sizes 64-83	3⅝	4⅛	4⅝	5⅛	5½	6	6½	6⅞	7⅜	7⅞	8¼	8¾	9¼	9⅝	10⅛	10⅝	11
"B" dim sizes 84-110	....	....	4 3/16	4 9/16	5	5⅜	5 13/16	6¼	6¾	7 1/16	7½	7⅞	8 5/16	8¾	9⅛	9 9/16	10

table continued

*NOTE: Total Travel equals Actual Travel plus 1 inch or 20% (whichever is greater), rounded up to the nearest ½ inch as applicable.

Constant supports are readily available for travel and loads not listed in this table. Dimensions and lug locations may vary from those shown on the following pages.

FIG. 10.2 CONTINUED

constant supports

load travel table *(continued from opposite page)*

load in pounds for total travel in inches

hanger size no.	total travel* in inches															
	12½	13	13½	14	14½	15	15½	16	16½	17	17½	18	18½	19	19½	20
64	6152	5915	5696	5492	5303	5126	4961	4806								
65	6432	6184	5955	5742	5544	5359	5187	5025								
66	7062	6790	6538	6304	6087	5884	5694	5517								
67	7690	7394	7120	6966	6629	6408	6201	6008								
68	8320	8000	7703	7428	7172	6933	6709	6500								
69	8843	8503	8188	7895	7623	7369	7131	6909								
70	9366	9005	8671	8361	8073	7804	7552	7317								
71	9888	9507	9155	8828	8523	8239	7973	7725								
72	10507	10103	9728	9380	9057	8755	8473	8209								
73	11126	10697	10301	9932	9590	9270	8971	8692								
74	11744	11292	10873	10484	10123	9786	9470	9175								
75	12416	11938	11496	11084	10703	10346	10012	9700								
76	13088	12584	12118	11684	11282	10906	10554	10225								
77	13760	13230	12740	12284	11861	11466	11096	10750								
78	14507	13949	13432	12951	12505	12088	11698	11334								
79	15254	14666	14123	13618	13149	12710	12300	11917								
80	16000	15384	14814	14284	13792	13332	12902	12500								
81	16800	16153	15555	14998	14482	14000	13547	13125								
82	17600	16922	16295	15712	15171	14665	14192	13750								
83	18400	17692	17036	16427	15861	15332	14837	14375								
84	19680	18922	18221	17569	16964	16398	15869	15375								
85	20960	20153	19406	18712	18068	17465	16902	16375								
86	22160	21307	20517	19783	19102	18465	17869	17313								
87	23360	22461	21628	20855	20136	19465	18837	18250								
88	24560	23614	22739	21926	21171	20465	19805	19188								
89	26400	25384	24443	23569	22757	21998	21288	20625								
90	29440	28307	27258	26283	25377	24531	23740	23000								
91	32240	31000	29850	28782	27791	26864	25998	25188								
92	35280	33922	32665	31496	30411	29397	28449	27563								
93	38800	37306	35924	34639	33446	32330	31287	30313								
94	42400	40768	39257	37853	36549	35330	34190	33125								
95	44200	42498	40924	39480	38100	36830	35642	34531	32119 33482	31175 32498	30285 31570	29442 30691	28647 29853	27894 29078	27179 28332	26500 27625
96	46000	44230	42590	41067	39652	38330	37093	35938	34845	33822	32856	31941	31080	30262	29486	28750
97	47800	45960	44257	42673	41204	39829	39545	37344	36209	35145	34141	33191	32295	31446	30640	29875
98	49600	47690	45923	44280	42755	41329	40000	38750	37572	36468	35427	34441	33511	32631	31794	31000
99	51600	49613	47775	46066	44479	42996	41609	40313	39067	37939	36855	35830	34862	33946	33076	32250
100	53600	51536	49627	47851	46203	44662	43221	41875	40602	39409	38284	37219	36214	35262	34358	33500
101	55600	53459	51479	49637	47927	46329	44834	43438	42117	40880	39712	38607	37565	36578	35640	34750
102	57600	56382	53330	51422	49651	47995	46447	45000	43632	42350	41141	39996	38916	37894	36922	36000
103	60200	57882	55738	53744	51892	50162	48544	47031	45602	44262	42998	41801	40673	39604	38586	37625
104	62800	60382	58145	56065	54134	52328	50640	49063	47571	46174	44855	43607	42429	41315	40255	39250
105	65400	62882	60552	58386	56375	54495	52737	51094	49541	48065	46712	45412	44186	43025	41921	40875
106	68000	65382	62960	60707	58616	56661	54834	53125	51510	50000	48569	47218	45943	44736	43588	42500
107	70960	68228	65700	63350	61168	59127	57220	55438	53752	52173	50683	49273	47942	46683	45485	44350
108	73920	71074	68441	65992	63719	61594	59607	57750	55994	54350	52797	51328	49942	48630	47383	46200
109	76060	74000	71255	68706	66340	64127	62059	60125	58297	56565	54959	53439	52000	50630	49331	48100
110	80000	76920	74070	71420	68960	66660	64510	62500	60600	58820	57140	55550	54050	52630	51280	50000
"B" dim sizes 64-83	11½	12	12⅜	12⅞	13⅜	13⅞	14¼	14¾	....	....	....	....	....	....	....	....
"B" dim sizes 84-110	10⅜	10 13/16	11 3/16	11⅝	12 1/16	12½	12⅞	13 5/16	13 11/16	14⅛	14 9/16	14 15/16	15⅜	15¾	16 3/16	16⅝

Constant supports are readily available for travel and loads not listed in this table. Dimensions and lug locations may vary from those shown on the following pages.

FIG. 10.2 CONTINUED

will be and if it will be in the appropriate range. Noting that the spring will be moving up (lesser displacement) as it goes from cold to hot, and the movement is 1 in., one can look at the displacement scale and determine that 1 in. down from an operating load setting of 3600 lb (1 in. on the load scale), the cold load would be 4400 lb (2 in. on the load scale), and it would remain in scale. The next step would be to check load variation, the difference between hot and cold settings divided by the hot setting, and make sure it is within acceptable limits (typically 25%).

Most springs are figure 268. A figure 82 spring is twice as stiff (essentially a half-spring). A figure 98 spring is twice as flexible as a figure 268 spring (essentially two springs stacked). A triple is three times as flexible and a quadruple is four times as flexible. Figure 82 springs are typically used when there are space limitations and the stiffer spring is an acceptable choice. The more flexible springs are used to accommodate greater movement.

When the load variation for variable-support springs is too great, a constant-effort spring can be used. Figure 10.2 provides a typical selection table for constant-effort springs. A spring size is selected based on the desired movement and supporting force. Note that the chart requires a margin be placed on the travel. The greater of 1 in. or 20% is added to the calculated travel, and then rounded up to the nearest ½ in.

If, for example, a constant support with a supporting force of 2100 lb and a calculated movement of 6.2 in. was required, a size 40 spring would be picked from the table in Fig. 10.2. This comes from looking in the column under total travel of 7.5 in. (6.2 in. plus 1 in., rounded up to the nearest ½ in.) for a spring size with a load greater than or equal to 2100 lb.

Insert 10.2   Stress Classification

The design-by-analysis rules of ASME BVPC Section VIII, Division 2, Part 5 are used to assess the results of detailed stress analysis of nozzles and attachments to pressure vessels. These rules include consideration of primary, secondary, and peak stresses. Primary stresses are load-controlled stresses and have essentially the same allowable limit for membrane stress as the basic allowable stress provided in Appendix A. A primary stress is one that is necessary for satisfying the laws of equilibrium of external and internal forces and moments.

Primary stresses include primary membrane, primary bending, and local primary stress. Primary membrane stress is limited to the basic allowable stress; it is the average of the stress through the wall thickness. Primary bending stress is rarely a consideration in these evaluations. Although through-wall bending is often present, it is a secondary stress. A typical example of a primary bending stress in a pressure vessel is in a flat head.

Local primary membrane stress is a primary membrane stress that is in a limited region. To be classified local, the meridional (longitudinal in a pipe) extent over which the stress intensity exceeds 1.1 times the basic allowable stress cannot exceed $\sqrt{Rt}$, where R is the midsurface radius of curvature measured normal to the surface from the axis of rotation and t is the minimum thickness in the region considered.

The limit for local primary membrane stress is 1.5 times the basic allowable stress. This higher stress can be used when the extent of the local stress region can be determined. This typically requires a detailed (e.g., finite-element) analysis.

Secondary stresses are deformation-controlled. The throughwall bending stresses in the vessel wall, calculated using WRC 107 (Wichman et al, 1979), WRC 297 (Mershon et al, 1984), Bednar (1986), finite-element analysis, or similar methods, are considered to be secondary stresses. Local yielding is considered to accommodate the imposed deformation, rather than result in gross failure. The combination of primary plus secondary stress is limited to three times the basic allowable stress (where the allowable stress is the average of the allowable stress at the temperature extremes of the condition being evaluated). At temperatures below the creep regime, this essentially limits the stress range to twice yield, or a shakedown limit (discussed in Section 8.1).

Peak stresses are highly localized and of concern in a fatigue analysis and limited in the ASME BVPC Section VIII, Division 2, by the Appendix 5 fatigue curves. However, these are not calculated in shell analysis, WRC 107 (Wichman et al, 1979), WRC 297 (Mershon et al, 1984), or Bednar (1986), although a stress concentration factor at the welds can be included. However, peak stresses are not generally considered in these evaluations and fatigue life calculations are generally not performed.

10.4 SPRING AND HANGER SUPPORTS

Specific requirements are provided in para. 121 for springs and hangers. To a great extent, these are per MSS SP-58. Spring selection is described in Insert 10.1, the B31.1 requirements are described below.

Load-carrying capacities for threaded rods are provided in Table 121.7.2(A). Note that this table is based on an allowable stress of 12,000 psi (82.7 MPa) reduced by 25%. This is more conservative than the allowable stress basis in ASME B31.1. For example, the allowable stress for A36 material in Appendix A is 14,500 psi (100 MPa). Higher allowable loads can be determined by calculation.

The minimum rod size for NPS 2 and smaller pipe is 3/8 in. (9.5 mm) diameter. For larger piping, the minimum rod size is 1/2 in. (12.5 mm).

Hangers for piping NPS 2 and larger are required to include means of adjustment, such as turnbuckles, that permit adjustment after erection while supporting the load. Additional specific requirements are provided in para. 121.4. Both variable and constant-effort spring supports are required to be designed and fabricated in accordance with MSS SP-58. The supporting force in variable spring hangers changes as the pipe moves up and/or down (e.g., from thermal expansion). It is recommended that the designer limit the load variation in the selected spring to 25% for the total travel that is expected from thermal movement. This is accomplished by selecting a more flexible spring support or by changing to a constant-effort support.

The limitation on load variation for constant-effort supports, per MSS SP-58 and para. 121.7.4(a), is 6%. This is a requirement for the hanger manufacturer to satisfy. They must design the constant-effort spring such that the load variation does not exceed 6% for the design travel range.

It is recommended that variable spring supports be provided with a means to indicate at all times the compression of the spring with respect to the approximate hot and cold positions of the piping systems (para. 121.7.3). An exception is provided for piping with temperatures not exceeding 250°F (120°C); however, it is still good practice to provide such an indicator. Additionally, it is also good practice to mark the design hot and cold positions of the spring on the load scale, so that the operation of the spring in the system can be easily visually checked. ASME B31.1 does not provide the same recommendation for constant-effort support springs, but it is also good practice to provide a travel indicator and to mark the design hot and cold positions. Note that the travel indicator on a constant-effort spring reads in portions of total travel and is not a direct measurement of travel (i.e., it is not a scale in inches).

10.5 FABRICATION OF SUPPORTS

ASME B31.1 refers to MSS SP-58 and para. 130 for fabrication of supports. It requires that welding, welding operators, and welding procedure specifications for standard pipe supports per MSS SP-58 be qualified in accordance with the requirements of ASME BPVC, Section IX. Other supports are required to follow the

ASME B31.1 requirements for welded joints, except that postweld heat treatment is only required if specified by the weld procedure, and only visual examination is required.

Welded attachments to the pressure boundary are required to comply with the ASME B31.1 requirements of Chapters V and VI, which cover fabrication (including welding, preheat, and postweld heat treatment), examination, and testing.

LOAD LIMITS FOR ATTACHED EQUIPMENT

11.1 OVERVIEW OF EQUIPMENT LOAD LIMITS

It is an ASME B31.1 requirement that piping systems be designed to comply with the load limits of equipment to which they are connected. This equipment can be machinery, pressure vessels, heat exchangers, as well as a myriad of other possibilities. The designer needs to be aware of the applicable load limits. Some are described herein as examples; however, this is not a complete list. The remainder of this chapter does not provide specific Code requirements; it provides general information on load limits for equipment.

11.2 PRESSURE VESSELS

There are several approaches to dealing with load limits for pressure vessels.

1. Specify the loads the nozzles are required to be able to accept in the vessel specifications and design the piping to comply with those limits.
2. Calculate loads on nozzles, and when they appear high by some rule of thumb, send them to the vessel manufacturer to confirm the nozzles are acceptable for the imposed loads.
3. Perform nozzle stress calculations to confirm the nozzles are acceptable for the loads, again typically when they exceed some rule-of-thumb value.

No rule of thumb works for every vessel, and a variety may well be in use. One of these is to be concerned when the total stress in the piping, including sustained and thermal loads at the vessel nozzle, exceeds S_h. With a little planning ahead, target allowable loads can be placed in the vessel specifications, which simplifies the entire process.

When a detailed stress calculation of the nozzle, with the imposed loads, is required, it is typical to turn to a design-by-analysis approach, per ASME BPVC, Section VIII, Division 2, Part 5. This is done whether the vessel is Division 1 or Division 2. Division 1 does not contain any rules for this evaluation. Stresses are calculated by hand calculations or the finite-element method.

A variety of hand calculation methods is available, based on work by Bijlaard (1954 to 1959). WRC 107 (Wichman et al, 1979) and WRC 297 (Mershon et al, 1984) contain procedures for analysis of stresses in nozzles due to external loads in cylinders and spheres. These are somewhat tedious to apply; the procedure is available in commercial programs, including some piping flexibility analysis programs. Bednar (1986) contains a more simplified approach, which is quicker to apply and is based on the same original theory. When calculating the stresses due to external loads one needs to keep in mind that these must be added to the membrane stresses in the vessel wall due to internal pressure. This is commonly forgotten. The pressure thrust load (pressure times inside area of the nozzle, in the absence of expansion joints) must be included in the nozzle load. As some flexibility analysis programs do not include this in reaction loads, it is sometimes missed.

Finite-element analysis of nozzles has become more common because a variety of finite-element programs have features that make it relatively easy to automatically create such models.

Once the stresses are calculated, they are typically compared to the primary, local primary, and secondary stress limits described in Part 5 of ASME BPVC, Section VIII, Division 2. These limits are described in Insert 10.2.

11.3   OTHER EQUIPMENT LOAD LIMITS

The load limits for other equipment must come from the equipment manufacturer or, if the equipment is constructed in accordance with a standard that provides acceptable load limits, as per those load limits. For equipment in accordance with those standards, the manufacturer is essentially required by the standard to design the equipment for certain minimum loads.

An alternative would be for the purchaser to specify minimum design loads in the purchase specification.

11.4   MEANS OF REDUCING LOADS ON EQUIPMENT

When faced with a situation where excessive loads are calculated to be exerted on equipment, typically the equipment has already been purchased, since the piping stress work is often done after equipment procurement. In such a case, there are a number of alternatives. Some of these are listed as follows in general order of preference:

1. If the loads are due to weight, add appropriate support.
2. Rerun the analysis using the hot modulus of elasticity to calculate the equipment loads.
3. If the loads are due to friction, consider using low-friction slide plates.
4. Add restraints to direct thermal expansion and resulting loads away from load-sensitive equipment.
5. Add flexibility to the piping system.
6. Consider using cold spring or expansion joints.

12

REQUIREMENTS FOR MATERIALS

12.1 OVERVIEW OF MATERIAL REQUIREMENTS

Requirements for materials are provided in Chapter III of ASME B31.1. Materials can be used for ASME B31.1 piping when at least one of six conditions exist. These are the following:

(1) when the material is listed in the allowable stress tables in Appendix A;

(2) when the material is listed in a Code Case that permits it;

(3) when the material is listed in a standard that is referenced in Table 126.1 and is not otherwise prohibited by ASME B31.1 (in this case, the material may only be used within the scope of and in the product form covered by the subject standard);

(4) the component is a flared, flareless, or compression-type tubing fitting that complies with the requirements of para. 115 (in this case, materials other than those listed in Appendix A {or Code Cases} may be used);

(5) the material is nonmetallic, in which case other conditions apply, as discussed in Chapter 16; or

(6) when the requirements for use of unlisted materials, described below, are satisfied.

In addition, for a material to be used for boiler external piping, it must be listed in the ASME BPVC, Section II, except under certain conditions. In general, it must be an SA, SB, or SFA specification listed in ASME BPVC, Section II. The exception is if the ASTM specification is equivalent to or more stringent than the ASME specification. The material manufacturer must certify, with evidence acceptable to the Authorized Inspector, that the requirements of the ASME specification are met by the material.

The allowable stress tables list the materials in terms of A and B designations (e.g., ASTM A-106). These are ASTM designations for the material. ASME BPVC, Section II replaces the initial letters — A and B — with SA and SB, respectively, when it lists the material standard. The SA designation identifies the material as ASME B&PV Code material in accordance with specifications listed in ASME BPVC, Section II, Part A, Ferrous Material Specifications. The SB designation identifies the material as ASME B&PV Code material in accordance with specifications listed in ASME BPVC, Section II, Part B, Nonferrous Material Specifications. The SFA designation identifies the material as ASME B&PV Code material in accordance with the specifications listed in ASME BPVC, Section II, Part C, Specifications for Welding Rods, Electrodes, and Filler Metals.

The SA and SB material specifications typically coincide with ASTM Specifications. For ASME B31.1, SA and SB material may be used interchangeably with A and B material (e.g., SA-106 with A106) for non-boiler external piping. As stated in the prior paragraph, SA and SB materials are required for boiler external piping, unless the ASTM specifications are equivalent or more stringent.

As of the 2001 edition of the Code, unlisted materials are permitted for use in the construction of non-boiler external piping. Essentially, the requirements for use of unlisted materials, described in para. 123.1.2, state that the material must be in accordance with a published specification covering chemistry, physical and mechanical properties, method and process of manufacture, heat treatment, and quality control; the allowable stresses are determined in accordance with Code rules; the material is qualified for the service temperature; the owner's acceptance is documented by the designer; and all other requirements of the Code are satisfied.

Selection of materials that are suitable to avoid deterioration in service is the responsibility of the designer and is not covered by the Code. Guidelines are provided in non-mandatory Appendix IV, Corrosion Control for ASME B31.1 Power Piping Systems.

12.2   TEMPERATURE LIMITS

ASME B31.1 does not permit application of materials above the maximum temperature for which allowable stresses are provided in Appendix A (or, where applicable, a listed standard for a component), except as provided in para. 122.6.2(G). This paragraph permits use of carbon steel above 800°F (427°C) for the discharge lines of pressure-relief devices under certain conditions.

Appendix A provides an allowable stress under a column −20°F to 100°F. This may be interpreted as limiting the minimum temperature to −20°F (−29°C), although the Code does not specifically prohibit applications below −20°F (−29°C). Power piping is generally considered to operate hot; cold applications simply are not addressed by the existing Code rules. In fact, risks of brittle fracture of carbon steel at temperatures above −20°F (−29°C) have led to changes in other Codes, such as ASME BPVC, Section VIII and ASME B31.3. Additional considerations, beyond what is provided in ASME B31.1, are appropriate for cold applications; ASME B31T may be referred to for guidance for such applications. Paragraph 124.1.2 was added in the 2007 edition; it requires consideration of brittle fracture at low service temperatures. Including ASME B31T, *Standard Toughness Requirements for Piping*, as a requirement by reference is being considered for future editions of ASME B31.1. This standard requires impact testing to qualify the material under various conditions.

12.3   MATERIAL LIMITATIONS

Specific limitations and requirements for materials are provided in para. 124. This includes steel, non-ferrous metals, non-metallic pipe, and iron. For iron, useful references to other paragraphs that limit the use of cast gray iron, malleable iron, and ductile iron are provided. In addition to specific limitations, the following general pressure–temperature limitations apply to iron materials:

(1) Cast Gray Iron (ASME SA-278 and ASTM A-278), maximum pressure of 250 psig [1,725 kPa (gage)], and maximum temperature of 450°F (232°C).

(2) Malleable Iron, maximum pressure of 350 psig [2,415 kPa (gage)], and maximum temperature of 450°F (232°C).

(3) Ductile (Nodular) Iron, maximum pressure of 350 psig [2,415 kPa (gage)], and maximum temperature of 450°F (232°C).

12.4 HOW TO USE THE ALLOWABLE STRESS TABLES IN APPENDIX A

The allowable stresses for use with the metallic materials are listed in Appendix A:

(1) Table A-1 for carbon steel;

(2) Table A-2 for low and intermediate alloy steel;

(3) Table A-3 for stainless steel;

(4) Table A-4 for nickel and high-nickel alloys;

(5) Table A-5 for cast iron;

(6) Table A-6 for copper and copper alloys;

(7) Table A-7 for aluminum and aluminum alloys; and

(8) Table A-9 for titanium and titanium alloys.

Table A-8 extends the allowable stress tables for stainless steel in Table A-3 to over 1,200°F (650°C).

Within each material group, the materials are grouped by product form (e.g., seamless pipe and tube, electric resistance welded pipe and tube), then by specification number within a product form, and then by alloy within a group of the same specification numbers.

The allowable stresses in the tables include weld joint efficiency and casting quality factor. As such, they are directly applicable to pressure design of straight pipe. For evaluation of longitudinal stresses from sustained loads, and for evaluation of thermal expansion stresses, the weld joint efficiency should not be included. For those evaluations, the allowable stress in the tables should be divided by the weld joint efficiency, E, or casting quality factor, F, listed in the table. The casting quality factor is assumed to be 0.80 in Appendix A; for cast steel components, it can be improved to as high as 1.0. See Section 3.3.

In addition to the designation of the material by alloy content and specification number, additional information is provided. This includes the P-number as well as the specified minimum tensile and yield strengths. The specified minimum strengths are from the Material Specifications.

The P-numbers are groupings of alloys for weld procedure qualification purposes; they group materials based on composition, weldability, and mechanical properties. These are assigned in the ASME BPVC, Section IX. The ASME B31.1 Section Committee has assigned P-numbers to some materials that are not covered in ASME BPVC, Section IX.

As of addenda a to the 2004 edition, when creep or creep rupture properties govern in setting the allowable stress, the allowable stress value is printed in italics.

Notes are typically provided. Prior to using a material, the notes should be reviewed. For example, Note (10) for Type 304 stainless steel in Table A-3 indicates that the allowable stresses that are listed for temperatures above 1000°F are only valid if the material has a carbon content of 0.04% or higher.

The General Notes provide specific guidance on features of the allowable stress tables. These notes have generally been covered in the prior discussions in this section.

The designer is permitted to linearly interpolate between temperatures for which allowable stresses are listed.

FABRICATION, ASSEMBLY, AND ERECTION

13.1 OVERVIEW OF CHAPTER V

Chapter V covers the ASME B31.1 rules for fabrication, assembly, and erection. It includes requirements for welding, details for specific types of welded joints, pre-heat, heat treatment, bending and forming, brazing and soldering, assembly, and erection.

13.2 GENERAL WELDING REQUIREMENTS

A variety of welding processes are used with piping. These include shielded-metal arc weld (SMAW), gas-tungsten arc weld (GTAW or TIG), gas-metal arc weld (GMAW or MIG), submerged arc weld (SAW), and flux-cored arc weld (FCAW). These are described in Insert 13.1.

Welding involves a welding procedure specification (WPS) that has been qualified by a procedure qualification test, which is documented in a procedure qualification record (PQR). Welders are required to pass a performance qualification test to be qualified to perform Code welding. This is documented in the welder qualification records (WQR).

Requirements for filler material, backing rings, consumable inserts, and end preparation are provided in paras. 127.2 and 127.3.

Filler metals, including consumable insert materials, are required to comply with the requirements of ASME BPVC, Section II, Part C. However, filler metals not incorporated in ASME BPVC, Section II, Part C may be used if they satisfy a procedure qualification test per ASME BPVC, Section IX. Unless otherwise specified by the designer, the completed weld is required to comply with the following:

(1) Nominal tensile strength of weld metal to equal or exceed that of the base metals being joined. If the base metals have different strengths, the weld tensile strength is required to exceed the lower part of the base metal tensile strengths.

(2) Nominal chemical analysis of the weld metal is required to be similar to the nominal chemical analysis of the base metal. If base metals being joined have different compositions, the weld chemistry is to be similar to either base metals or an intermediate composition.

(3) As an exception to (2), austenitic steels may be used to join ferritic steels; the weld metal is required to have an austenitic composition.

(4) For non-ferrous materials, the weld metal is required to be that as recommended by the material manufacturer or by an industry association for that metal.

Tack welds may be left in place, provided they are not cracked. However, they must be made by a qualified welder using the same or equivalent electrode and WPS as the first pass of the weld. See para. 127.4.1(C).

Insert 13.1   Arc Welding Processes

Shielded Metal Arc Welding (SMAW)

Principles of Operation

Shielded metal arc welding (SMAW) is the most widely used of all arc welding processes (Fig. 13.1). It is a manual process, which employs the heat of an arc to melt the base metal and the tip of a consumable covered electrode. The bare end of the electrode is clamped in an electrode holder, which is connected to one terminal of a welding power source by a welding lead (cable). The base metal (work) is connected by a cable to the other terminal of the power source. The arc is initiated by touching the electrode tip against the work and then withdrawing it slightly. The heat of the arc melts the base metal in the immediate area, the electrode metal core, and the electrode covering. As the electrode is consumed, shielding of the arc and the weld metal may be provided by gases formed or by slag produced during decomposition of the covering, depending on the electrode type. Also, as the electrode melts, tiny droplets of molten metal are transferred to the molten weld pool by the arc stream. The welder moves the electrode progressively along the joint, and the deposited filler metal and the molten base metal coalesce to form the weld.

Process Capabilities and Limitations

The SMAW process is suitable for most of the commonly used metals and alloys. However, other processes, such as gas tungsten arc welding (GTAW) and gas metal arc (GMAW) welding, are much preferred for some metals, such as aluminum and aluminum alloys and copper and some copper alloys, and for certain applications, such as root pass welding of groove joints in pipe. The equipment required for SMAW is relatively portable and the process can often be used in areas of limited accessibility.

Because SMAW electrodes are produced in individual straight lengths, they can be consumed only to some minimum length. When this occurs, welding must stop while the electrode is replaced. This results in a loss of arc time; hence, the overall weld metal deposition rate tends to be lower than with a continuous electrode process. However, the deposition rate with SMAW is higher than with GTAW. With SMAW the slag which forms on the weld bead surface usually must be removed from the end

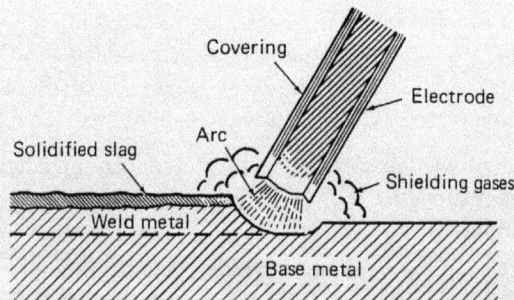

FIG. 13.1
SHIELDED METAL ARC WELDING
(COURTESY OF THE JAMES F. LINCOLN FOUNDATION)

of the weld bead before continuing with a new electrode. For many applications, the completed weld surface must be cleaned of slag or oxides, which could contaminate the process stream or cause corrosion. In SMAW, the weld quality is highly dependent on the skill of the welder.

Gas Tungsten Arc Welding (GTAW)

Principles of Operation

Gas tungsten arc welding (GTAW) uses a special torch, which contains a non-consumable tungsten or tungsten alloy electrode and provides a concentric flow of inert gas to shield the electrode, arc, and molten metal from the atmosphere (Fig. 13.2). The process is sometimes called TIG, for "tungsten inert gas." The heat for welding is produced by an electric arc between the non-consumable electrode and the part to be welded. Filler metal is added by feeding a bare rod or wire into the zone of the arc and the molten weld metal, which is protected by the inert atmosphere. The shielding atmosphere consists of argon, helium, or mixtures thereof. Alternating current is preferred for manual welding of aluminum and magnesium, whereas direct current, electrode negative, is preferred for welding most other materials. A high-frequency pilot arc is usually used for arc starting in order to keep from damaging the electrode or contaminating the weld pool or the workpiece. A weld is produced by applying the arc so that the abutting workpieces and the added filler metal are melted and joined together as the weld metal solidifies.

GTAW equipment is portable, and the process is applicable to most metals in a wide range of thickness and in all welding positions. The process can be used to weld all types of joint geometries and overlays in plate, sheet, pipe, tubing, and structural shapes. It is especially suitable for welding sections of pipe less than 10 mm (3/8 in.) in thickness and also 25.4 mm to 152.4 mm (1 in. to 6 in.) in diameter. Thicker sections can be welded, but it is usually less expensive to use a consumable electrode process, which affords a higher deposition rate. For some applications the process can be used for welding by fusion alone without the addition of filler metal. For welding pipe the combination of GTAW for the root pass followed by either SMAW or GMAW is particularly advantageous. The GTAW process provides a root pass surface which is smooth and uniform on the inside of the pipe, whereas the fill and

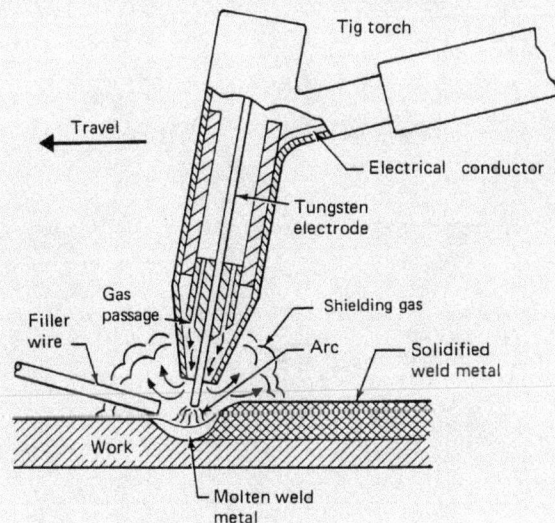

FIG. 13.2
GAS TUNGSTEN ARC WELDING
(COURTESY OF THE JAMES F. LINCOLN FOUNDATION)

cap passes are made with a more economical process. High-quality welds can be produced with the GTAW process when adequately trained welders and proper procedures are used.

For GTA welding it is essential to carefully clean the joint surfaces and use clean filler metal in order to avoid weld defects. The GTAW process is slower than consumable electrode processes. Argon and helium shielding gases are relatively expensive. Improper technique which causes transfer of tungsten into the weld or exposure of the hot filler rod to air can result in unsatisfactory welds.

Gas Metal Arc Welding (GMAW)

Principles of Operation

Gas metal arc welding, popularly known as MIG welding, uses a continuous solid-wire consumable electrode for filler metal and an externally supplied gas or gas mixture for shielding (Fig. 13.3). The shielding gas—helium, argon, carbon dioxide, or mixtures thereof—protects the molten metal from reacting with constituents of the atmosphere.

Process Capabilities and Limitations

Gas metal arc welding is operated in semiautomatic, machine, and automatic modes. It is utilized particularly in high-production welding operations. All commercially important metals, such as carbon steel, stainless steel, aluminum, and copper, can be welded with this process in all positions by choosing the appropriate shielding gas, electrode, and welding conditions.

Flux-Cored Arc Welding (FCAW)

Principles of Operation

Flux-cored arc welding uses a continuous tubular wire consumable filler metal. The filler metals are designed principally for joining carbon and low-alloy steels and stainless steels. However, some nickel-base

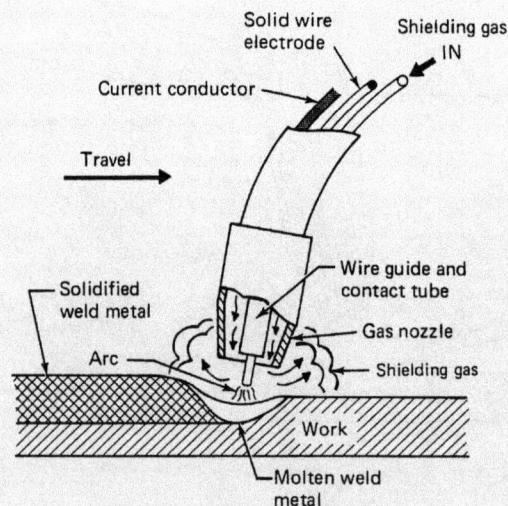

FIG. 13.3
GAS METAL ARC WELDING (COURTESY OF THE JAMES F. LINCOLN FOUNDATION)

filler metals of this type are also available. The flux core may contain minerals, ferroalloys, and ingredients that provide shielding gases, deoxidizers, and slag-forming materials. There are two basic types of cored electrodes, those designed to be used with an additional external shielding gas (Fig. 13.4), and those of self-shielding type, designed for use without external shielding gas. For carbon and low-alloy steels the external shielding gas is usually carbon dioxide or a mixture of argon and carbon dioxide. For stainless steels, argon and argon–oxygen or argon–carbon dioxide mixtures are often used. Self-shielded FCAW electrodes are designed to generate protective shielding gases from core ingredients, similar to the generation of gases by SMAW electrodes. Self-shielded electrodes do not require external shielding.

Process Capabilities and Limitations

Flux-cored arc welding can be operated in either automatic or semi-automatic modes. Weld metal can be deposited at higher rates, and the welds can be larger and better contoured than those made with solid electrodes (GMAW), regardless of the shielding gas. FCAW lends itself to high-production welding applications, as in a manufacturing plant or fabrication shop.

Some classifications of FCAW electrodes of both the externally shielded and self-shielded types are designed for single-pass welding only. Such electrodes should not be used for multipass welding, because there is a risk of weld cracking. For low-temperature service applications, care should be exercised in selecting a classification of FCAW electrode which is designed and tested to ensure good low-temperature notch ductility.

Submerged Arc Welding (SAW)

Principles of Operation

Submerged arc welding uses an arc (or arcs) maintained between a bare electrode and the work (Fig. 13.5). The feature that distinguishes submerged arc welding from other arc welding processes is

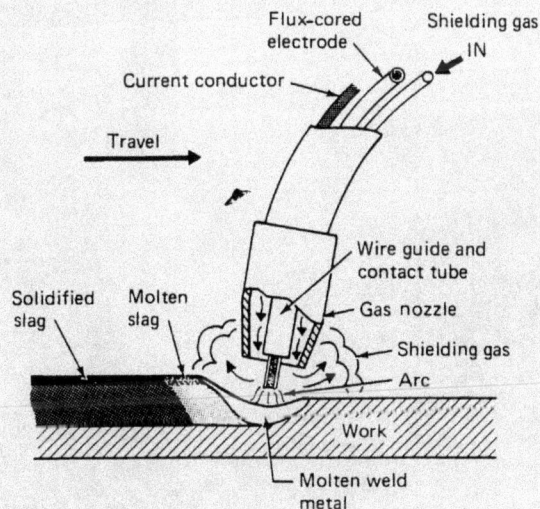

FIG. 13.4
GAS-SHIELDED FLUXED CORED ARC WELDING
(COURTESY OF THE JAMES F. LINCOLN FOUNDATION)

the blanket of granular, fusible material that covers the welding area. By common usage the material is termed a flux, although it performs several important functions in addition to those strictly associated with a fluxing agent. Flux plays a central role in achieving the high deposition rates and weld quality that characterize the SAW process in joining and surfacing applications.

Process Capabilities and Limitations

With proper selection of equipment, SAW can be applied to a wide range of industrial applications. The high quality of welds, the high deposition rates, the deep penetration, and the adaptability to automatic operation make the process particularly suitable for fabrication of large weldments. It is used extensively in ship building, railroad car fabrication, pressure vessel fabrication, pipe manufacturing, and the fabrication of structural members where long welds are required.

SAW is not suitable for all metals and alloys. It is widely used for welding carbon steels, low-alloy structural steels, and stainless steels.

Submerged arc welding can be used for welding butt joints in the flat position, making fillet welds in the flat and horizontal positions, and surfacing in the flat position. With special tooling and fixturing, lap and butt joints can be welded in the horizontal position. The use of this process for welding piping systems is limited to fabrication shops where the pipe can be rolled.

Insert 13.2   Brazing Process

Description of Process.   Brazing includes a group of metal-joining processes which produce coalescence of materials by heating them to a suitable temperature and by using a filler metal which melts at a temperature above 427°C (800°F) and below the temperature at which the base metal starts to melt. The brazing filler metal is distributed between closely fitted surfaces of the joint by capillary attraction.

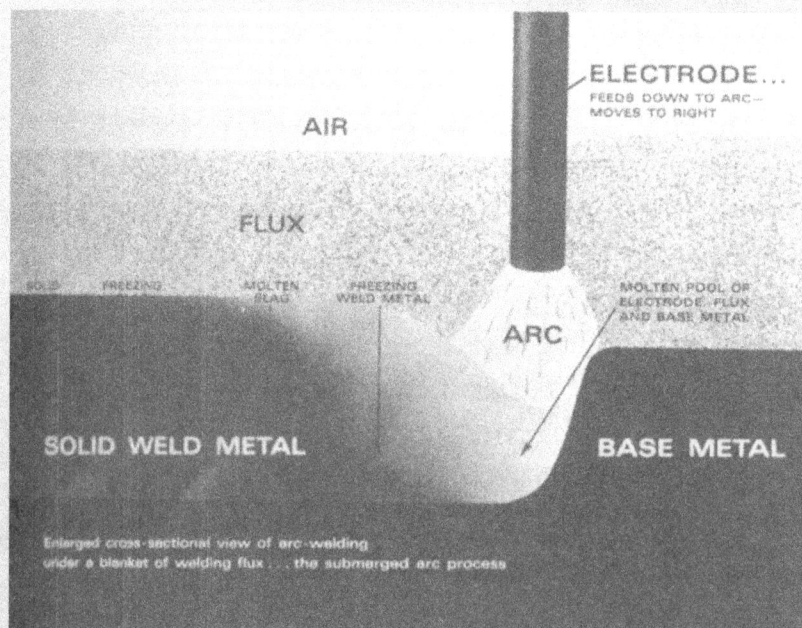

FIG. 13.5
SUBMERGED ARC WELDING (COURTESY OF THE JAMES F. LINCOLN FOUNDATION)

Various types of brazing are defined according to the method of heating employed, for example, dip brazing in a molten metal or salt bath, furnace brazing, induction brazing, infrared brazing, resistance brazing, and torch brazing. Torch brazing is the most commonly used brazing process for process plant construction and maintenance operations.

Torch Brazing: Principles of Operation.   Torch brazing is accomplished by heating with one or more torches using a gas fuel (acetylene, propane, city gas, etc.) burned with air, compressed air, or oxygen. Brazing filler metal may be preplaced at the joint in the form of rings, washers, strips, powder, etc., or its may be fed in wire or rod form by hand. In any case, proper precleaning and fluxing are essential.

Capabilities and Limitations

Brazing can be employed for joining a wide variety of metals and alloys, and appropriate brazing filler metals are available for various applications. To achieve a good joint, the parts must be properly cleaned and must be protected during the heating process, usually by use of a flux in torch brazing, in order to prevent excessive oxidation. The parts must be designed and properly aligned to afford capillary flow of the filler metal. Heat must be applied so as to provide the required temperature range and uniform temperature distribution in the parts.

13.3   WELDING PROCEDURE SPECIFICATION

Welds are conducted in accordance with welding procedure specifications (WPS). A WPS is a written qualified welding procedure prepared to provide direction for making production welds to specified requirements. The WPS or other documents may be used to provide direction to the welder or welding operator to ensure compliance with Code or other specification requirements. The WPS references one or more supporting Procedure Qualification Record(s) (PQRs). ASME BPVC, Section IX, QW-482, gives a suggested format for Welding Procedure Specifications (WPS).

The WPS consist of a number of items, called variables, that define the application of a welding process or combination of processes. Examples of variables are welding process, base metal type and thickness, filler metal type, welding position, pre-heat, post-heat, electrical characteristics, and type of shielding gas. Variables are divided into three types: essential, non-essential, and supplementary essential.

An essential variable is one that, if changed to a value that is outside the limits permitted by the original procedure qualification test, requires a new procedure qualification test.

Non-essential variables include items in the WPS that may have to be changed to satisfy a particular welding application but do not affect the properties of the weld. An example is a change in the groove design. A change in a non-essential variable requires revision or amendment of the WPS or a new WPS; however, requalification testing is not required.

A supplementary essential variable is one that comes into consideration when a particular fabrication Code, such as ASME BPVC, Section VIII, or ASME B31.1, for example, requires supplementary testing in addition to the customary tensile and bend tests required for procedure qualification (e.g., impact testing for low temperature service).

ASME BPVC, Section IX, Article II, lists the essential, non-essential, and supplementary essential variables for welding procedure specifications and procedure qualification testing for each welding process used for joining and for special applications such as hard-facing overlay and corrosion-resistant overlay.

Standard Welding Procedure Specifications listed in ASME BPVC, Section IX, Appendix E may be used within the limitations established by Article V of that Code.

13.4   WELDING PROCEDURE QUALIFICATION RECORD

The purpose of procedure qualification testing is to determine or demonstrate that the weldment proposed for construction is capable of providing the required properties for its intended application.

It is pre-supposed that the welder or welding operator performing the welding procedure qualification test is a skilled worker. That is, the welding procedure qualification test establishes the properties of the weldment, not the skill of the welder or welding operator.

A procedure qualification record (PQR) includes a record of the welding data used to weld a prescribed test coupon in accordance with a WPS and the results of specified mechanical and other tests. The completed PQR must document all essential and, when required, supplementary essential variables of ASME BPVC, Section IX. ASME BPVC, Section IX, QW-483, is a suggested format for a procedure qualification record (PQR).

If any changes are made in the WPS that involve essential variables or, when required, supplementary essential variables, a new procedure qualification test and a new PQR are required to document the changes and support a revision of the WPS or issuance of a new WPS.

The details of preparing a weld test assembly or assemblies, testing of the weld, and test results required for procedure qualification are given in ASME BPVC, Section IX. Customary tests for qualification of joint welding procedures are tensile and bend tests.

13.5   WELDER PERFORMANCE QUALIFICATION

The manufacturer or contractor is responsible for conducting tests to qualify welders and welding operators in accordance with qualified welding procedure specifications, which the organization employs in the construction of weldments built in accordance with the ASME Code. The purpose of welder and welding operator qualification tests is to ensure that the manufacturer or contractor has determined that his or her welder(s) and welding operator(s) using his or her procedures are capable of developing the minimum requirements specified for an acceptable weldment. Performance qualification tests are intended to determine the ability of welders and welding operators to make sound welds.

ASME BPVC, Section IX lists and defines essential variables for each welding process for performance qualification of welders and welding operators. It should be noted that the essential variables for performance qualification are in many cases quite different from the variables for procedure qualification, discussed in Section. 13.4. For the SMAW process, for example, the essential variables for performance qualification are the following: elimination of backing, change in pipe diameter, change in base material P-number, change in filler metal F-number, change in thickness of the weld deposit, the addition of welding positions beyond that originally qualified, and a change in direction of weld progression for vertical welding. A change in an essential variable for performance qualification, beyond prescribed limits, requires requalification testing and the issuance of a new qualification record.

The details of preparing the weld test assembly or assemblies, testing of the weld, and the test results required for welder and welding operator performance qualification are given in ASME BPVC Section IX. Customary tests for performance qualification are visual and bend tests. However, for some base materials and welding processes there are also alternate provisions for examination of the weld(s) by radiography in lieu of bend tests.

ASME BPVC, Section IX, QW-484, is a suggested format for recording welder/welding operator performance qualifications (WPQ). A welder or welding operator performance qualification for a specific process expires when he or she has not welded with that process during a period of 6 months or longer.

Within a 6-month period before the expiration of a performance qualification, a welder who welds using a manual or semi-automatic process maintains all of his or her qualifications for manual or semi-automatic welding with that process. For example, assuming that a welder is qualified for manual SMAW welding of both carbon steel and stainless steel but welds only carbon steel with SMAW during a 6-month period, his or her qualification for SMAW of stainless steel as well as for SMAW of carbon steel is maintained. This points to the importance of keeping timely records of the welding process(es) used during a welder's production welding assignments.

When there is a specific reason to question the ability of a welder or welding operator to make welds that meet the specification, his or her performance qualifications supporting that welding can be revoked. Other qualifications that are not in question remain in effect.

Renewal of performance qualification that has expired may be made for any process by welding a single test coupon of either plate or pipe, of any material, thickness or diameter, in any position, and by testing that coupon by bending or by radiography [if the latter is permitted by ASME BPVC, Section IX for the process and material involved; (see para. QW-304)]. Alternatively, where radiography is a permissible method of examination (QW-304), the renewal of qualification may be done on production work.

13.6   PRE-HEATING

Pre-heating requirements are provided in para. 131 of ASME B31.1. Pre-heating is used, along with heat treatment, to minimize the detrimental effects of high temperature and severe thermal gradients in welding and to drive out hydrogen that could cause weld cracking. Excessive rates of cooling after the weld is made cause adverse microstructural phases in some alloys that result in high hardness, reduced ductility, and reduced fracture toughness. Pre-heating the surrounding material serves to slow the rate of cooling, thereby preventing the formation of these adverse phases. It also drives off moisture, which can introduce hydrogen into the weld. The presence of hydrogen in the weld metal can result in a variety of weld-cracking problems.

The pre-heat requirements, which are applicable to all types of welding, including tack welds, repair welds, and seal welds of threaded joints, are provided in para. 131. The requirements for pre-heat are a function of the P-number of the base material, the nominal wall thickness, and the specified minimum tensile strength of the base metal. The minimum pre-heat temperature is 50°F (10°C). When welding materials with different thicknesses, the requirements are based on the greater of the thicknesses of the parts being joined. When welding dissimilar metals, the higher pre-heat temperature should be used. Except for tack welds, the pre-heat zone is required to extend at least 3 in. (75 mm) or 1.5 times the base metal thickness, whichever is greater, beyond each edge of the weld. The temperature in this zone must be at or greater than the minimum required pre-heat temperature. For tack welds, the zone is 1 in. (25 mm) on each side of the weld. The temperature must be obtained before—and maintained during— the welding.

The pre-heat is required to be maintained until PWHT is performed for P-Nos. 3, 4, 5A, 5B, 6, and 15E materials, except when the conditions of para. 131.6.1 are satisfied.

13.7   HEAT TREATMENT

Post-weld heat treatment is performed to temper the weldment, relax residual stresses, and remove hydrogen. The consequential benefits are avoidance of hydrogen-induced cracking and improved ductility, toughness, corrosion resistance, and dimensional stability.

Heat treatment requirements are provided in para. 132 of ASME B31.1. The Code requires heat treatment after certain welding, bending, and forming operations. Post-weld heat treatment serves to improve the metallurgy (changing the adverse phases mentioned in pre-heat) in some materials and relieve residual stresses caused by shrinkage of the weld material.

Specific requirements for post-weld heat treatment are provided in Table 132. The portion of Table 132 covering P-1, carbon steel, is provided herein as Table 13.1. This table specifies the heat treatment time and temperature based on the P-number, material chemistry, wall thickness, and specified minimum tensile

TABLE 13.1
POSTWELD HEAT TREATMENT (ASME B31.1, PART OF TABLE 132)

P-Number From Mandatory Appendix A	Holding Temperature Range, °F (°C)	Holding Time Based on Nominal Thickness	
		Up to 2 in. (50 mm)	Over 2 in. (50 mm)
P-No. 1 Gr. Nos. 1, 2, 3	1,100 (600) to 1,200 (650)	1 hr/in. (25 mm), 15 min minimum	2 hr plus 15 min for each additional inch (25 mm) over 2 in. (50 mm)

GENERAL NOTES:

(a) PWHT of P-No. 1 materials is not mandatory, provided that all of the following conditions are met:

　(1) the nominal thickness, as defined in para. 132.4.1, is $3/4$ in. (19.0 mm) or less

　(2) a minimum preheat of 200°F (95°C) is applied when the nominal material thickness of either of the base metals exceeds 1 in.(25.0 mm)

(b) PWHT of low hardenability P-No. 1 materials with a nominal material thickness, as defined in para. 132.4.3, over $3/4$ in. (19.0 mm) butnot more than $1\frac{1}{2}$ in. (38 mm) is not mandatory, provided all of the following conditions are met:

　(1) the carbon equivalent, *CE*, is 0.50, using the formula

$$CEp \quad C + (Mn + Si)/6 + (Cr + Mo + V)/5 + (Ni + Cu)/15$$

The maximum chemical composition limit from the material specification or actual values from a chemical anal ysis or material test report shall be used in computing *CE*. If analysis for the last twoterms is not available, 0.1% may be substituted for those two terms as follows:

$$CEp \quad C + (Mn + Si)/6 + 0.1$$

　(2) a minimum preheat of 250°F (121°C) is applied

　(3) the maximum weld deposit thickness of each weld pass shall not exceed $1/4$ in. (6 mm)

(c) When it is impractical to PWHT at the temperature range specified in Table 132, it is permissible to perform the PWHT of this material at lower temperatures for longer periods of time in accordance with Table 132.1.

strength of the base material. A lower heat treatment temperature is permitted for carbon and low-alloy steel if the holding time is increased. This is provided in Table 132.1 or Table 13.2 herein.

The upper temperature provided in Table 13.1 (B31.1 Table 132) may be exceeded as long as the temperature does not exceed the lower critical temperature per Table 13.3. If two materials with different P-numbers are to be joined by welding, the higher temperature requirement shall be used. However, when a non-pressure part is joined to a pressure part, the maximum heat treatment temperature for the pressure part is not permitted to be exceeded.

Furnace heat treatment is preferred; however, local heat treatment is permitted. Local heat treatment is required to be a full circumferential band around the entire component, with the weld located in the center of the band. This is to minimize the stresses caused by differential thermal expansion caused by the heat treatment. For nozzles and attachment welds, the band must encircle the header (for nozzles) or pipe (for attachment welds). The width of the band is required to be at least three times the wall thickness at the weld of the thickest part being joined. For nozzles or attachments, the band is required to extend beyond the nozzle or attachments, the band is required to extend beyond the nozzle or attachment weld at least two times the thickness of the pipe to which the nozzle or thickness is welded.

The maximum rate of heating and cooling at temperatures above 600°F (315°C) is specified in para. 132.5. For materials with a maximum thickness less than 2 in. (50 mm), the maximum rate is 600°F (315°C) per hour. The rate is reduced for thicker components; the maximum rate is 600°F (315°C) per hour divided by half the maximum thickness in inches.

Requirements for heat treatment after bending or forming are stated in para. 129.3. These depend on the material, diameter, thickness, and whether the material is hot or cold bent or formed. When material is bent or formed at temperatures more than 100°F (56°C) below the critical temperatures, it is considered to be cold bent or formed. At temperatures above that, it is assumed to be hot bent or formed. The ap-

TABLE 13.2
ALTERNATE POSTWELD HEAT TREATMENT REQUIREMENTS
FOR CARBON AND LOW ALLOY STEELS (ASME B31.1,
TABLE 132.1)

Decrease in Temperatures Below Minimum Specified Temperature, °F(°C)	Minimum Holding Time at Decreaed Temperature, hr [Note (1)]
50 (28)	2
100 (56)	4
150 (84) [Note (2)]	10
200 (112) [Note (2)]	20

GENERAL NOTE: Postweld heat treatment at lower temperatures for longer periods of time, in accordance with this Table, shall be used only where permitted in Table 132.

NOTES:

(1) Times shown apply to thicknesses up to 1 in. (25 mm). Add 15 min/in. (15 min/25 mm) of thickness for thicknesses greater than 1 in. (25 mm).

(2) A decrease of more than 100°F (56°C) below the minimum specified temperature is allowable only for P-No. 1, Gr. Nos. 1 and 2 materials.

proximate lower critical temperatures are provided in Table 129.3.2 of ASME B31.1, provided herein as Table 13.3.

For P-No. 1 materials, post-weld heat treatment is required when both the thickness is greater than 3/4 in. (19 mm) and the bending or forming operation is performed at a temperature less than 1650°F (900°C). For other ferritic alloy steel, heat treatment is required if the nominal pipe size is NPS 4 (DN 100) or larger, or if the nominal thickness is 1/2 in. (13 mm) or greater. This heat treatment is required to be a full anneal, normalize and temper, or tempering heat treatment as specified by the designer if the component was hot bent or

TABLE 13.3
APPROXIMATE LOWER CRITICAL TEMPERATURES
(ASME B31.1, TABLE 129.3.1)

Material	Approximate Lower Critical °F (°C) [Note (1)]
Carbon steel (P-No. 1)	1,340 (725)
Carbon–molybdenum steel (P-No. 3)	1,350 (730)
1Cr–$\frac{1}{2}$Mo (P-No. 4, Gr. No. 1)	1,375 (745)
1$\frac{1}{4}$Cr–$\frac{1}{2}$Mo (P-No. 4, Gr. No. 1)	1,430 (775)
2$\frac{1}{4}$Cr–1Mo, 3Cr–1Mo (P-No. 5A)	1,480 (805)
5Cr–$\frac{1}{2}$Mo (P-No. 5B, Gr. No. 1)	1,505 (820)
9Cr	1,475 (800)
9Cr–1Mo–V, 9Cr–2W (P-No. 15E)	1,470 (800)

NOTE:

(1) These values are intended for guidance only. The user may apply values obtained for the specific material in lieu of these values.

formed. Otherwise, the normal heat treatment per Table 132 is required. Requirements for post-bending or post-forming heat treatment for austenitic alloys are provided in para. 129.3.4. Post-bending or post-forming heat treatment is not required for other alloys.

While not referenced by ASME B31.1, AWS D10.10 provides useful guidelines for local heating of welds in piping.

13.8   GOVERNING THICKNESS FOR HEAT TREATMENT

When using Table 132, the thickness to be used is the lesser of the thickness of the weld and the thickness of the thicker of the materials being joined by welding. For full penetration butt welds, the thickness is the thicker of the two components, measured at the joint, that are being joined by welding. For example, if a pipe is welded to a heavier wall valve but the valve thickness is tapered to the pipe thickness at the welded joint, the governing thickness will be the greater of the valve thickness at the end of the taper at the weld joint (presumably the nominal pipe wall thickness) or the pipe thickness. Branch connections and fillet weld joints are special cases.

Paragraph 132.4.2 provides the following definitions for nominal thickness used in Table 132:

(1) groove welds (girth and longitudinal): the thicker of the two abutting ends after weld preparation, including ID machining;

(2) partial penetration welds: the depth of the weld groove;

(3) fillet welds: the throat thickness of the weld; and

(4) material repair welds: the depth of the cavity to be repaired.

For branch connections, the dimension is essentially the thickness of the groove weld plus the throat thickness of the cover fillet weld. Specific requirements are provided in para. 132.4.2(E).

13.9   PIPE BENDS

Pipe may be hot or cold bent. For cold bending of ferritic materials, the temperature must be below the transformation range. For hot bending, the temperature must be done above the transformation range.

The thickness after bending must comply with the requirements of para. 102.4.5.

When pipe is bent, it tends to ovalize (also termed flattening). Paragraph 104.2.1 provides limitations on flattening (the difference between the maximum and minimum diameters at any cross section) and buckling for some bends by reference to PFI ES-24. However, 104.2.1(C) indicates that alternative limits may be specified by design.

See Section. 13.7 herein for heat treatment after bending.

13.10   BRAZING

Brazing procedures, brazers, and brazing operators are required to be qualified in accordance with ASME BPVC Section IX, Part QB, of the ASME Boiler and Pressure Vessel Code. Insert 13.2 provides general information on the brazing process.

Qualification of the Brazing Procedure Specification and Brazers or Brazing Operators by others (than the employer) is permitted under certain conditions, as specified in para. 128.5.3. Requirements for records and identification symbols for brazers and brazing operators are provided in para. 128.6.

Aside from these requirements, general good practice requirements for brazing are specified in para. 128 of ASME B31.1.

13.11   BOLTED JOINTS

Proper assembly of bolted joints is essential to avoid leakage during service. Information of flange bolting is provided in Appendix S of ASME BPVC, Section VIII, Division 1. Guidelines for bolted joint assembly are provided in ASME PCC-1.

ASME B31.1 provides some good practice with respect to flange bolt-up in para. 135.3. Paragraph 135.3.4 requires that the bolts extend completely through their threads. Flanged joints are required to be fitted up so that the gasket contact surfaces bear uniformly on the gasket and for the joint to be made up so that the bolt stress is applied relatively uniformly to the gasket.

13.12   WELDED JOINT DETAILS

Welded joint details, including socket weld joints, socket weld and slip-on flanges, and branch connections are provided in Chapter V.

Girth butt welds are covered in para. 127.4.2. A fundamental requirement is that the finished weld has at least the minimum required thickness, per the design rules of Chapter II of ASME B31.1, for the particular size and wall of the pipe used. While partial penetration welds are not prohibited, para. 127.4.2(B.3) requires that when welding pipe to pipe, the surface of the weld shall, as a minimum, be flush with the outer surface of the weld pipe. Some undercuts and concavities are permitted per para. 127.4.2, and the reinforcement (extra thickness) is limited per Table 127.4.2 in ASME B31.1 based on the metal thickness and design temperature.

When components with different outside diameters or thicknesses are joined, the welding end of the thicker component must be trimmed to transition to the thickness of the thinner component. Alternatively, the transition may be accomplished with the weld metal. Figure 127.4.2 in ASME B31.1 provides the envelope for the welding-end transition.

Standard details for slip-on and socket welding flange attachment welds are provided in Fig. 127.4.4(B), provided herein as Fig. 13.6. A couple points worth noting are the fillet weld size for a slip on flange or socket weld flange, which is 1.4 times the nominal pipe wall thickness or the thickness of the hub, whichever is less; and the small gap shown between the flange face and the toe of the inside fillet for slip-on flanges. The small gap is intended to avoid damage to the flange face from welding. It indicates a gap, but there is no specific limit.

The question arose whether a specific limit to the gap between the fillet weld and flange face was appropriate. Studies, including finite element analysis (Becht et al, 1992) and earlier Markl fatigue testing, indicated that it essentially did not matter how much the pipe was inserted into the flange. Insertion by an amount equal to the hub height was optimal for fatigue life, but there was not a significant difference. To minimize future confusion, inclusion of minimum insertion depth has been recommended and may be specified in a future edition.

The required fillet weld size for socket welds other than socket weld flanges is specified in Fig. 127.4.4(C), provided herein as Fig. 13.7. There are a couple points worth mentioning in this figure. The fillet weld size for a socket weld other than a flange is 1.09 times (1.09 is 1.25 times 0.875, where 0.875 considers a nominal 12.5% mill tolerance on the pipe) the nominal pipe wall thickness, or the thickness of the hub (whichever is less).

A second issue with this figure that has caused considerable controversy is the 1/16 in. (2 mm) approximate gap before welding indicated on the figure. This is a requirement for a gap before welding so that weld shrinkage will be less likely to cause small cracks in the root of the fillet weld. Whether such cracks cause problems is questionable, and fatigue testing has shown that socket welds that are welded after jamming the pipe into the socket have longer fatigue lives than ones welded with a gap. There is no requirement for a gap after welding, and weld shrinkage can close a gap that was present prior to welding.

Some owners require random radiographic examination to ensure proper socket welding practice. One of the points that is checked for is the presence of a gap. One argument for this is that it is not possible to determine if there was a gap prior to welding unless there is a gap shown by radiography. If this is desired, it should be specified (the requirement that there be a gap after welding) as an additional requirement of the engineering design. It is not a Code requirement.

Attachment Welds

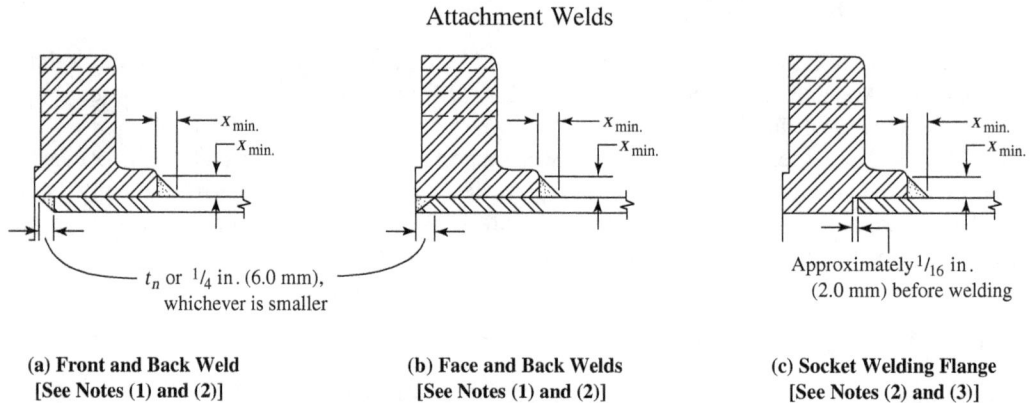

(a) Front and Back Weld
[See Notes (1) and (2)]

(b) Face and Back Welds
[See Notes (1) and (2)]

(c) Socket Welding Flange
[See Notes (2) and (3)]

t_n p   nominal pipe wall thickness
$x_{min.}$ p $1.4\,t_n$ or thickness of the hub, whichever is smaller

NOTES:
(1) Refer to para. 122.1.1(F) for limitations of use.
(2) Refer to para. 104.5.1 for limitations of use.
(3) Refer to para. 122.1.1(H) for limitations of use.

FIG. 13.6
**WELDING DETAILS FOR SLIP-ON AND SOCKET-WELDING FLANGES; SOME
ACCEPTABLE TYPES OF FLANGE ATTACHMENT WELDS (ASME B31.1, FIG. 127.4.4(B))**

Acceptable details for welded branch connections are provided in Fig. 127.4.8(D), provided herein as Fig. 13.8. ASME B31.1 does not include calculations for required weld sizes for these connections; rather, the minimum weld sizes are specified in this figure. Acceptable details for integrally reinforced outlet fittings are provided in Fig. 127.4.8(E), provided herein as Fig. 13.9. The two primary points on this recently added figure are intended to illustrate are the required weld size and the detailing to show a cover weld providing

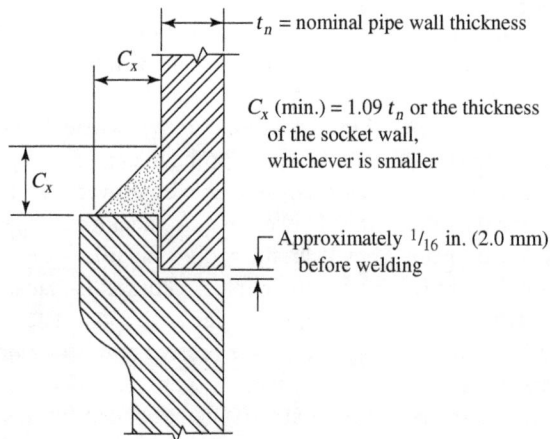

FIG. 13.7
**MINIMUM WELDING DIMENSIONS REQUIRED FOR SOCKET WELDING COMPONENTS
OTHER THAN FLANGES (ASME B31.1, FIG. 127.4.4 (C))**

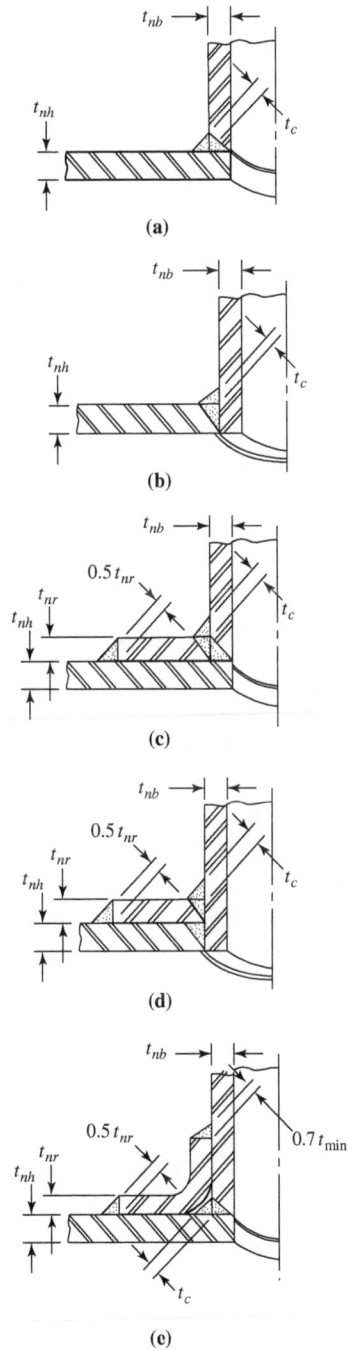

t_{nb}

t_{nh} t_c

(a)

t_{nb}

t_{nh} t_c

(b)

t_{nb}

$0.5\,t_{nr}$

t_{nr}

t_{nh} t_c

(c)

t_{nb}

$0.5\,t_{nr}$

t_{nr}

t_{nh} t_c

(d)

t_{nb}

$0.5\,t_{nr}$

$0.7\,t_{min}$

t_{nr}

t_{nh}

t_c

(e)

GENERAL NOTE: Weld dimensions may be larger than the
minimum values shown here.

FIG. 13.8
SOME ACCEPTABLE TYPES OF WELDED BRANCH ATTACHMENT DETAILS SHOWING
MINIMUM ACCEPTABLE WELDS (ASME B31.1, FIG. 127.4.8. (D))

GENERAL NOTES;
(a) Welds shall be in accordance with para. 127.4.8(C).
(b) Weld attachment details for branch fittings that do not match the schedule or weight designation of the run pipe as defined by MSS SP-97 Table 1 shall be designed to meet the requirements in paras. 104.3.1 and 104.7.2.
(c) The stress intensification factors as required by paras. 104.8 and 119.7.3, for the fittings represented by drawings (b-1), (b-2), (c-1), and (c-2), should be obtained from the fitting manufacturer.

NOTES;
(1) When the fitting manufacturer has not provided a visible scribe line on the branch fitting, the weld line shall be the edge of the first bevel on the branch fitting adjacent to the run pipe.
(2) The minimum cover weld throat thickness, t_c, applies when the angle between the branch fitting groove weld face and the run pipe surface is less than 135 deg. For areas where the angle between the groove weld face and the run pipe surface is 135 deg or greater, the cover weld may transition to nothing.
(3) Cover weld shall provide a smooth transition to the run pipe.
(4) t_{nb} shall be measured at the longitudinal centerline of the branch fitting. When t_{nb} in the crotch area does not equal t_{nb} in the heel area, the thicker of the two shall govern in determining the heat treatment in accordance with para. 132.4.

a radius between the weld and run pipe. Without the radius, a notch can be left on the outer perimeter of the weld that substantially reduces the fatigue life (increases the stress intensification factor) or the outlet fitting, in particular with out-plane bending.

13.13   MISCELLANEOUS ASSEMBLY REQUIREMENTS

Threaded joints should generally be lubricated with a suitable thread compound or lubricant suitable for the service conditions and that does not react unfavorably with either the service fluid or the piping material. However, if the joint is intended to be seal welded, ASME B31.1 states that the joint should not be made up with thread compound or lubricant. The material can result in a poor quality seal weld. Note, as discussed in the design section, that seal welds are not considered to contribute to the joint strength. Also, per para. 127.4.5, the seal weld is required to completely cover the threads. This results in improved fatigue performance.

Various good practice requirements are provided in para. 135 for assembly of straight-threaded joints, tubing joints, caulked joints, and packed joints.

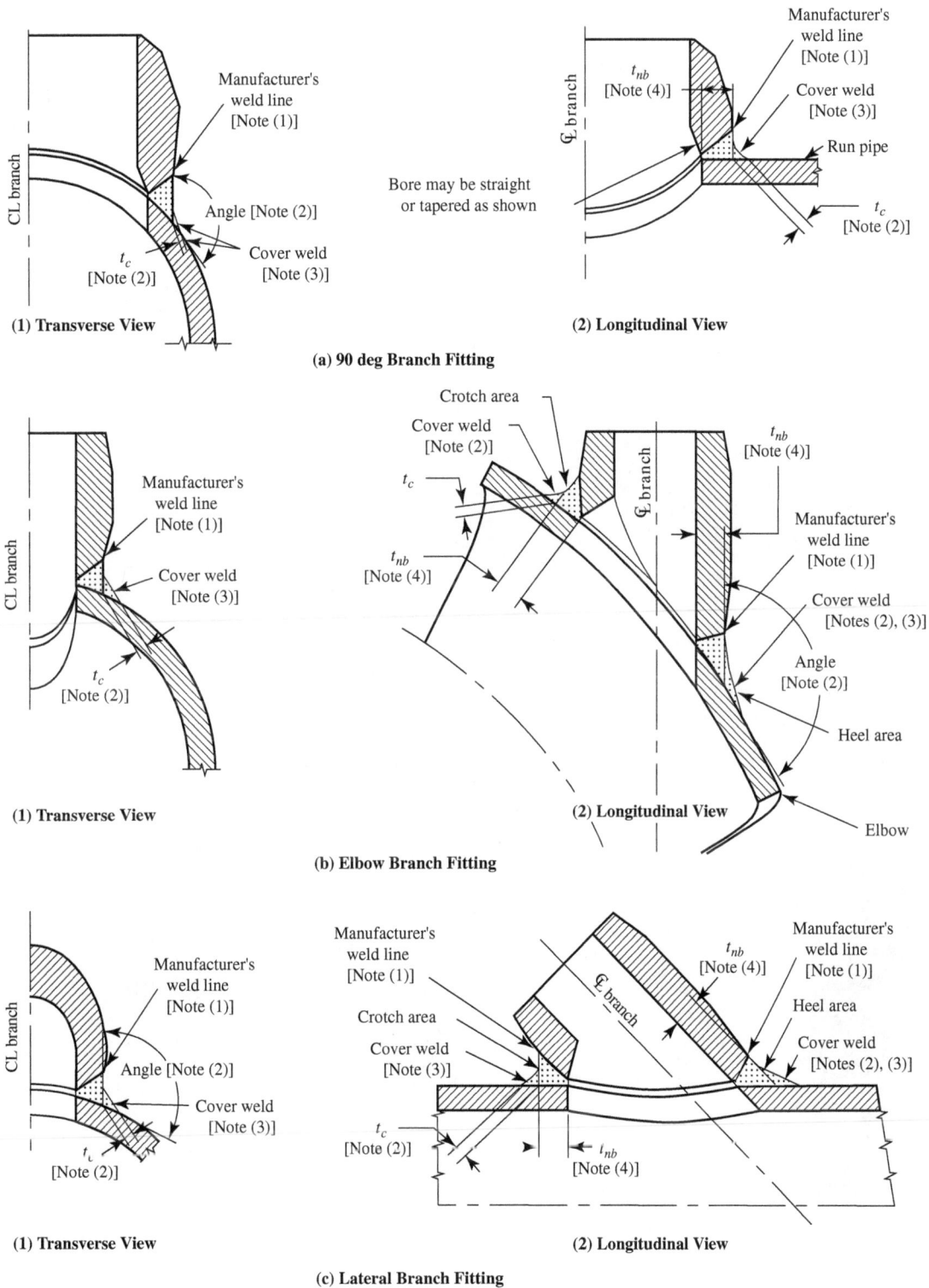

(1) Transverse View

(2) Longitudinal View

(a) 90 deg Branch Fitting

(1) Transverse View

(2) Longitudinal View

(b) Elbow Branch Fitting

(1) Transverse View

(2) Longitudinal View

(c) Lateral Branch Fitting

FIG. 13.9
SOME ACCEPTABLE DETAILS FOR INTEGRALLY REINFORCED OUTLET FITTINGS
(ASME B31.1, FIG. 127.4.8(E))

EXAMINATION

14.1 OVERVIEW OF EXAMINATION REQUIREMENTS

ASME B31.1 requires that examination of the piping be performed by the piping manufacturer, fabricator, erector, or a party authorized by the owner as a quality control function. These examinations include visual observations and non-destructive examination, such as radiography, ultrasonic, eddy-current, liquid-penetrant, and magnetic-particle methods.

Inspection is the responsibility of the owner and may be performed by employees of the owner or a party authorized by the owner except in the case of boiler external piping, which requires inspection by an Authorized Inspector. The Inspector is responsible for ensuring, before the initial operation, compliance with the engineering design as well as with the material, fabrication, assembly, examination, and test requirements of ASME B31.1. Note that the process of inspection does not relieve the manufacturer, fabricator, or erector of his or her responsibilities for complying with the Code. ASME B31.1 does not specify qualifications for the Inspector.

Boiler external piping must be inspected by an Authorized Inspector. The requirements are provided in ASME BPVC, Section I. Qualifications for the Authorized Inspector are described in Section 1.5.4 herein; duties are summarized in ASME BPVC, Section I, para. PG-90. They include verifying that the manufacturer or assembler has a valid ASME Certificate of Authorization; monitoring compliance with the accepted Quality Control program; verifying that the Certificate Holder has the necessary Code books, addenda, and Code Cases to cover the work being performed; reviewing a selected number of the manufacturer's design calculations to verify compliance with ASME BPVC, Section I; witnessing and approving proof tests to establish Maximum Allowable Working Pressure; verifying that the Certificate Holder has sufficient material control; verifying that the Certificate Holder's controls provide a positive means of identification to maintain traceability of materials; verifying that the Certificate Holder's personnel are examining cut edges before welding; verifying that all aspects of welding conform to the Code requirements; providing acceptance of the method and extent of welded repairs and verifying that only qualified welding procedures, welders, and welding operators are used; verifying that all required heat treatments have been performed and are properly documented; verifying that required non-destructive examinations and tests have been performed by qualified personnel and that the results are properly documented; performing the required inspections and witnessing hydrostatic tests; verifying that the responsible representative has signed the Data Report and that it is correct before signing; and verifying that the item to be stamped is in compliance with the requirements of the Code and the nameplate, if used, is properly attached.

Personnel who perform non-destructive examination of welds are required to be qualified and certified for the examination method they use. The employer of the examiner is responsible for establishing the program to qualify the examiners. Minimum aspects of the program are specified in para. 136.1, including instruction, on-the-job training, eye examination, and written or oral examination. If the examiner has not performed examinations using a specific method for a period of one year or more, recertification is required. Additionally, substantial changes in procedures require recertification.

As an alternative to the program described above, ASME Section V, Article 1 may be used to qualify personnel. Also, AWS QC1, Qualification and Certification of Welding Inspectors, may be used for establishing qualifications of individuals performing visual examination of welds.

Requirements for the examination processes are described in ASME BPVC, Section V, with limited exceptions and additions. ASME BPVC, Section V is referenced by ASME B31.1. The required degree of examination and the acceptance criteria for the examinations are provided in Chapter VI of ASME B31.1.

Depending on pressure and temperature conditions, ASME B31.1 either requires 100% examination by radiography, magnetic-particle, ultrasonic, or liquid-penetrant methods, or by a visual examination. If only visual examination is required, the piping is judged to be acceptable if it meets the visual examination requirements of para. 136.4.2 and the pressure test requirements of the Code.

The examinations are required to take place after post-weld heat treatment unless otherwise directed by the engineering design for P-Nos 3, 4, 5A, 5B and 15E material welds.

It is not the intent of ASME B31.1 that the examination will ensure that the constructed piping system will be free of defects, even ones that are rejectable if found. If additional examinations are performed that reveal defects, it is beyond the Code requirements, as the system already complied with the Code, and the issue of whether to repair these defects and who is to pay for such repairs is a purely contractual issue. The defect can be left in and the piping system will still comply with Code. Of course, it would be judicious to at least perform a fitness-for-service evaluation of the defect if it is intended to be left unrepaired.

14.2   REQUIRED EXAMINATION

The required examination depends upon the type of weld, pressure, temperature, pipe diameter, and thickness. These requirements are spelled out in Table 136.4 (Table 14.1 herein). The requirements tend to require more examination on high-energy systems (higher temperature and pressure). Fluid hazards such as toxicity and flammability are not considered.

Piping systems with design temperatures over 750°F (400°C) are subject to the most stringent examination, with all welds other than 1/4 in. (6 mm) and smaller fillet welds attaching non-pressure-retaining parts subject to radiographic (RT), ultrasonic (UT), magnetic-particle (MT), or liquid-penetrant (PT) examination in addition to visual examination (VT).

The next lower category with respect to non-destructive examination is for temperatures between 350°F (175°C) and 750°F (400°C), inclusive, with pressures over 1025 psig [7,100 kPa (gage)]. For this category, fewer of the welds require examination other than VT, and the examination requirements depend on the wall thickness.

The final category is all other systems for which only visual examination is required.

Table 136.4.1 (Table 14.2 herein) summarizes the types of examination that are used and the types of imperfections that are detected by each method.

14.3   VISUAL EXAMINATION

Visual examination is covered in para. 136.4.2 and means using the unaided eye (except for corrective lenses) to inspect the exterior and readily accessible internal surface areas of piping assemblies or components. It does not include nor require remote examination such as by use of boroscopes. Visual examination is used to confirm that the requirements of the design and the WPS are satisfied. This should include a check of materials and components for conformance to specifications and freedom from defects; fabrication including welds; assembly of threaded, bolted, and other joints; piping during erection; and piping after erection. Requirements for visual examination can be found in the ASME BPVC, Section V, Article 9. The following indications in welds that are visually examined, per para. 136.4.2, are unacceptable:

TABLE 14.1
MANDATORY MINIMUM NONDESTRUCTIVE EXAMINATIONS FOR PRESSURE WELDS OR WELDS TO PRESSURE-RETAINING COMPONENTS (ASME B31.1, TABLE 136.4)

Type Weld	Piping Design Conditions and Nondestructive Examination		
	Temperatures Over 750°F (400°C) Pressures	Temperatures Between 350°F (175°C) and 750°F (400°C) Inclusive, With All Pressures Over 1,025 psig [7 100 kPa (gage)]	All Others
Butt welds (girth and longitudinal) [Note (1)]	RT or UT for over NPS 2. MT or PT for NPS 2 and less [Note (2)].	RT or UT for over NPS 2, $\frac{3}{4}$ with thickness over $\frac{3}{4}$ in. (19.0 mm). VT for all sizes with thickness $\frac{3}{4}$ in (19.0 mm) or less.	Visual for all sizes and thicknesses
Welded branch connections (size indicated is branch size) [Notes (3) and (4)]	RT or UT for over NPS 4, MT or PT for NPS 4 and less [Note (2)].	RT or UT for branch over NPS 4 and thickness of branch over $\frac{3}{4}$ in. (19.0 mm) MT or PT for branch NPS 4 and less with thickness of branch over $\frac{3}{4}$ in. (19 mm) VT for all sizes with branch thickness $\frac{3}{4}$ in. (19.0 mm) or less	VT for all sizes and thickness
Fillet, socket, attachment, and seal welds	PT or MT for all sizes and thickness [Note (5)]		VT for all sizes and thickness

GENERAL NOTES:
(a) All welds shall be given a visual examination in addition to the type of specific nondestructive examination specified.
(b) NPS — nominal pipe size.
(c) RT — radiographic examination; UT — ultrasonic examination; MT — magnetic particle examination; PT — liquid penetrant examination; VT — visual examination.
(d) For nondestructive examinations of the pressure retaining component, refer to the standards listed in Table 126.1 or manufacturing specifications.
(e) Acceptance standards for nondestructive examinations performed are as follows: MT — see para. 136.4.3; PT — see para. 136.4.4; VT — see para. 136.4.2; RT — see para. 136.4.5; UT — see para. 136.4.6.
(f) All longitudinal welds and spiral welds in pipe intended for sustained operation in the creep range (see paras. 104.1.1 and 123.4, and Table 102.4.7) must receive and pass a 100% volumetric examination (RT or UT) per the applicable material specification or in accordance with para. 136.4.5 or 136.4.6.

NOTES:
(1) The thickness of butt welds is defined as the thicker of the two abutting ends after end preparation.
(2) RT may be used as an alternative to PT or MT when it is performed in accordance with para. 136.4.5.
(3) RT or UT of branch welds shall be performed before any nonintegral reinforcing material is applied.
(4) In lieu of volumetric examination (RT, UT) of welded branch connections when required above, surface examination (PT, MT) is acceptable and, when used, shall be performed at the lesser of one-half of the weld thickness or each $\frac{1}{2}$ in. (12.5 mm) of weld thickness and all accessible final weld surfaces.
(5) Fillet welds not exceeding $\frac{1}{4}$ in. (6 mm) throat thickness which are used for the permanent attachment of nonpressure retaining parts are exempt from the PT or MT requirements of the above Table.

TABLE 14.2
WELD IMPERFECTIONS INDICATED BY VARIOUS TYPES OF EXAMINATION
(ASME B31.1, TABLE 136.4.1)

Imperfection	Visual	Magnetic Particle	Liquid Penetrant	Radiography	Ultrasonic
Crack — surface	X [Note (1)]	X [Note (1)]	X [Note (1)]	X	X
Crack — internal	. . .	. . .	. . .	X	X
Undercut — surface	X [Note (1)]	X [Note (1)]	X [Note (1)]	X	. . .
Weld reinforcement	X [Note (1)]	. . .	. . .	X	. . .
Porosity	X [Notes (1), (2)]	X [Notes (1), (2)]	X [Notes (1), (2)]	X	. . .
Slag inclusion	X [Note (2)]	X [Note (2)]	X [Note (2)]	X	X
Lack of fusion (on surface)	X [Notes (1), (2)]	X [Notes (1), (2)]	X [Notes (1), (2)]	X	X
Incomplete penetration	X [Note (3)]	X [Note (3)]	X [Note (3)]	X	X

NOTES:
(1) Applies when the outside surface is accessible for examination and/or when the inside surface is readily accessible.
(2) Discontinuities are detectable when they are open to the surface.
(3) Applies *only* when the inside surface is readily accessible.

(1) cracks on the external surface;

(2) undercut on the surface that is greater than 1/32 in. (1.0 mm) deep, or encroaches on the minimum required section thickness;

(3) weld reinforcement that exceeds the permitted reinforcement in Table 127.4.2 (this depends on the wall thickness and design temperature);

(4) lack of fusion on the surface;

(5) incomplete penetration (when the inside surface is readily accessible);

(6) any other linear indications greater than 3/16 in. (5.0 mm) long; and

(7) surface porosity with rounded indications greater than 1/16 in. (2.0 mm) or less edge to edge in any direction (rounded indications are indications that are circular or elliptical with their length less than three times their width).

There are several interpretations with respect to the required access for visual inspection, concerning whether internal surfaces require inspection.

Interpretation 25-7, Questions (2) and (4), address the matter:

Question (2): Is it a requirement of the ASME B31.1 Code that the fabrication sequence shall be arranged so as to provide for making the maximum number of joint inside surfaces "readily accessible" for examination for incomplete penetration?

Reply (2): No.

Question (4): Is it required that the required visual examination occur "before, during, or after the manufacture, fabrication, assembly, or test," as stated in the ASME B31.1 Code, para. 136.4.2?

Reply (4): No. Visual examinations may occur whenever the examiner wishes to be present. However, it is required that the completed weld receive a visual examination that verifies that the completed weld meets the acceptance criteria of para. 136.4.2.

14.4   RADIOGRAPHIC EXAMINATION

Radiographic examination means using X-ray or gamma-ray radiation to produce a picture of the subject part, including subsurface features, on radiographic film for subsequent interpretation. It is a volumetric examination procedure, which provides a means of detecting defects that are not observable on the surface of the material. Requirements for radiographic examination of welds are provided in the ASME BPVC, Section V, Article 2, except the requirements of T-274 are used as a guide, but not for the rejection of radiographs unless the geometrical unsharpness exceeds 0.07 in. (2.0 mm). T-274 specifies limits to geometric unsharpness, depending on material thickness, which are only required when referenced by a Code section (e.g., ASME B31.1).

The acceptance standards for radiographic examination are provided in para. 136.4.5. In the acceptance criteria, t is the thickness of the weld or, when two members having different thicknesses are welded together, the thinner portion of the weld. The following indications are unacceptable:

(1) Any type of crack or zone of incomplete fusion or penetration.

(2) Any other elongated indication that has a length greater than one of the following:

    (a) 1/4 in (6.0 mm) for t up to 3/4 in. (19.0 mm);

    (b) 1/3t for t from 3/4 in. (19.0 mm) to 2 to 1/4 in. (57.0 mm), inclusive; and

    (c) 3/4 in. (19.0 mm) for t greater than 2 to 1/4 in. (57.0 mm).

(3) Any group of indications in a line that have an aggregate length greater than t in a length of 12t, except where the distance between the successive indications exceeds 6L, where L is the longest indication in the group.

(4) Porosity in excess of that shown as acceptable in Appendix A-250 of ASME BPVC, Section I.

(5) Root concavity when there is an abrupt change in density as indicated on the radiograph.

14.5   ULTRASONIC EXAMINATION

Ultrasonic examination means detecting defects using high-frequency sound impulses. The defects are detected by the reflection of sound waves from them. It is also a volumetric examination method, which can be used to detect subsurface defects. It is used in ASME B31.1 as an alternative to radiography for weld examination. The requirements for ultrasonic examination of welds are provided in the ASME BPVC, Section V, Article 4. Furthermore, para. 136.4.6 provides additional requirements for the equipment (it must be capable of recording the UT data unless physical obstructions prevent the use of such a system and the use of manual UT is approved by the owner) and personnel qualification.

The acceptance standards for UT examination are provided in para. 136.4.6. When an ultrasonic indication greater than 20% of the reference level is found, the examiner is required to characterize the shape, identity, and location of the discontinuity. The following discontinuities are unacceptable:

(1) discontinuities evaluated as being cracks, lack of fusion, or incomplete penetration; and

(2) other discontinuities if the indication exceeds the reference level and their length exceeds a thickness dependent limit. This limit is the same as that for radiography for "other elongated indications" listed in the prior section.

14.6   LIQUID-PENETRANT EXAMINATION

Liquid-penetrant examination means detecting surface defects by spreading a liquid dye penetrant on the surface, removing the dye after sufficient time has passed for the dye to penetrate into any surface defect, and applying a thin coat of developer to the surface, which draws the dye from defects. The defects are observable by the contrast between the color of the dye penetrant and the color of the developer. The indication may well be larger than the discontinuity that the liquid penetrant has penetrated; however, it is the size of the indication, not the size of the defect, that is evaluated in accordance with the acceptance standards.

The examiner must distinguish between relevant and non-relevant indications. Indications may arise from surface imperfections such as machining marks, which are not considered defects and are thus non-relevant. Furthermore, indications that are 1/16 in. (2.0 mm) or less are also considered to be non-relevant. ASME B31.1 provides a discussion of this.

Liquid-penetrant examination is used for girth welds in piping NPS 2 (DN 50) and less; branch connections NPS 4 (DN 100) and less; and fillet, socket, attachment, and seal welds when a non-destructive examination other than VT is required (MT may also be used). In addition, PT is used to detect surface defects. The requirements for PT of welds are provided in ASME BPVC, Section V, Article 6.

Paragraph 136.4.4 covers both the evaluation of indications and their acceptance criteria. The following indications are not acceptable if they are relevant:

(1) any cracks or linear indications (a linear indication has a length three or more times the width);

(2) rounded (indications other than linear) indications with dimensions greater than 3/16 in. (5.0 mm);

(3) four or more rounded indications in a line separated by 1/16 in. (2.0 mm) or less edge to edge; and

(4) ten or more rounded indications in any 6 sq. in. (3870 mm^2) of surface with the major dimension of this area not to exceed 6 in. (150 mm) with the area taken in the most unfavorable location relative to the indications being evaluated.

14.7   MAGNETIC-PARTICLE EXAMINATION

Magnetic-particle examination employs either electric coils wound around the part or prods to create a magnetic field. A magnetic powder is applied to the surface and defects are revealed by patterns the powder forms in response to the magnetic field disturbances caused by defects. This technique reveals surface and shallow subsurface defects. As such, it can provide more information than PT. However, its use is limited to magnetic materials. Magnetic-particle examination is used for girth welds in piping NPS 2 (DN 50) and less; branch connections NPS 4 (DN 100) and less; and fillet, socket, attachment, and seal welds when a non-destructive examination other than VT is required (PT may also be used). The requirements for MT of welds and components other than castings are provided in the ASME BPVC, Section V, Article 7.

Guidance on the evaluations of indications and the acceptance criteria for them are provided in para. 136.4.3. The same considerations regarding relevant and non-relevant indications apply for MT as for PT. The acceptance criteria for MT are the same as for PT.

PRESSURE TESTING

15.1 OVERVIEW OF PRESSURE TEST REQUIREMENTS

ASME B31.1 requires leak testing of all piping systems other than lines open to the atmosphere, such as vents or drains downstream of the last shut-off valve. For boiler external piping, a hydrostatic test in accordance with PG-99 of ASME BPVC Section I, conducted in the presence of the Authorized Inspector, is required. For non-boiler external piping, the following options are available:

(1) hydrostatic testing,

(2) pneumatic testing,

(3) mass-spectrometer testing, and

(4) initial service testing.

The required hydrostatic test of PG-99 is similar to the ASME B31.1 hydrotest. It requires a 1.5 times design pressure hydrotest. The test pressure is then permitted to be reduced to the maximum allowable working pressure, and the boiler examined. The maximum stress during the hydrotest is not permitted to exceed 90% of the yield strength of the material at test temperature.

The maximum stress during the test is limited by para. 102.3.3(B). This paragraph limits the circumferential (hoop) stress and the longitudinal stress (from weight, pressure, and other loads during the test, but not including occasional loads such as wind and earthquake) to 90% of the yield strength (0.2% offset) at test temperature.

The leak test is required to be conducted after post-weld heat treatment, non-destructive examination, and other fabrication, assembly, and erection activities required to provide the systems being tested with pressure-retaining capability have been completed. If repairs or additions are made following the leak test, the affected piping must generally be re-tested unless the repairs or additions comply with the limitations in para. 137.8.

All joints, except those previously tested, are generally required to be left un-insulated and exposed for the leak test. ASME B31.1 permits painting of the joints prior to the test; however, paint can effectively seal small leaks to extremely high pressures and would of course render the mass-spectrometer testing option relatively useless for detecting leaks through painted welds. Para. 137.2.1 permits leaving insulation on the system during the leak test, by prior agreement, if the hold time under pressure is extended to check for possible leakage through the insulation barrier.

Expansion joints are to be provided with temporary restraints if required for the additional pressure load under test, or they are to be isolated during the system test. This is per para. 137.2.3. Note, however, that the means, such as pipe anchors and expansion joint hardware, which resist the pressure thrust forces resulting from the presence of an expansion joint in the system, are essential to maintaining the system pressure integrity. Thus, these restraints should be designed to be as capable as the piping for withstanding a 1.5 times hydrostatic test. Unless there are very specific reasons not to do so, it is better to include the expansion joint and its pressure-thrust-restraining hardware as part of the system pressure test. If it would fail in this pressure test, it should probably not be put in operation.

Piping containing toxic fluids are required to either be pneumatically tested, or hydrostatically tested with an additional mass spectrometer test [para. 122.8.2(H)].

15.2   HYDROSTATIC TESTING

A hydrostatic test is generally the preferred alternative because it is conducted at a higher pressure, which has beneficial effects such as crack blunting and warm pre-stressing, and entails substantially less risk than a pneumatic test. These reduce the risk of crack growth and brittle fracture after the hydrostatic test when the pipe is placed in service. The minimum test pressure is generally 1.5 times the design pressure. However, the test pressure may be limited to a lower value, as it is not permitted to exceed the maximum allowable test pressure of any non-isolated components, such as vessels, pumps, or valves, nor to exceed the stress limits mentioned in 15.1. There is no temperature correction factor for systems that will operate at higher temperatures.

The springs should generally be left with their travel stops in place through the hydrostatic test. Furthermore, if the line normally contains vapor or a fluid with a lower density than water, the need for supplemental temporary supports must be considered because of the higher fluid weight than normal operation.

The pressure is required to be held for at least 10 minutes, then it may be reduced to the design pressure and held for as long as necessary to conduct examinations for leakage. All joints and connections must be visually inspected for leakage. The acceptance criteria are that there be no evidence of weeping or leaking, with the exception of possible localized instances at pump or valve packing. Note, however, para. 137.2.1, discussed above, permits this examination to be made from the outside of the insulation under certain conditions. There is no provision in ASME B31.1 for substituting a monitoring of pressure decay for the 100% visual examination of the pipe joints during the hydrostatic test.

15.3   PNEUMATIC TESTING

A pneumatic test is considered to potentially entail a significant hazard from the amount of stored energy in the compressed gas. A rupture can result in an explosive release of this energy. For example, an explosion of 200 ft (60 m) of NPS 36 (DN 1600) line containing air at 500 psi (3,500 kPa) can create a blast wave roughly equivalent to 80 lbs (35 kg) of TNT. The hazard is proportional to both the volume and the pressure. Guidelines for pneumatic testing are contained in ASME PCC-2.

As a result, pneumatic testing is not permitted unless the owner either specifies it or permits it as an alternative. It is recommended by ASME B31.1 that pneumatic testing only be performed when one of the following conditions exist:

(1) when the piping systems are so designed that they cannot be filled with water, or

(2) when the piping systems are to be used in services where traces of the testing medium cannot be tolerated.

Because of this concern, the pneumatic test may be conducted at a lower pressure than a hydrostatic test. It is permitted to be between 1.2 and 1.5 times the system design pressure, inclusive. The 1.5 factor is an upper limit, rather than a lower limit. Similar to hydrostatic tests, the test pressure is not permitted to exceed the maximum allowable test pressure of any non-isolated component and must also satisfy the stress limits of para. 102.3.3(B) during test.

A preliminary pneumatic test, not to exceed 25 psig [175 kPa (gage)], may be performed to locate major leaks prior to conducting other leak testing.

Specific precautions are required. The pressure is to first be gradually increased to not more than one-half of the test pressure and then increased in steps not to exceed one-tenth the test pressure until the required test pressure is reached. The test pressure must be maintained for a minimum time of 10 minutes, and then it may be reduced to the lesser of the design pressure or 100 psig [700 kPa (gage)] and held for as long as necessary to conduct the examination for leakage.

Examination for leakage is required to be by bubble test or equivalent examination methods at all joints and connections. The acceptance criterion is that there be no evidence of leakage other than possible localized instances at pump or valve packing.

15.4   MASS-SPECTROMETER TESTING

An mass-spectrometer leak test is permitted when specified by the owner. This leak test provides a greater sensitivity to leaks than the hydrostatic or pneumatic test, although it does not provide the benefits such as crack blunting provided by the hydrostatic test mentioned above.

ASME B31.1 requires that the test be conducted in accordance with the instructions of the manufacturer of the test equipment. The equipment is required to be calibrated against a reference leak, which cannot exceed the maximum permissible leak rate. Note that methods for performing this type of test are described in the ASME BPVC, Section V, Article 10. The acceptance criterion is that the leak rate not exceed the maximum permissible leakage from the system.

Halide testing remains listed in ASME B31.1 as an option but would generally be prohibited by environmental regulations.

15.5   INITIAL SERVICE TESTING

When specified by the owner, ASME B31.1 permits an initial service leak test in lieu of other leak tests such as hydrostatic or pneumatic when other types of tests are not practical or when tightness is demonstrable because of the nature of the service. Note that it is the responsibility of the owner to make this determination. Examples of systems where initial leak testing may be an appropriate selection are provided in para. 137.7.1. These examples are the following:

(1) piping where shut-off valves are not available for isolating a line and where temporary closures are impractical; and

(2) piping where, during the course of checking out pumps, compressors, or other equipment, ample opportunity is afforded for examination for leakage prior to full-scale operation.

Note that flange joints with temporary blanks for other leak tests are also subject to the initial service leak test. See para. 137.2.5.

In this test, the system is gradually brought up to the normal operating pressure during initial operation and held for at least 10 minutes. All joints and connections must be examined for leakage. The acceptance criterion is that there be no evidence of weeping or leaking except for localized instances at pump or valve packing.

15.6   RE-TESTING AFTER REPAIR OR ADDITIONS

For boiler external piping, the requirements of ASME BPVC, Section I apply. Thus, PW-54.2 applies for repairs to pressure parts and PW-54.3 applies for non-pressure parts. For pressure parts, after repair, the part must be re-tested. If it fails the re-test, the Inspector may permit supplementary repairs or permanently reject the part. For non-pressure parts, re-testing is not required under limited conditions (e.g., P-1 material; stud weld or fillet weld of limited size-with pre-heat in larger thicknesses; and completed weld inspection by the Authorized Inspector). Otherwise, a re-test is also required. For non-boiler external piping, the affected area of the piping is required to be re-tested after repair.

Additional welding to the system, limited to the following, is permitted without re-test (this is an addition, not a repair). Seal welds and attachment of lugs, brackets, insulation supports, name-plates, or other non-pressure-retaining attachments may be made after the pressure test under the following conditions:

(1) fillet welds do not exceed 3/8 in. (10.0 mm);

(2) full penetration welded attachments do not exceed the nominal thickness of the pressure-retaining member or 1/2 in. (12.0 mm), whichever is less;

(3) pre-heat is per para. 131 and examination is per Table 136.4 (these are the normal requirements for pre-heat and examination of welded joints); and

(4) seal welds are examined for leakage after system start-up.

NON-METALLIC PIPING

16.1 ORGANIZATION AND SCOPE

ASME B31.1 provides rules for non-metallic pipe in paras. 105.3 and 124.9. For some services, non-mandatory rules are provided in Appendix III. The mandatory rules are found in the aforementioned paragraphs. Only the following service applications are permitted:

(1) plastic pipe containing water and non-flammable liquids, within manufacturer's limitations, but not to exceed 150 psi (1000 kPa) pressure and 140°F (60°C) temperature;

(2) reinforced thermosetting resin (RTR) pipe may be used in buried flammable and combustible liquid service;

(3) polyethylene pipe may be used in buried flammable liquid and gas service [for natural gas service, per para. 122.8.1(B.4), the following limitations apply: a maximum temperature of 140°F (60°C), minimum temperature of −20°F (−30°C), and maximum pressure of 100 psi (690 kPa)]; heat fusion joints are required, and Appendix III becomes mandatory;

(4) reinforced-concrete pipe may be used in accordance with the specifications listed in Table 126.1 for water service;

(5) metallic piping lined with non-metals for fluids that would corrode or be contaminated by unprotected metal. Per para. 122.9, such piping is required to be qualified as unlisted components.

Non-mandatory rules for non-metallic piping in some services are located in Appendix III of ASME B31.1. The scope is limited to plastic and elastomer-based piping materials, with or without fabric or fibrous material added for pressure reinforcement. The services within the scope of Appendix III include the following:

(1) water service;

(2) non-flammable and non-toxic liquid, dry material, and slurry systems;

(3) RTR pipe in buried flammable and combustible liquid service as described above;

(4) polyethylene pipe in buried flammable and combustible liquid and gas service as described above; and

(5) metallic piping lined with non-metals

The behavior of non-metallic piping is more complex than metallic piping, and the design criteria are significantly less well developed. As a result, the designers are left to their best judgment in many circumstances. For example, while a formal flexibility analysis is required, no methods are provided for doing so.

Non-metallic piping systems per Appendix III are prohibited from use in confined space where toxic gases could be produced and accumulate, either from combustion of the piping materials or from exposure to flame or elevated temperatures from fire.

The remainder of this chapter covers the non-mandatory Appendix III rules.

16.2 DESIGN CONDITIONS

Chapter II requirements with respect to design pressure and temperatures are generally applicable.

16.3 ALLOWABLE STRESS

Various plastic materials have different, established methods of determining allowable stresses. Some limited allowable stress values are provided in Tables III-4.2.1, III-4.2.2, and III-4.2.3 for thermoplastic, laminated reinforced thermosetting resin (RTR), and machine-made RTR plastic pipe, respectively. For the most part, allowable stresses or pressure ratings must be determined from tests performed by the manufacturer.

The methods of determining the allowable stresses in thermoplastics and reinforced thermosetting resins are provided in ASTM specifications as follows:

(1) Thermoplastic: hydrostatic design stress (HDS) is determined in accordance with ASTM D 2837. Note that the strength is determined based on time-dependent properties [long-term tests extrapolated to longer design times (100,000 hours and 50 years)]. This is because creep is significant for this material even at ambient temperature. Furthermore, the strength of this material is highly sensitive to temperature.

(2) Reinforced Thermosetting Resin (laminated): design stress (DS) is taken as one-tenth of the minimum tensile strengths specified in Table 1 of ASTM C 582. This is also called hand layup. The strength of RTR is not particularly temperature sensitive in the range of application, so this allowable stress is considered to be valid from −29°C (−20°F) through 82°C (180°F).

(3) Reinforced Thermosetting Resin (filament wound and centrifugally cast): the hydrostatic design stress (HDS) used in design is the hydrostatic design basis stress (HDBS) times a service factor, F (often taken as 0.5), which is selected in accordance with ASTM D 2992. The HDBS is determined in accordance with ASTM D 2992. The HDBS is determined from long-term testing.

16.4 PRESSURE DESIGN

The philosophy of the Chapter II with respect to metallic piping applies to non-metallic piping. The primary differences are that the Table of listed components is Table III-4.1.1 rather than 126.1, and there are substantially fewer pressure design equations provided in Appendix III.

Listed components with established ratings are accepted at those ratings. Listed components without established ratings, but with allowable stresses listed, can be rated using the pressure design rules of III-2.2; however, these are very limited. Otherwise (listed components without allowable stresses or unlisted components), components must be rated per para. III-2.2.9.

The variations permitted in the base Code (para. 102.2.4) are not permitted for non-metallic piping.

The equations that are available for sizing non-metallic components are very limited in Appendix III. It includes straight pipe, bends, flanges, and blind flanges. Furthermore, the use of the referenced flange design method (per ASME BPVC, Section VIII, Division 1, Appendix 2) is questionable for many non-metallics. As a result, for pressure design, most non-metallic piping components must be either per a listed standard (i.e., listed in Table III-4.1.1) or qualified per III-2.2.9.

Paragraph III-2.2.9, pressure design of unlisted components and joints, differs from the rules for metallic pipe in that neither experimental stress analysis nor numerical analysis (e.g., finite element) are listed as acceptable alternatives for qualifying components. The two methods that are considered acceptable for substantiating the pressure design are extensive successful service experience under comparable design conditions with similarly proportioned components made of the same or similar material, or a performance test.

The performance test must include the effects of time, since failure of non-metallic components can be time dependent.

For straight pipe, equations (2) for thermoplastic, (3) for RTR-laminated pipe, and (4) for RTR-machine-made pipe (filament wound and RPM centrifugally cast pipe) are provided in Appendix III. They are of the following form:

$$t = \frac{D}{\dfrac{2SF}{P} + 1}$$

(2, 3, 4)

where:

    D = outside diameter of pipe

    F = service factor, which is only used for machine-made (filament-wound and centrifugally cast) pipe

    P = internal design gage pressure

    S = design stress from applicable table

    t = pressure design thickness

16.5   LIMITATIONS ON COMPONENTS AND JOINTS

Requirements for non-metallic piping components and joints are covered in III-2.3 and III-2.4. There are not any specific component limitations. Requirements for joints include bonded joints, flanged joints, expanded or rolled joints (not permitted), threaded joints, caulked joints, and proprietary joints.

16.6   FLEXIBILITY AND SUPPORT

Rules for piping flexibility and support for non-metallic piping are provided in para. III-2.5. Appendix III does not provide detailed rules for evaluation of non-metallic piping systems for thermal expansion. However, it requires a formal flexibility analysis when the following exemptions from formal flexibility analysis are not met:

(1) duplicates, or replaces without significant change, a system operating with a successful service record;

(2) can readily be judged adequate by comparison with previously analyzed systems; and

(3) is laid out with a conservative margin of inherent flexibility, or employs joining methods or expansion-joint devices, or a combination of these methods, in accordance with manufacturer's instructions.

As in metallic piping, a formal analysis is not necessarily a computer analysis. It can be any appropriate method, including charts and simplified calculations. The objectives in the design of a piping system for thermal expansion are the same as for metallic piping systems. Specifically, they are to prevent the following:

(1) failure of piping or supports from overstrain or fatigue;

(2) leakage at joints; or

(3) detrimental stresses or distortion in piping or in connected equipment (pumps, for example) resulting from excessive thrusts and moments in the piping (para. III-2.5.4).

One of the significant differences from metallic systems is that fully restrained designs are commonly used. That is, systems where the thermal expansion is offset by elastic compression/extension of the piping between axial restraints. This is possible because of the relatively low elastic modulus of plastic piping. The resulting loads are generally reasonable for the design of structural anchors. Note, however, that in performing a computer flexibility analysis of such systems, the axial load component of thermal expansion stress must be included. See Section 9.1 for a discussion of stresses from axial loads in flexibility analysis. Other significant differences from metallic piping include the following:

(1) Most RTR and RPM systems are non-isotropic. That is, the material properties are different in different directions, depending on the orientation of the reinforcing fibers.

(2) Axial extension of the pipe from internal pressure can be significant and should be considered. Note that fiber-wound RTR pipe can either extend or contract from internal pressure, depending on the orientation of the reinforcing fibers.

(3) Plastic materials creep at ambient temperature. For example, a plastic pipe that is fully restrained and compressed as it heats, can experience compressive creep strain during operation. When it cools back to ambient temperature, this can result in tension in the pipe and a load reversal on the restraints.

(4) In plastic piping, particularly RTR systems, the limiting component is often a fitting or joint. For such systems, the results of the flexibility analysis can be an evaluation of the loads versus the allowable loads on components, rather than a comparison of stress with allowable stress.

(5) Material properties, even for nominally the same material, are often manufacturer-specific. Thus, the design of plastic systems generally requires interaction and consultation with the manufacturer of the pipe and information on the resin. This is particularly so for RTR and RPM piping, which also includes the consideration of the fiber reinforcing.

(6) Stress intensification factors have not been developed for non-metallic piping. For many non-metallic components (RTR in particular), the design is manufacturer-specific. Thus, the development of industry-standard stress-intensification factors is problematic.

In general, design of RTR and RPM piping considers the material to be brittle. Thus, there is essentially no difference between stresses from thermal expansion and those from weight or pressure, and an allowable stress that is comparable to that permitted for pressure is commonly used as an allowable for the total weight plus pressure plus thermal expansion stress. For laminated RTR, the allowable for pressure stress is one-tenth of the tensile strength, and a commonly used allowable for longitudinal stress from combined loads is one-fifth of the tensile strength.

While the behavior of thermoplastics is generally not brittle so that the shakedown concepts of metallic piping may be applicable, this technology is not developed. Thus, the allowable for longitudinal stresses from combined loads is often taken conservatively as the allowable stress for internal pressure.

Some methods have been developed for evaluation of plastic piping. A literature survey is provided in WRC 415 (Short et al, 1996).

With respect to support, Appendix III highlights some specific concerns for non-metallic piping in para. III-2.6. In non-metallic piping that has limited ductility, avoidance of point loads can be critical to system performance. While local loads may be accommodated in ductile systems by local plastic deformation, such loads can result in brittle fracture in materials that are brittle.

Another consideration is that deformation can accumulate over time because of creep. Thus, support spacing must be sufficiently close to avoid excessive long-term sagging from creep.

16.7   MATERIALS

Limitations with respect to application of materials based on the fluid are discussed in Section 16.1.

The maximum design temperature for a material listed in the allowable stress tables is not permitted to exceed the maximum temperature listed in the table. Furthermore, listed materials are not permitted to be used below the minimum temperature specified in the allowable stress tables.

16.8   FABRICATION, ASSEMBLY, AND ERECTION

One of the key elements to successful construction of a plastic piping system is the joints. Appendix III requires a formal process of developing, documenting, and qualifying bonding procedures and personnel performing the bonding. The joints in plastic (RTR, RPM, and thermoplastic) piping are called bonds. The requirements are similar to the requirements for qualification of welds and welders.

The first step is to have a documented bonding procedure specification (BPS). The specification must document the procedures for making the joint, as set forth in para. III-5.1.2. This procedure must be qualified by a bonding procedure qualification test. Once it is so qualified, it may be used by personnel to bond non-metallic ASME B31.1 piping systems. Those bonders, however, must also be qualified to perform the work.

The bonders are qualified by performance qualification testing. The qualification test for the bonding procedure and the bonder are the same. They must fabricate and assemble, including at least one pipe-to-pipe joint and one pipe-to-fitting joint, and pressure test it. There are two options for the pressure test. The first is to perform a hydrostatic pressure test on the assembly to a pressure of the maximum of 150 psi (1000 kPa) or 1.5 times the maximum allowable pressure of the assembly. The second option is to cut three coupons containing the joint and bending the strips using a procedure to be defined in the bonding procedure specification. In the bend test option, the test strips shall not break when bent a minimum of 90 deg., at ambient temperature, over an inside bend radius of 1.5 times the nominal diameter of the tested pipe.

To qualify the bonding procedure specification (BPS), at least one of each joint type covered by the BPS must be included in the test(s). With respect to size, if the largest joint in 110 mm (NPS 4) or smaller, the test assembly is required to be the largest size to be joined. If the largest pipe to be joined is greater than 110 mm (NPS 4), the size needs to be greater than 110 mm (NPS 4) or 25% of the largest pipe to be joined, whichever is greater.

The same as for welding, the employer of the bonder is responsible for performing the bonding procedure qualification test, qualifying bonders, and maintaining records of the specifications and test. Under certain circumstances, as described in ASME B31.1 including approval of the designer, use of bonding procedure specifications qualified by others and bonders qualified by others is permitted.

If a bonder or bonding operator has not used a specific bonding process for a period of six months, re-qualification is required.

Again, similar to welding, the bonds that are made are required to be identified with a symbol that indicates which joints are made by which bonder. As an alternative, appropriate records that provide this information may be used instead of physically marking each joint.

General requirements are provided in para. III-5.1.3 for solvent-cemented joints in thermoplastic piping and heat-fusion joints in thermoplastic piping, and in para. III-5.1.4 for adhesive joints in RTR, and butt-and-wrapped joints in RTR piping.

Other requirements for fabrication of plastic piping are provided in paras. III-5.2 through III-5.4. These are not comprehensive requirements; they address some specific considerations that apply to plastic piping.

Insert 16.1   Bonding Processes

Thermosetting Resin Piping

A thermosetting plastic material is one that is polymerized (cured) by application of heat or chemical means and is thus changed into an infusible or insoluble final product. Fiberglass pipe is an example of a reinforced thermosetting resin (RTR) pipe. Unlike thermoplastic materials, they cannot be readily reshaped by heating them to their melting range once they are cured. The thermosetting plastic products will usually degrade once they are heated to such temperatures. The two types of thermosetting resin for piping are (1) RTR, which is composed of a fibrous reinforcement material such as glass or carbon fiber in a thermosetting plastic resin such as epoxy, polyester, furan, or phenolic, and (2) reinforced plastic mortar (RPM), which also includes a filler material such as sand. Bonding processes for thermoplastic pipe include adhesive joints and various types of butt-and-wrap joints.

Adhesive Joints

Principles of Operation

Adhesive joining for RTR piping uses adhesive to join the pipes (Fig. 16.1). The pipe joint usually has a bell-and-spigot or a tapered-joint configuration. Depending on the adhesive used and the piping materials, the joint may require curing by one of several methods. These include using chemical heat wraps, applying heat to the pipe joint by other means, or allowing the joint to cure naturally.

Process Limitations

The application of the adhesive to the surfaces to be joined and subsequent assembly of the surfaces should result in a continuous bond. These joints require the pipe to overlap in order to provide a sufficient area of glued surfaces. As a result, in assembling a piping system, the piping may need to have sufficient flexibility for the final pipe joint or joints to be made. Strength of the glued joint may be less than that of the pipe, although the strength of a properly made tapered joint can be as high as that of the pipe.

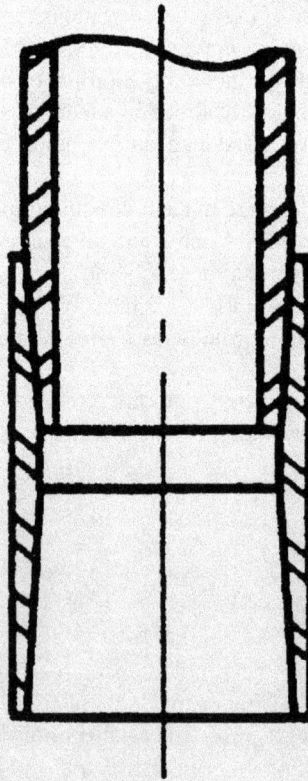

FIG. 16.1
FULLY TAPERED THERMOSETTING ADHESIVE JOINT (ASME B31.3, FIG. A328.5.6)

Butt-and-Wrap Joints

Principles of Operation

Butt-and-wrap pipe joints may be applied to a variety of RTR piping (Fig. 16.2). This joining technique can be applied to pipe joints by butting two pipes end to end or where the joints have a bell-and spigot arrangement. In either case, layers of resin-impregnated glass fiber cloth (or other reinforcing fiber consistent with the pipe) overwrap the pipe joint area. The cloths are applied in layers on the joint, building it up. Ideally, the finished joint should be capable of withstanding internal pressure, longitudinal force, and bending moments.

Cuts of the pipe should always be sealed so the fiber reinforcement can be protected from contacting the service fluid. Branch connections may be made using similar hand layup techniques.

Process Limitations

The strength of the pipe joint may be designed such that it equals or exceeds that of the pipe. However, the joints can also be a weak link in the system, with butt-and-wrap strength (using square-ended pipes butted together) is sometimes considered to be one-half the strength of the pipe.

Thermoplastic Piping

Thermoplastic piping is typically unreinforced and composed of thermoplastic material such as polyethylene (PE), high-density polyethylene (HDPE), polypropylene (PP), ethylene-chlorotrifluoroethylene (ECTFE), ethylene-tetrafluoroethylene (ETFE), poly(vinyl chloride) (PVC), chlorinated poly(vinyl chloride) (CPVC), acrylonitrile–butadiene–styrene (ABS), and poly(vinylidene fluoride) (PVDF). These materials soften when heated and can be shaped and fused in the heated state and cooled without

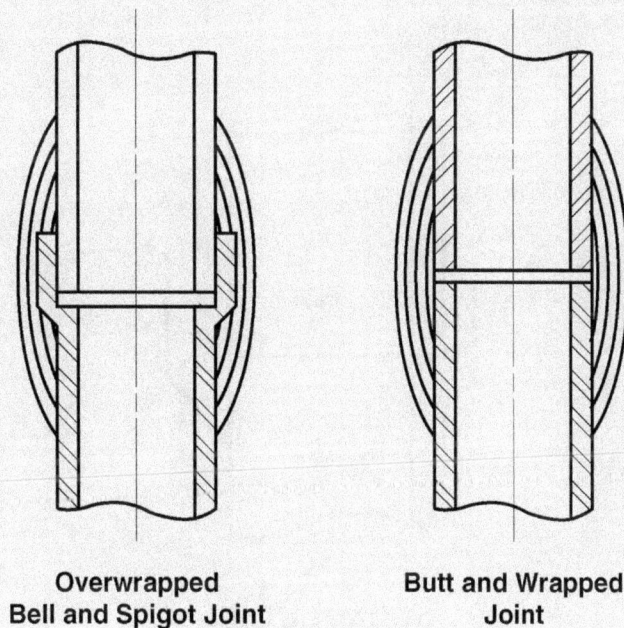

Overwrapped
Bell and Spigot Joint

Butt and Wrapped
Joint

FIG. 16.2
THERMOSETTING WRAPPED JOINTS (ASME B31.3, FIG. A328.5.7)

degradation of the material. As a result, fusion bonding of thermoplastic is a possible and often preferred option for bonding.

Solvent-Cement Joining

Principles of Operation

Solvent-cement joints are used on some common types of thermoplastic piping materials, including PVC, CPVC, and ABS (Fig. 16.3). Joining methods are covered by ASTM standards. Usually, the pipe and socket fitting to be joined have a slight interference fit. Proper joining relies on softening and fusing of the pipe and fitting material and should result in a bond that is stronger than the pipe. Both parts of the joint to be made are first cleaned by wiping with an appropriate cleaning agent. The joint to be made is prepared by applying a suitable primer before the solvent cement is applied. The primers are described in an ASTM specification that applies to the material being solvent-cemented. Primer and cement are applied to both surfaces to be joined according to recommended procedures. A continuous bond must be produced, with a small bead of excess cement appearing on the outer limit of the joint, when the fully wetted male end is inserted into the fully wetted female socket.

Process Limitations

Some materials, such as PP, cannot be suitably joined using solvent-cement techniques. Further, a good solvent-cement joint must have a slight interference fit between the outside diameter of the pipe and the inside diameter of the mating socket. The diametrical clearance between the pipe and the entrance of the fitting socket should never exceed 0.04 in. With a large gap, heavier cement, which has a greater film thickness to bridge the gap, is used. However, such joints can end up relying on the shear strength of the glue rather than a bonding of the mating plastic parts, which is significantly weaker than proper solvent-cement joints with interference fits.

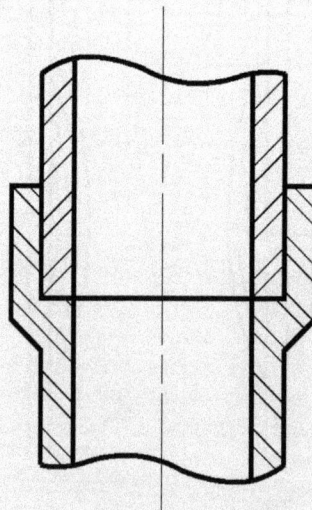

Socket Joint

FIG. 16.3
THERMOPLASTIC SOLVENT-CEMENTED JOINT (ASME B31.3, FIG. A328.5.3)

Hot Gas Welding

Principles of Operation

Hot gas welding is used exclusively for thermoplastic materials; it is not suitable for thermosetting resin materials (Fig. 16.4). It employs hot air or inert gas to melt the base material and tip of the welding rod. The melted welding rod fuses with the base material to form a weld joint. Although it is common practice to use partial-penetration welds, it is essential to use full-penetration welds to properly develop the joint strength, which will be weaker than the base material in any case.

Extrusion welding is similar to hot gas welding except that the weldment is extruded from a screw-driven nozzle. The equipment required for extrusion welding is much larger than that for hot gas welding. The nozzle must be large enough to hold an extrusion screw and a motor must be attached to drive the screw. As a result, extrusion welding requires a great deal of space to conduct the welding; approximately 3 ft of clearance in all directions is the usual requirement.

The advantage of extrusion welding over hot gas welding is the application of a thick weld bead in a single pass (usually the equivalent to 5 to 12 passes of hot gas welding). The result is a substantially lower degree of residual stress and lesser effects of oxidation.

Process Limitations

Proper joining by this technique is more difficult for some materials, such as PVC, and should not be used for those materials except in special circumstances. In those circumstance, special precautions, such as use of an inert gas such as nitrogen for the hot gas, may be necessary.

Due to a number of factors, including imperfections, residual stress, and oxidation, the strength of hot-gas-welded joints and extrusion-welded joints is lower than that of the base material. According to German Standard DVS 2203, the long-term strength of a hot-gas-welded joint is 40% of that of the base material for HDPE, PP, PVC, and PVDF. The long-term strength of extrusion welds is, according to the same standard, 60% of that of the base material for HDPE and PP. The relatively common practice of making partial-penetration welds using hot gas welding techniques results in substantial further reduction of joint strength.

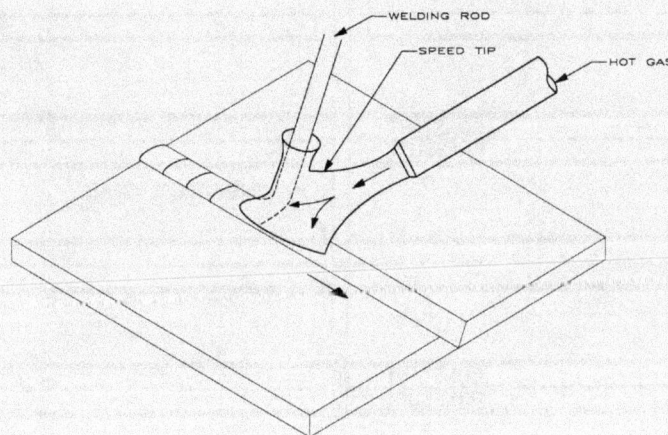

FIG. 16.4
HOT GAS WELDING

Heat Fusion

Principles of Operation

Heat fusion relies on using a suitable heating element to bring the mating surfaces up to the melting temperature and then forcing the two surfaces together to cause them to flow and fuse. It is the most common techniques for bonding all polyolefins, such as HDPE and PE, and fluoropolymers, such as PVDF, ECTFE, and ETFE. Available techniques include heat-element butt fusion, heat element socket fusion, electrical resistance fusion, and branch fabrication techniques such as sidewall fusion. In heat-element butt fusion, the mating pipe ends are forced against a heating element with a specific pressure to bring the mating surfaces up to melting temperature (Fig. 16.5). The heating element is removed and then the mating pipe ends are forced against each other with a specified pressure for a specified duration of time.

In heat-element socket fusion, a heating element is inserted into the socket and the pipe end is inserted into a heating element (Fig. 16.6). After the mating surfaces are brought up to melting temperature, the pipe is inserted into the socket, again with a specified force and duration. There is an interference fit and the pipe fuses to the socket.

In electric resistance socket fusion, electric heating element embedded in the socket fitting wall brings the surfaces up to melting temperature, again accomplishing the same objective (Fig. 16.7).

In sidewall (saddle) fusion techniques (Fig. 16.7), a hole is cut in the sidewall of the pipe with a bevel and the branch pipe is prepared with a mating bevel. A heating element is used to heat the end of the branch and mating surfaces in the run pipe. Joining is similar to other heat fusion techniques. Other heat fusion techniques based on the same principles are also used in branch and reinforcement bonding.

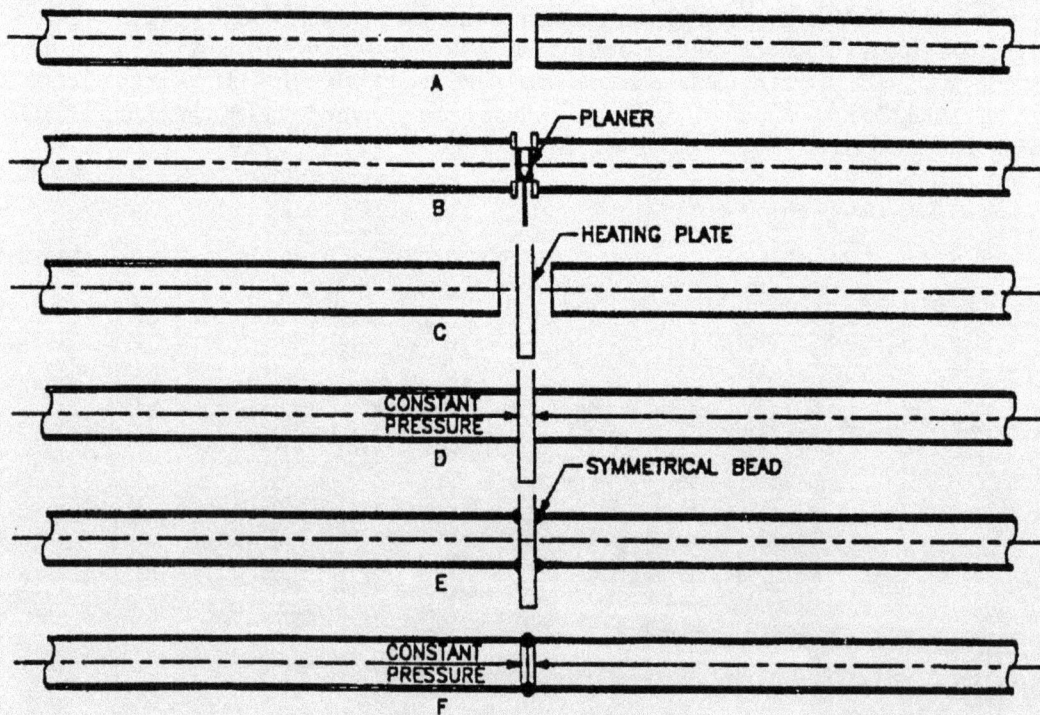

FIG. 16.5
STEPS FOR HEAT-ELEMENT BUTT FUSION (COURTESY OF CHRIS ZIU)

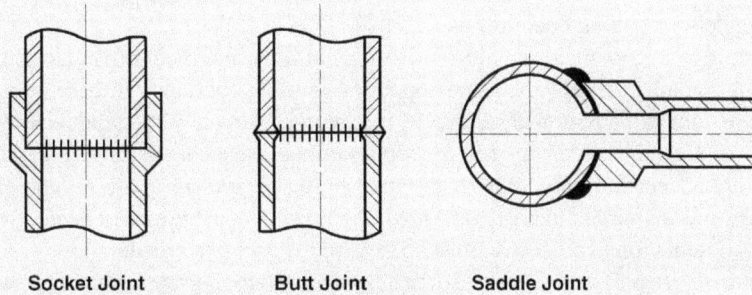

Socket Joint   Butt Joint   Saddle Joint

FIG. 16.6
THERMOPLASTIC HEAT FUSION JOINTS (ASME B31.3, FIG. A328.5.4)

Process Limitations

Heat fusion provides superior joints to hot gas and extrusion welding. However, special equipment is required to heat the surfaces and draw the mating components together with the appropriate force (pressure on the surfaces being bonded). The equipment can be large, so sufficient working space is required. Also, the mating parts must be physically separated to provide space for the heating element and then pressed together. As such, sufficient pipe flexibility or other means must be provided to make the final joint in piping systems between anchors or other fixed points.

According to German Standard DVS 2203, the long-term strength of a properly made heat fusion butt joint is 80% of that of the base material for HDPE and PP and 60% of that for PVC and PVDF.

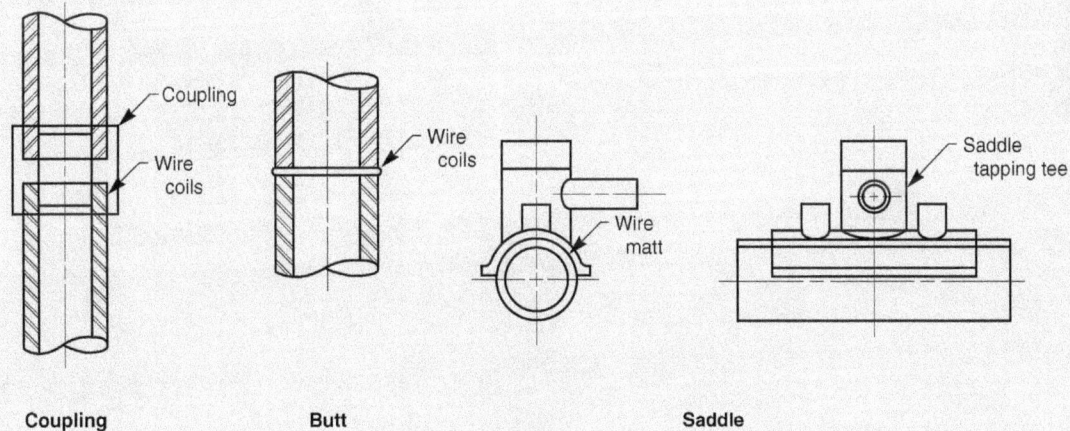

Coupling   Butt   Saddle

FIG. 16.7
THERMOPLASTIC ELECTROFUSION JOINTS (ASME B31.3, FIG. A328.5.5)

16.9   EXAMINATION AND TESTING

The non-destructive examination techniques for non-metallic piping are not nearly as well developed as for metallic piping. As a result, the only technique that is used is visual examination. Unlike the rules for metallic piping in ASME B31.1, progressive examination techniques are adapted in para. III-6.2. This is a random examination technique that is intended to result in a certain level of quality. If items that fail the examination are found, more items are examined. The basic visual examination requirement is 5%, including, for bonds, the work of each bonder or bonding operator.

If a defective item is found, two additional examinations of the same type are required to be made of the same kind of item (if a bond, others by the same bonder or bonding operator). If these items pass, the items represented by the random examination are accepted (pending passing any required leak testing). If the additional examination finds another defective item, two additional items of the same type are required to be examined. If these are acceptable, the items represented by the examination are accepted. If any of these examinations fail, the entire group of items represented by the random examination is required to be replaced or fully examined. All items found defective must be repaired and re-examined.

Leak tests, if required, are performed in accordance with the Code rules described in Chapter 15.

POST-CONSTRUCTION

ASME B31.1 provides Chapter VI, Operation and Maintenance, as well as non-mandatory appendices IV (Corrosion Control for ASME B31.1 Power Piping Systems) and V (Recommended Practice for Operation, Maintenance, and Modification of Power Piping Systems) dealing with post-construction issues such as inspection and repair. Post-construction deals with issues after the piping has been put in service.

Chapter VII of ASME B31.1 provides requirements for operation and maintenance of covered piping systems (CPS). Covered piping systems are defined (see definitions in para. 100.2) as at a minimum the following in an electric power generating station.

- NPS 4 and larger main steam, hot reheat steam, cold reheat steam, and boiler feedwater piping systems
- NPS 4 and larger piping in other systems that operate above 750°F (400°C) or above 1,025 psi (7100 kPa)

In addition, the Operating Company may include other piping systems determined to be hazardous considering both probability and consequences of failure.

Chapter VII includes the following elements.

- Specific requirements for written operation and maintenance procedures which document not only how it should be operated but also how it has been operated and maintained.
- A program for condition assessment with the results of the assessments documented
- As built and as modified or repaired piping drawings
- A program for periodic walkdowns with requirements to evaluate things such as unexpected piping position changes and significant vibration.
- A program for data collection
- An evaluation of high-priority areas operating in the creep regime

Guidance in ASME B31.1 on post-construction issues is provided in non-mandatory Appendices IV and V. Appendix IV covers Corrosion Control for ASME B31.1 Power Piping Systems. The Foreword to that appendix states it provides minimum requirements; however, the appendix is non-mandatory.

Appendix V provides Recommended Practice for Operation, Maintenance, and Modification of Power Piping Systems. It requires the following:

(1) procedures for assuring that personnel performing maintenance are qualified by training or experience for their tasks

(2) documented operating and maintenance programs

(3) documentation related to design and system construction details

(4) documented procedures for operation and maintenance

(5) material records

(6) records as to repairs, testing (e.g. wall thickness measurements), NDE reports, and failures

(7) documented failure investigations with respect to all material failures in critical piping systems

(8) documentation with respect to repair or replacement of failed components

(9) special inspection programs for materials with known adverse performance history

(10) periodic documented visual surveys of critical piping, including piping position, hanger position, vibration, interferences, etc.

In addition to the B31.1 provision, other standards should be considered for application, as follows.

ASME PCC-3 Inspection Planning Using Risk Based Methods — The post-construction inspection guidance provided in ASME B31.1 are prescriptive whereas the current state of the art in inspection planning uses risk based methods. In risk-based inspection planning, the specific deterioration mechanisms relevant to the component are considered, and the probability of failure is multiplied by the consequence of failure to determine risk. Inspection strategies are evaluated relative to their effectiveness in reducing risk. This is best done on a cost benefit basis. Other maintenance strategies can be considered in the same manner.

ASME PCC-2, *Repair of Pressure Equipment and Piping* — The repair guidance in ASME B31.1 is very limited. PCC-2 provides many recognized and generally accepted practices for many repair methods. The 2011 edition provides 28 articles covering repairs using welding, mechanical repairs such as clamps, non-metallic repairs such as composite wraps, and examination and testing. The repair method articles include relevant design, fabrication, examination, and testing practices.

The standard has been written in modular form. Each repair article is essentially standalone, with the exception that one section on general requirements applies to all. The following articles are in the 2011 edition; the article on field heat treating of vessels is new to the 2011 edition. As can be seen, a wide variety of repair methods is covered, and the committee continues to develop additional articles as well as improve existing ones.

Welded Repairs

- Butt-Welded Insert Plates in Pressure Components
- External Weld Overlay to Repair Internal Thinning
- Seal-Welded Threaded Connections and Seal Weld Repairs
- Welded Leak Repair Box
- Full Encirclement Steel Sleeves for Piping
- Fillet Welded Patches with Reinforcing Plug Welds
- Alternatives to Traditional Welding Preheat
- Alternatives to Post-weld Heat Treatment
- In-Service Welding onto Carbon Steel Pressure Components or Pipelines
- Weld Buildup, Weld Overlay and Clad Restoration
- Fillet Welded Patches
- Threaded or Welded Plug Repairs
- Field Heat Treating of Vessels

Mechanical Repairs

- Replacement of Pressure Components
- Freeze Plugs
- Damaged Threads in Tapped Holes
- Flaw Excavation and Weld Repair
- Flange Refinishing
- Mechanical Clamp Repair
- Pipe Straightening or Alignment Bending
- Damaged Anchors in Concrete (Post-Installed Mechanical Anchors)

- Hot and Half Bolting Removal Procedures
- Inspection and Repair of Shell and Tube Heat Exchangers

Non-Metallic and Bonded Repairs

- Non-Metallic Composite Wrap Systems for Piping and Pipework: High Risk Applications
- Non-Metallic Composite Wrap Systems for Pipe: Low Risk Applications
- Non-Metallic Internal Lining for Pipe-Sprayed Form for Buried Piping

Examination and Testing

- Pressure and Tightness Testing of Piping and Equipment
- Nondestructive Examination in Lieu of Pressure Testing for Repairs and Alterations

API 579/ASME FFS-1, *Fitness for Service* — ASME B31.1 provides very limited information on performing fitness for service work. The recommendations are limited to references some methods for evaluating wall thinning, in the references of Appendix IV. API 579/ASME FFS-1 provides comprehensive methods for performing fitness for service assessments for a variety of deterioration, including general wall thinning, local thinning, cracks, creep damage, and geometric defects such as dents and gouges. The standard is organized into parts, each covering assessment of a different type of damage, as follows.

- Part 1 — Introduction
- Part 2 — Fitness-for Service Engineering Assessment Procedure
- Part 3 — Assessment of Existing Equipment for Brittle Fracture
- Part 4 — Assessment of General Metal Loss
- Part 5 — Assessment of Local Metal Loss
- Part 6 — Assessment of Pitting Corrosion
- Part 7 — Assessment of Hydrogen Blisters and Hydrogen Damage Associated with HIC and SOHIC
- Part 8 — Assessment of Weld Misalignment and Shell Distortions
- Part 9 — Assessment of Crack-Like Flaws
- Part 10 — Assessment of Components Operating in the Creep Range
- Part 11 — Assessment of Fire Damage
- Part 12 — Assessment of Dents, Gouges, and Dent-Gouge Combinations
- Part 13 — Assessment of Laminations

Appendix I

Properties of Pipe and Pressure Ratings of Listed Piping Components

I-1 THREAD AND GROOVE DEPTHS

The following table provides the nominal thread depth, h, from Table 2 of ANSI/ASME B1.20.1—1983. It is applicable for pressure design of pipe with American Standard Taper Pipe Thread (NPT). It is 0.8 times the thread pitch. The thread depth must be considered as a mechanical allowance in the pressure design of straight pipe in accordance with the base Code.

Nominal Pipe Size	Height of Thread, h (in.)
$^{1}/_{16}$	0.02963
$^{1}/_{8}$	0.02963
$^{1}/_{4}$	0.04444
$^{3}/_{8}$	0.04444
$^{1}/_{2}$	0.05714
$^{3}/_{4}$	0.05714
1	0.06957
1 $^{1}/_{4}$	0.06957
1 $^{1}/_{2}$	0.06957
2	0.06957
2 $^{1}/_{2}$	0.100000
3	0.100000
3 $^{1}/_{2}$	0.100000
4	0.100000
5	0.100000
6	0.100000
8	0.100000
10	0.100000
12	0.100000
14 O.D.	0.100000
16 O.D.	0.100000
18 O.D.	0.100000
20 O.D.	0.100000
24 O.D.	0.100000

The following table provides standard cut groove depths for IPS pipe from Victaulic®. Roll grooving does not remove metal from the pipe and therefore need not be considered in the allowances in pressure design. The cut groove depth must be considered in the allowances in pressure design.

Nominal Size (in.)	Groove Depth, D (in.)	Minimum Allowable Pipe Wall Thickness, T (in.)
$^3/_4$	0.056	0.113
1	0.063	0.133
1 $^1/_4$	0.063	0.140
1 $^1/_2$	0.063	0.145
2	0.063	0.154
2 $^1/_2$	0.078	0.188
3 O.D.	0.078	0.188
3	0.078	0.188
3 $^1/_2$	0.083	0.188
4	0.083	0.203
4 $^1/_4$ O.D.	0.083	0.203
4 $^1/_2$	0.083	0.203
5 $^1/_4$ O.D.	0.083	0.203
5 $^1/_2$ O.D.	0.083	0.203
5	0.084	0.203
6 O.D.	0.085	0.219
6 $^1/_4$ O.D.	0.109	0.249
6 $^1/_2$ O.D.	0.085	0.219
6	0.085	0.219
8 O.D.	0.092	0.238
8	0.092	0.238
10 O.D.	0.094	0.250
10	0.094	0.250
12 O.D.	0.109	0.279
12	0.109	0.279
14 O.D.	0.109	0.281
15 O.D.	0.109	0.312
16 O.D.	0.109	0.312
18 O.D.	0.109	0.312
20 O.D.	0.109	0.312
22 O.D.	0.172	0.375
24 O.D.	0.172	0.375
30 O.D.	0.250	0.625

The groove depth, D, is the nominal groove depth which is required to be uniform. The thickness, T, is the minimum pipe wall thickness that may be cut grooved. Note that the pressure rating of the fitting, provided by the manufacturer, may govern the pressure design. Also note that the pressure rating of Victaulic® couplings on roll and cut grooved standard wall pipes are the same.

I-2   PROPERTIES OF PIPE

The following table provides properties for nominal pipe. The schedule numbers are provided as follows.

   a. Pipe schedule numbers per ASME B36.10, Welded and Seamless Wrought Steel Pipe.
   b. Pipe nominal wall thickness designation per ASME B36.10.
   c. Pipe schedule numbers per ASME B36.19, Stainless Steel Pipe.

The weight per foot of pipe is based on carbon steel.

PROPERTIES OF PIPE
(Courtesy of Anvil International)

The following formulas are used in the computation of the values shown in the table:

$$\text{†weight of pipe per foot (pounds)} = 10.6802t\,(D - t)$$
$$\text{weight of water per foot (pounds)} = 0.3405d^2$$
$$\text{square feet outside surface per foot} = 0.2618D$$
$$\text{square feet inside surface per foot} = 0.2618d$$
$$\text{inside area (square inches)} = 0.785d^2$$
$$\text{area of metal (square inches)} = 0.785\,(D^2 - d^2)$$
$$\text{moment of inertia (inches}^4) = 0.0491\,(D^4 - d^4)$$
$$= A_m R_g^2$$
$$\text{section modulus (inches}^3) = \frac{0.0982\,(D^4 - d^4)}{D}$$
$$\text{radius of gyration (inches)} = 0.25\sqrt{D^2 + d^2}$$

A_m = area of metal (square inches)
d = inside diameter (inches)
D = outside diameter (inches)
R_g = radius of gyration (inches)
t = pipe wall thickness (inches)

† The ferritic steels may be about 5% less, and the austenitic stainless steels about 2% greater than the values shown in this table which are based on weights for carbon steel.

*schedule numbers

Standard weight pipe and schedule 40 are the same in all sizes through 10-inch; from 12-inch through 24-inch, standard weight pipe has a wall thickness of ⅜-inch.

Extra strong weight pipe and schedule 80 are the same in all sizes through 8-inch; from 8-inch through 24-inch, extra strong weight pipe has a wall thickness of ½-inch.

Double extra strong weight pipe has no corresponding schedule number.

a: ANSI B36.10 steel pipe schedule numbers

b: ANSI B36.10 steel pipe nominal wall thickness designation

c: ANSI B36.19 stainless steel pipe schedule numbers

nominal pipe size outside diameter, in.	schedule number*			wall thick-ness, in.	inside diam-eter, in.	inside area, sq. in.	metal area, sq. in.	sq ft outside surface, per ft	sq ft inside surface, per ft	weight per ft, lb†	weight of water per ft, lb	moment of inertia, in.⁴	section modu-lus, in.³	radius gyra-tion, in.
	a	b	c											
⅛ 0.405	—	—	10S	0.049	0.307	0.0740	0.0548	0.106	0.0804	0.186	0.0321	0.00088	0.00437	0.1271
	40	Std	40S	0.068	0.269	0.0568	0.0720	0.106	0.0705	0.245	0.0246	0.00106	0.00525	0.1215
	80	XS	80S	0.095	0.215	0.0364	0.0925	0.106	0.0563	0.315	0.0157	0.00122	0.00600	0.1146
¼ 0.540	—	—	10S	0.065	0.410	0.1320	0.0970	0.141	0.1073	0.330	0.0572	0.00279	0.01032	0.1694
	40	Std	40S	0.088	0.364	0.1041	0.1250	0.141	0.0955	0.425	0.0451	0.00331	0.01230	0.1628
	80	XS	80S	0.119	0.302	0.0716	0.1574	0.141	0.0794	0.535	0.0310	0.00378	0.01395	0.1547
⅜ 0.675	—	—	10S	0.065	0.545	0.2333	0.1246	0.177	0.1427	0.423	0.1011	0.00586	0.01737	0.2169
	40	Std	40S	0.091	0.493	0.1910	0.1670	0.177	0.1295	0.568	0.0827	0.00730	0.02160	0.2090
	80	XS	80S	0.126	0.423	0.1405	0.2173	0.177	0.1106	0.739	0.0609	0.00862	0.02554	0.1991
½ 0.840	—	—	5S	0.065	0.710	0.3959	0.1583	0.220	0.1859	0.538	0.171	0.0120	0.0285	0.2750
	—	—	10S	0.083	0.674	0.357	0.1974	0.220	0.1765	0.671	0.1547	0.01431	0.0341	0.2692
	40	Std	40S	0.109	0.622	0.304	0.2503	0.220	0.1628	0.851	0.1316	0.01710	0.0407	0.2613
	80	XS	80S	0.147	0.546	0.2340	0.320	0.220	0.1433	1.088	0.1013	0.02010	0.0478	0.2505
	160	—	—	0.187	0.466	0.1706	0.383	0.220	0.1220	1.304	0.0740	0.02213	0.0527	0.2402
	—	XXS	—	0.294	0.252	0.0499	0.504	0.220	0.0660	1.714	0.0216	0.02425	0.0577	0.2192
¾ 1.050	—	—	5S	0.065	0.920	0.665	0.2011	0.275	0.2409	0.684	0.2882	0.02451	0.0467	0.349
	—	—	10S	0.083	0.884	0.614	0.2521	0.275	0.2314	0.857	0.2661	0.02970	0.0566	0.343
	40	Std	40S	0.113	0.824	0.533	0.333	0.275	0.2157	1.131	0.2301	0.0370	0.0706	0.334
	80	XS	80S	0.154	0.742	0.432	0.435	0.275	0.1943	1.474	0.1875	0.0448	0.0853	0.321
	160	—	—	0.218	0.614	0.2961	0.570	0.275	0.1607	1.937	0.1284	0.0527	0.1004	0.304
	—	XXS	—	0.308	0.434	0.1479	0.718	0.275	0.1137	2.441	0.0641	0.0579	0.1104	0.2840
1 1.315	—	—	5S	0.065	1.185	1.103	0.2553	0.344	0.310	0.868	0.478	0.0500	0.0760	0.443
	—	—	10S	0.109	1.097	0.945	0.413	0.344	0.2872	1.404	0.409	0.0757	0.1151	0.428
	40	Std	40S	0.133	1.049	0.864	0.494	0.344	0.2746	1.679	0.374	0.0874	0.1329	0.421
	80	XS	80S	0.179	0.957	0.719	0.639	0.344	0.2520	2.172	0.311	0.1056	0.1606	0.407
	160	—	—	0.250	0.815	0.522	0.836	0.344	0.2134	2.844	0.2261	0.1252	0.1903	0.387
	—	XXS	—	0.358	0.599	0.2818	1.076	0.344	0.1570	3.659	0.1221	0.1405	0.2137	0.361
1¼ 1.660	—	—	5S	0.065	1.530	1.839	0.326	0.434	0.401	1.107	0.797	0.1038	0.1250	0.564
	—	—	10S	0.109	1.442	1.633	0.531	0.434	0.378	1.805	0.707	0.1605	0.1934	0.550
	40	Std	40S	0.140	1.380	1.496	0.669	0.434	0.361	2.273	0.648	0.1948	0.2346	0.540
	80	XS	80S	0.191	1.278	1.283	0.881	0.434	0.335	2.997	0.555	0.2418	0.2913	0.524
	160	—	—	0.250	1.160	1.057	1.107	0.434	0.304	3.765	0.4580	0.2839	0.342	0.506
	—	XXS	—	0.382	0.896	0.0631	1.534	0.434	0.2346	5.214	0.2732	0.341	0.411	0.472
1½ 1.900	—	—	5S	0.065	1.770	2.461	0.375	0.497	0.463	1.274	1.067	0.1580	0.1663	0.649
	—	—	10S	0.109	1.682	2.222	0.613	0.497	0.440	2.085	0.962	0.2469	0.2599	0.634

PROPERTIES OF PIPE (CONTINUED)
(Courtesy of Anvil International)

nominal pipe size *outside diameter,* in.	schedule number* a	b	c	wall thickness, in.	inside diameter, in.	inside area, sq. in.	metal area, sq. in.	sq ft outside surface, per ft	sq ft inside surface, per ft	weight per ft, lb†	weight of water per ft, lb	moment of inertia, in.⁴	section modulus, in.³	radius gyration, in.
1½ 1.900	40	Std	40S	0.145	1.610	2.036	0.799	0.497	0.421	2.718	0.882	0.310	0.326	0.623
	80	XS	80S	0.200	1.500	1.767	1.068	0.497	0.393	3.631	0.765	0.391	0.412	0.605
	160	—	—	0.281	1.338	1.406	1.429	0.497	0.350	4.859	0.608	0.483	0.508	0.581
	—	XXS	—	0.400	1.100	0.950	1.885	0.497	0.288	6.408	0.412	0.568	0.598	0.549
	—	—	—	0.525	0.850	0.567	2.267	0.497	0.223	7.710	0.246	0.6140	0.6470	0.5200
	—	—	—	0.650	0.600	0.283	2.551	0.497	0.157	8.678	0.123	0.6340	0.6670	0.4980
2 2.375	—	—	5S	0.065	2.245	3.96	0.472	0.622	0.588	1.604	1.716	0.315	0.2652	0.817
	—	—	10S	0.109	2.157	3.65	0.776	0.622	0.565	2.638	1.582	0.499	0.420	0.802
	40	Std	40S	0.154	2.067	3.36	1.075	0.622	0.541	3.653	1.455	0.666	0.561	0.787
	80	XS	80S	0.218	1.939	2.953	1.477	0.622	0.508	5.022	1.280	0.868	0.731	0.766
	160	—	—	0.343	1.689	2.240	2.190	0.622	0.442	7.444	0.971	1.163	0.979	0.729
	—	XXS	—	0.436	1.503	1.774	2.656	0.622	0.393	9.029	0.769	1.312	1.104	0.703
	—	—	—	0.562	1.251	1.229	3.199	0.622	0.328	10.882	0.533	1.442	1.2140	0.6710
	—	—	—	0.687	1.001	0.787	3.641	0.622	0.262	12.385	0.341	1.5130	1.2740	0.6440
2½ 2.875	—	—	5S	0.083	2.709	5.76	0.728	0.753	0.709	2.475	2.499	0.710	0.494	0.988
	—	—	10S	0.120	2.635	5.45	1.039	0.753	0.690	3.531	2.361	0.988	0.687	0.975
	40	Std	40S	0.203	2.469	4.79	1.704	0.753	0.646	5.793	2.076	1.530	1.064	0.947
	80	XS	80S	0.276	2.323	4.24	2.254	0.753	0.608	7.661	1.837	1.925	1.339	0.924
	160	—	—	0.375	2.125	3.55	2.945	0.753	0.556	10.01	1.535	2.353	1.637	0.894
	—	XXS	—	0.552	1.771	2.464	4.03	0.753	0.464	13.70	1.067	2.872	1.998	0.844
	—	—	—	0.675	1.525	1.826	4.663	0.753	0.399	15.860	0.792	3.0890	2.1490	0.8140
	—	—	—	0.800	1.275	1.276	5.212	0.753	0.334	17.729	0.554	3.2250	2.2430	0.7860
3 3.500	—	—	5S	0.083	3.334	8.73	0.891	0.916	0.873	3.03	3.78	1.301	0.744	1.208
	—	—	10S	0.120	3.260	8.35	1.274	0.916	0.853	4.33	3.61	1.822	1.041	1.196
	40	Std	40S	0.216	3.068	7.39	2.228	0.916	0.803	7.58	3.20	3.02	1.724	1.164
	80	XS	80S	0.300	2.900	6.61	3.02	0.916	0.759	10.25	2.864	3.90	2.226	1.136
	160	—	—	0.437	2.626	5.42	4.21	0.916	0.687	14.32	2.348	5.03	2.876	1.094
	—	XXS	—	0.600	2.300	4.15	5.47	0.916	0.602	18.58	1.801	5.99	3.43	1.047
	—	—	—	0.725	2.050	3.299	6.317	0.916	0.537	21.487	1.431	6.5010	3.7150	1.0140
	—	—	—	0.850	1.800	2.543	7.073	0.916	0.471	24.057	1.103	6.8530	3.9160	0.9840
3½ 4.000	—	—	5S	0.083	3.834	11.55	1.021	1.047	1.004	3.47	5.01	1.960	0.980	1.385
	—	—	10S	0.120	3.760	11.10	1.463	1.047	0.984	4.97	4.81	2.756	1.378	1.372
	40	Std	40S	0.226	3.548	9.89	2.680	1.047	0.929	9.11	4.28	4.79	2.394	1.337
	80	XS	80S	0.318	3.364	8.89	3.68	1.047	0.881	12.51	3.85	6.28	3.14	1.307
	—	XXS	—	0.636	2.728	5.845	6.721	1.047	0.716	22.850	2.530	9.8480	4.9240	1.2100
4 4.500	—	—	5S	0.083	4.334	14.75	1.152	1.178	1.135	3.92	6.40	2.811	1.249	1.562
	—	—	10S	0.120	4.260	14.25	1.651	1.178	1.115	5.61	6.17	3.96	1.762	1.549
	—	—	—	0.188	4.124	13.357	2.547	1.178	1.082	8.560	5.800	5.8500	2.6000	1.5250
	40	Std	40S	0.237	4.026	12.73	3.17	1.178	1.054	10.79	5.51	7.23	3.21	1.510
	80	XS	80S	0.337	3.826	11.50	4.41	1.178	1.002	14.98	4.98	9.61	4.27	1.477
	120	—	—	0.437	3.626	10.33	5.58	1.178	0.949	18.96	4.48	11.65	5.18	1.445
	—	—	—	0.500	3.500	9.621	6.283	1.178	0.916	21.360	4.160	12.7710	5.6760	1.4250
	160	—	—	0.531	3.438	9.28	6.62	1.178	0.900	22.51	4.02	13.27	5.90	1.416
	—	XXS	—	0.674	3.152	7.80	8.10	1.178	0.825	27.54	3.38	15.29	6.79	1.374
	—	—	—	0.800	2.900	6.602	9.294	1.178	0.759	31.613	2.864	16.6610	7.4050	1.3380
	—	—	—	0.925	2.650	5.513	10.384	1.178	0.694	35.318	2.391	17.7130	7.8720	1.3060
5 5.563	—	—	5S	0.109	5.345	22.44	1.868	1.456	1.399	6.35	9.73	6.95	2.498	1.929
	—	—	10S	0.134	5.295	22.02	2.285	1.456	1.386	7.77	9.53	8.43	3.03	1.920
	40	Std	40S	0.258	5.047	20.01	4.30	1.456	1.321	14.62	8.66	15.17	5.45	1.878
	80	XS	80S	0.375	4.813	18.19	6.11	1.456	1.260	20.78	7.89	20.68	7.43	1.839
	120	—	—	0.500	4.563	16.35	7.95	1.456	1.195	27.04	7.09	25.74	9.25	1.799
	160	—	—	0.625	4.313	14.61	9.70	1.456	1.129	32.96	6.33	30.0	10.80	1.760
	—	XXS	—	0.750	4.063	12.97	11.34	1.456	1.064	38.55	5.62	33.6	12.10	1.722
	—	—	—	0.875	3.813	11.413	12.880	1.456	0.998	43.810	4.951	36.6450	13.1750	1.6860
	—	—	—	1.000	3.563	9.966	14.328	1.456	0.933	47.734	4.232	39.1110	14.0610	1.6520

PROPERTIES OF PIPE (CONTINUED)
(Courtesy of Anvil International)

nominal pipe size outside diameter, in.	schedule number* a	b	c	wall thickness, in.	inside diameter, in.	inside area, sq. in.	metal area, sq. in.	sq ft outside surface, per ft	sq ft inside surface, per ft	weight per ft, lb†	weight of water per ft, lb	moment of inertia, in.⁴	section modulus, in.³	radius gyration, in.
6 6.625	—	—	5S	0.109	6.407	32.2	2.231	1.734	1.677	5.37	13.98	11.85	3.58	2.304
	—	—	10S	0.134	6.357	31.7	2.733	1.734	1.664	9.29	13.74	14.40	4.35	2.295
	—	—	—	0.129	6.187	30.100	4.410	1.734	1.620	15.020	13.100	22.6600	6.8400	2.2700
	40	Std	40S	0.280	6.065	28.89	5.58	1.734	1.588	18.97	12.51	28.14	8.50	2.245
	80	XS	80S	0.432	5.761	26.07	8.40	1.734	1.508	28.57	11.29	40.5	12.23	2.195
	120	—	—	0.562	5.501	23.77	10.70	1.734	1.440	36.39	10.30	49.6	14.98	2.153
	160	—	—	0.718	5.189	21.15	13.33	1.734	1.358	45.30	9.16	59.0	17.81	2.104
	—	XXS	—	0.864	4.897	18.83	15.64	1.734	1.282	53.16	8.17	66.3	20.03	2.060
	—	—	—	1.000	4.625	16.792	17.662	1.734	1.211	60.076	7.284	72.1190	21.7720	2.0200
	—	—	—	1.125	4.375	15.025	19.429	1.734	1.145	66.084	6.517	76.5970	23.1240	1.9850
8 8.625	—	—	5S	0.109	8.407	55.5	2.916	2.258	2.201	9.91	24.07	26.45	6.13	3.01
	—	—	10S	0.148	8.329	54.5	3.94	2.258	2.180	13.40	23.59	35.4	8.21	3.00
	—	—	—	0.219	8.187	52.630	5.800	2.258	2.150	19.640	22.900	51.3200	11.9000	2.9700
	20	—	—	0.250	8.125	51.8	6.58	2.258	2.127	22.36	22.48	57.7	13.39	2.962
	30	—	—	0.277	8.071	51.2	7.26	2.258	2.113	24.70	22.18	63.4	14.69	2.953
	40	Std	40S	0.322	7.981	50.0	8.40	2.258	2.089	28.55	21.69	72.5	16.81	2.938
	60	—	—	0.406	7.813	47.9	10.48	2.258	2.045	35.64	20.79	88.8	20.58	2.909
	80	XS	80S	0.500	7.625	45.7	12.76	2.258	1.996	43.39	19.80	105.7	24.52	2.878
8 8.625	100	—	—	0.593	7.439	43.5	14.96	2.258	1.948	50.87	18.84	121.4	28.14	2.847
	120	—	—	0.718	7.189	40.6	17.84	2.258	1.882	60.63	17.60	140.6	32.6	2.807
	140	—	—	0.812	7.001	38.5	19.93	2.258	1.833	67.76	16.69	153.8	35.7	2.777
	160	—	—	0.906	6.813	36.5	21.97	2.258	1.784	74.69	16.09	165.9	38.5	2.748
	—	—	—	1.000	6.625	34.454	23.942	2.258	1.734	81.437	14.945	177.1320	41.0740	2.7190
	—	—	—	1.125	6.375	31.903	26.494	2.258	1.669	90.114	13.838	190.6210	44.2020	2.6810
10 10.750	—	—	5S	0.134	10.482	86.3	4.52	2.815	2.744	15.15	37.4	63.7	11.85	3.75
	—	—	10S	0.165	10.420	85.3	5.49	2.815	2.728	18.70	36.9	76.9	14.30	3.74
	—	—	—	0.219	10.312	83.52	7.24	2.815	2.70	24.63	36.2	100.46	18.69	3.72
	20	—	—	0.250	10.250	82.5	8.26	2.815	2.683	28.04	35.8	113.7	21.16	3.71
	30	—	—	0.307	10.136	80.7	10.07	2.815	2.654	34.24	35.0	137.5	25.57	3.69
	40	Std	40S	0.365	10.020	78.9	11.91	2.815	2.623	40.48	34.1	160.8	29.90	3.67
	60	XS	80S	0.500	9.750	74.7	16.10	2.815	2.553	54.74	32.3	212.0	39.4	3.63
	80	—	—	0.593	9.564	71.8	18.92	2.815	2.504	64.33	31.1	244.9	45.6	3.60
	100	—	—	0.718	9.314	68.1	22.63	2.815	2.438	76.93	29.5	286.2	53.2	3.56
	120	—	—	0.843	9.064	64.5	26.24	2.815	2.373	89.20	28.0	324	60.3	3.52
	—	—	—	0.875	9.000	63.62	27.14	2.815	2.36	92.28	27.6	333.46	62.04	3.50
	140	—	—	1.000	8.750	60.1	30.6	2.815	2.291	104.13	26.1	368	68.4	3.47
	160	—	—	1.125	8.500	56.7	34.0	2.815	2.225	115.65	24.6	399	74.3	3.43
	—	—	—	1.250	8.250	53.45	37.31	2.815	2.16	126.82	23.2	428.17	79.66	3.39
	—	—	—	1.500	7.750	47.15	43.57	2.815	2.03	148.19	20.5	478.59	89.04	3.31
12 12.750	—	—	5S	0.156	12.438	121.4	6.17	3.34	3.26	20.99	52.7	122.2	19.20	4.45
	—	—	10S	0.180	12.390	120.6	7.11	3.34	3.24	24.20	52.2	140.5	22.03	4.44
	20	—	—	0.250	12.250	117.9	9.84	3.34	3.21	33.38	51.1	191.9	30.1	4.42
	30	—	—	0.330	12.090	114.8	12.88	3.34	3.17	43.77	49.7	248.5	39.0	4.39
	—	Std	40S	0.375	12.000	113.1	14.58	3.34	3.14	49.56	49.0	279.3	43.8	4.38
	40	—	—	0.406	11.938	111.9	15.74	3.34	3.13	53.53	48.5	300	47.1	4.37
	—	XS	80S	0.500	11.750	108.4	19.24	3.34	3.08	65.42	47.0	362	56.7	4.33
	60	—	—	0.562	11.626	106.2	21.52	3.34	3.04	73.16	46.0	401	62.8	4.31
	80	—	—	0.687	11.376	101.6	26.04	3.34	2.978	88.51	44.0	475	74.5	4.27
	—	—	—	0.750	11.250	99.40	28.27	3.34	2.94	98.2	43.1	510.7	80.1	4.25
	100	—	—	0.843	11.064	96.1	31.5	3.34	2.897	107.20	41.6	562	88.1	4.22
	—	—	—	0.875	11.000	95.00	32.64	3.34	2.88	110.9	41.1	578.5	90.7	4.21
	120	—	—	1.000	10.750	90.8	36.9	3.34	2.814	125.49	39.3	642	100.7	4.17
	140	—	—	1.125	10.500	86.6	41.1	3.34	2.749	139.68	37.5	701	109.9	4.13
	—	—	—	1.250	10.250	82.50	45.16	3.34	2.68	153.6	35.8	755.5	118.5	4.09
	160	—	—	1.312	10.126	80.5	47.1	3.34	2.651	160.27	34.9	781	122.6	4.07

PROPERTIES OF PIPE (CONTINUED)
(Courtesy of Anvil International)

nominal pipe size outside diameter, in.	schedule number*			wall thick-ness, in.	inside diam-eter, in.	inside area, sq. in.	metal area, sq. in.	sq ft outside surface, per ft	sq ft inside surface, per ft	weight per ft, lb†	weight of water per ft, lb	moment of inertia, in.⁴	section modu-lus, in.³	radius gyra-tion, in.
	a	b	c											
14 14.000	—	—	5S	0.156	13.688	147.20	6.78	3.67	3.58	23.0	63.7	162.6	23.2	4.90
	—	—	10S	0.188	13.624	145.80	8.16	3.67	3.57	27.7	63.1	194.6	27.8	4.88
	—	—	—	0.210	13.580	144.80	9.10	3.67	3.55	30.9	62.8	216.2	30.9	4.87
	—	—	—	0.219	13.562	144.50	9.48	3.67	3.55	32.2	62.6	225.1	32.2	4.87
	10	—	—	0.250	13.500	143.1	10.80	3.67	3.53	36.71	62.1	255.4	36.5	4.86
	—	—	—	0.281	13.438	141.80	12.11	3.67	3.52	41.2	61.5	285.2	40.7	4.85
	20	—	—	0.312	13.376	140.5	13.42	3.67	3.50	45.68	60.9	314	44.9	4.84
	—	—	—	0.344	13.312	139.20	14.76	3.67	3.48	50.2	60.3	344.3	49.2	4.83
	30	Std	—	0.375	13.250	137.9	16.05	3.67	3.47	54.57	59.7	373	53.3	4.82
	40	—	—	0.437	13.126	135.3	18.62	3.67	3.44	63.37	58.7	429	61.2	4.80
	—	—	—	0.469	13.062	134.00	19.94	3.67	3.42	67.8	58.0	456.8	65.3	4.79
	—	XS	—	0.500	13.000	132.7	21.21	3.67	3.40	72.09	57.5	484	69.1	4.78
	60	—	—	0.593	12.814	129.0	24.98	3.67	3.35	84.91	55.9	562	80.3	4.74
	—	—	—	0.625	12.750	127.7	26.26	3.67	3.34	89.28	55.3	589	84.1	4.73
	80	—	—	0.750	12.500	122.7	31.2	3.67	3.27	106.13	53.2	687	98.2	4.69
	100	—	—	0.937	12.126	115.5	38.5	3.67	3.17	130.73	50.0	825	117.8	4.63
	120	—	—	1.093	11.814	109.6	44.3	3.67	3.09	150.67	47.5	930	132.8	4.58
	140	—	—	1.250	11.500	103.9	50.1	3.67	3.01	170.22	45.0	1028	146.8	4.53
	160	—	—	1.406	11.188	98.3	55.6	3.67	2.929	189.12	42.6	1117	159.6	4.48
16 16.000	—	—	5S	0.165	15.670	192.90	8.21	4.19	4.10	28	83.5	257	32.2	5.60
	—	—	10S	0.188	15.624	191.70	9.34	4.19	4.09	32	83.0	292	36.5	5.59
	10	—	—	0.250	15.500	188.7	12.37	4.19	4.06	42.05	81.8	384	48.0	5.57
	20	—	—	0.312	15.376	185.7	15.38	4.19	4.03	52.36	80.5	473	59.2	5.55
	30	Std	—	0.375	15.250	182.6	18.41	4.19	3.99	62.58	79.1	562	70.3	5.53
	40	XS	—	0.500	15.000	176.7	24.35	4.19	3.93	82.77	76.5	732	91.5	5.48
	60	—	—	0.656	14.688	169.4	31.6	4.19	3.85	107.50	73.4	933	116.6	5.43
	80	—	—	0.843	14.314	160.9	40.1	4.19	3.75	136.46	69.7	1157	144.6	5.37
	100	—	—	1.031	13.938	152.6	48.5	4.19	3.65	164.83	66.1	1365	170.6	5.30
	120	—	—	1.218	13.564	144.5	56.6	4.19	3.55	192.29	62.6	1556	194.5	5.24
	140	—	—	1.437	13.126	135.3	65.7	4.19	3.44	223.64	58.6	1760	220.0	5.17
	160	—	—	1.593	12.814	129.0	72.1	4.19	3.35	245.11	55.9	1894	236.7	5.12
18 18.000	—	—	5S	0.165	17.670	245.20	9.24	4.71	4.63	31	106.2	368	40.8	6.31
	—	—	10S	0.188	17.624	243.90	10.52	4.71	4.61	36	105.7	417	46.4	6.30
	10	—	—	0.250	17.500	240.5	13.94	4.71	4.58	47.39	104.3	549	61.0	6.28
	20	—	—	0.312	17.376	237.1	17.34	4.71	4.55	59.03	102.8	678	75.5	6.25
	—	Std	—	0.375	17.250	233.7	20.76	4.71	4.52	70.59	101.2	807	89.6	6.23
	30	—	—	0.437	17.126	230.4	24.11	4.71	4.48	82.06	99.9	931	103.4	6.21
	—	XS	—	0.500	17.00	227.0	27.49	4.71	4.45	93.45	98.4	1053	117.0	6.19
	40	—	—	0.562	16.876	223.7	30.8	4.71	4.42	104.75	97.0	1172	130.2	6.17
	60	—	—	0.750	16.500	213.8	40.6	4.71	4.32	138.17	92.7	1515	168.3	6.10
	80	—	—	0.937	16.126	204.2	50.2	4.71	4.22	170.75	88.5	1834	203.8	6.04
	100	—	—	1.156	15.688	193.3	61.2	4.71	4.11	207.96	83.7	2180	242.2	5.97
	120	—	—	1.375	15.250	182.6	71.8	4.71	3.99	244.14	79.2	2499	277.6	5.90
	140	—	—	1.562	14.876	173.8	80.7	4.71	3.89	274.23	75.3	2750	306	5.84
	160	—	—	1.781	14.438	163.7	90.7	4.71	3.78	308.51	71.0	3020	336	5.77
20 20,000	—	—	5S	0.188	19.634	302.40	11.70	5.24	5.14	40	131.0	574	57.4	7.00
	—	—	10S	0.218	19.564	300.60	13.55	5.24	5.12	46	130.2	663	66.3	6.99
	10	—	—	0.250	19.500	298.60	15.51	5.24	5.11	52.73	129.5	757	75.7	6.98
	20	Std	—	0.375	19.250	291.0	23.12	5.24	5.04	78.60	126.0	1114	111.4	6.94
	30	XS	—	0.500	19.000	283.5	30.6	5.24	4.97	104.13	122.8	1457	145.7	6.90
	40	—	—	0.593	18.814	278.0	36.2	5.24	4.93	122.91	120.4	1704	170.4	6.86
	60	—	—	0.812	18.376	265.2	48.9	5.24	4.81	166.40	115.0	2257	225.7	6.79
	—	—	—	0.875	18.250	261.6	52.6	5.24	4.78	178.73	113.4	2409	240.9	6.77
	80	—	—	1.031	17.938	252.7	61.4	5.24	4.70	208.87	109.4	2772	277.2	6.72
	100	—	—	1.281	17.438	238.8	75.3	5.24	4.57	256.10	103.4	3320	332	6.63

PROPERTIES OF PIPE (CONTINUED)
(Courtesy of Anvil International)

nominal pipe size outside diameter, in.	schedule number*			wall thick-ness, in.	inside diam-eter, in.	inside area, sq. in.	metal area, sq. in.	sq ft outside surface, per ft	sq ft inside surface, per ft	weight per ft, lb†	weight of water per ft, lb	moment of inertia, in.⁴	section modul-lus, in.³	radius gyra-tion, in.
	a	b	c											
20 20.000	120	—	—	1.500	17.000	227.0	87.2	5.24	4.45	296.37	98.3	3760	376	6.56
	140	—	—	1.750	16.500	213.8	100.3	5.24	4.32	341.10	92.6	4220	422	6.48
	160	—	—	1.968	16.064	202.7	111.5	5.24	4.21	379.01	87.9	4590	459	6.41
22 22.000	—	—	5S	0.188	21.624	367.3	12.88	5.76	5.66	44	159.1	766	69.7	7.71
	—	—	10S	0.218	21.564	365.2	14.92	5.76	5.65	51	158.2	885	80.4	7.70
	10	—	—	0.250	21.500	363.1	17.18	5.76	5.63	58	157.4	1010	91.8	7.69
	20	Std	—	0.375	21.250	354.7	25.48	5.76	5.56	87	153.7	1490	135.4	7.65
	30	XS	—	0.500	21.000	346.4	33.77	5.76	5.50	115	150.2	1953	177.5	7.61
	—	—	—	0.625	20.750	338.2	41.97	5.76	5.43	143	146.6	2400	218.2	7.56
	—	—	—	0.750	20.500	330.1	50.07	5.76	5.37	170	143.1	2829	257.2	7.52
	60	—	—	0.875	20.250	322.1	58.07	5.76	5.30	197	139.6	3245	295.0	7.47
	80	—	—	1.125	19.750	306.4	73.78	5.76	5.17	251	132.8	4029	366.3	7.39
	100	—	—	1.375	19.250	291.0	89.09	5.76	5.04	303	126.2	4758	432.6	7.31
	120	—	—	1.625	18.750	276.1	104.02	5.76	4.91	354	119.6	5432	493.8	7.23
	140	—	—	1.875	18.250	261.6	118.55	5.76	4.78	403	113.3	6054	550.3	7.15
	160	—	—	2.125	17.750	247.4	132.68	5.76	4.65	451	107.2	6626	602.4	7.07
24 24.000	10	—	—	0.250	23.500	434	18.65	6.28	6.15	63.41	188.0	1316	109.6	8.40
	20	Std	—	0.375	23.250	425	27.83	6.28	6.09	94.62	183.8	1943	161.9	8.35
	—	XS	—	0.500	23.000	415	36.9	6.28	6.02	125.49	180.1	2550	212.5	8.31
	30	—	—	0.562	22.876	411	41.4	6.28	5.99	140.80	178.1	2840	237.0	8.29
	—	—	—	0.625	22.750	406	45.9	6.28	5.96	156.03	176.2	3140	261.4	8.27
	40	—	—	0.687	22.626	402	50.3	6.28	5.92	171.17	174.3	3420	285.2	8.25
	—	—	—	0.750	22.500	398	54.8	6.28	5.89	186.24	172.4	3710	309	8.22
	—	—	5S	0.218	23.564	436.1	16.29	6.28	6.17	55	188.9	1152	96.0	8.41
	—	—	—	0.875	22.250	388.6	63.54	6.28	5.83	216	168.6	4256	354.7	8.18
	60	—	—	0.968	22.064	382	70.0	6.28	5.78	238.11	165.8	4650	388	8.15
	80	—	—	1.218	21.564	365	87.2	6.28	5.65	296.36	158.3	5670	473	8.07
	100	—	—	1.531	20.938	344	108.1	6.28	5.48	367.40	149.3	6850	571	7.96
	120	—	—	1.812	20.376	326	126.3	6.28	5.33	429.39	141.4	7830	652	7.87
	140	—	—	2.062	19.876	310	142.1	6.28	5.20	483.13	134.5	8630	719	7.79
	160	—	—	2.343	19.314	293	159.4	6.28	5.06	541.94	127.0	9460	788	7.70
26 26.000	—	—	—	0.250	25.500	510.7	19.85	6.81	6.68	67	221.4	1646	126.6	9.10
	10	—	—	0.312	25.376	505.8	25.18	6.81	6.64	86	219.2	2076	159.7	9.08
	—	Std	—	0.375	25.250	500.7	30.19	6.81	6.61	103	217.1	2478	190.6	9.06
	20	XS	—	0.500	25.000	490.9	40.06	6.81	6.54	136	212.8	3259	250.7	9.02
	—	—	—	0.625	24.750	481.1	49.82	6.81	6.48	169	208.6	4013	308.7	8.98
	—	—	—	0.750	24.500	471.4	59.49	6.81	6.41	202	204.4	4744	364.9	8.93
	—	—	—	0.875	24.250	461.9	69.07	6.81	6.35	235	200.2	5458	419.9	8.89
	—	—	—	1.000	24.000	452.4	78.54	6.81	6.28	267	196.1	6149	473.0	8.85
	—	—	—	1.125	23.750	443.0	87.91	6.81	6.22	299	192.1	6813	524.1	8.80
28 28.000	—	—	—	0.250	27.500	594.0	21.80	7.33	7.20	74	257.3	2098	149.8	9.81
	10	—	—	0.312	27.376	588.6	27.14	7.33	7.17	92	255.0	2601	185.8	9.79
	—	Std	—	0.375	27.250	583.2	32.54	7.33	7.13	111	252.6	3105	221.8	9.77
	20	XS	—	0.500	27.000	572.6	43.20	7.33	7.07	147	248.0	4085	291.8	9.72
	30	—	—	0.625	26.750	562.0	53.75	7.33	7.00	183	243.4	5038	359.8	9.68
	—	—	—	0.750	26.500	551.6	64.21	7.33	6.94	218	238.9	5964	426.0	9.64
	—	—	—	0.875	26.250	541.2	74.56	7.38	6.87	253	234.4	6865	490.3	9.60
	—	—	—	1.000	26.000	530.9	84.82	7.33	6.81	288	230.0	7740	552.8	9.55
	—	—	—	1.125	25.750	520.8	94.08	7.33	6.74	323	225.6	8590	613.6	9.51
30 30.000	—	—	5S	0.250	29.500	683.4	23.37	7.85	7.72	79	296.3	2585	172.3	10.52
	10	—	10S	0.312	29.376	677.8	29.19	7.85	7.69	99	293.7	3201	213.4	10.50
	—	Std	—	0.375	29.250	672.0	34.90	7.85	7.66	119	291.2	3823	254.8	10.48
	20	XS	—	0.500	29.000	660.5	46.34	7.85	7.59	158	286.2	5033	335.5	10.43
	30	—	—	0.625	28.750	649.2	57.68	7.85	7.53	196	281.3	6213	414.2	10.39

PROPERTIES OF PIPE (CONTINUED)
(Courtesy of Anvil International)

nominal pipe size *outside diameter,* in.	schedule number*			wall thickness, in.	inside diameter, in.	inside area, sq. in.	metal area, sq. in.	sq ft outside surface, per ft	sq ft inside surface, per ft	weight per ft, lb†	weight of water per ft, lb	moment of inertia, in.⁴	section modulus, in.³	radius gyration, in.
	a	b	c											
30 30.000	40	—	—	0.750	28.500	637.9	68.92	7.85	7.46	234	276.6	7371	491.4	10.34
	—	—	—	0.875	28.250	620.7	80.06	7.85	7.39	272	271.8	8494	566.2	10.30
	—	—	—	1.000	28.000	615.7	91.11	7.85	7.33	310	267.0	9591	639.4	10.26
	—	—	—	1.125	27.750	604.7	102.05	7.85	7.26	347	262.2	10653	710.2	10.22
32 32.000	—	—	—	0.250	31.500	779.2	24.93	8.38	8.25	85	337.8	3141	196.3	11.22
	10	—	—	0.312	31.376	773.2	31.02	8.38	8.21	106	335.2	3891	243.2	11.20
	—	Std	—	0.375	31.250	766.9	37.25	8.38	8.18	127	332.5	4656	291.0	11.18
	20	XS	—	0.500	31.000	754.7	49.48	8.38	8.11	168	327.2	6140	383.8	11.14
	30	—	—	0.625	30.750	742.5	61.59	8.38	8.05	209	321.9	7578	473.6	11.09
	40	—	—	0.688	30.624	736.6	67.68	8.38	8.02	230	319.0	8298	518.6	11.07
	—	—	—	0.750	30.500	730.5	73.63	8.38	7.98	250	316.7	8990	561.9	11.05
	—	—	—	0.875	30.250	718.3	85.52	8.38	7.92	291	311.6	10372	648.2	11.01
	—	—	—	1.000	30.000	706.8	97.38	8.38	7.85	331	306.4	11680	730.0	10.95
	—	—	—	1.125	29.750	694.7	109.0	8.38	7.79	371	301.3	13023	814.0	10.92
34 34.000	—	—	—	0.250	33.500	881.2	26.50	8.90	8.77	90	382.0	3773	221.9	11.93
	10	—	—	0.312	33.376	874.9	32.99	8.90	8.74	112	379.3	4680	275.3	11.91
	—	Std	—	0.375	33.250	867.8	39.61	8.90	8.70	135	376.2	5597	329.2	11.89
	20	XS	—	0.500	33.000	855.3	52.62	8.90	8.64	179	370.8	7385	434.4	11.85
	30	—	—	0.625	32.750	841.9	65.53	8.90	8.57	223	365.0	9124	536.7	11.80
	40		—	0.688	32.624	835.9	72.00	8.90	8.54	245	362.1	9992	587.8	11.78
	—	—	—	0.750	32.500	829.3	78.34	8.90	8.51	266	359.5	10829	637.0	11.76
	—	—	—	0.875	32.250	816.4	91.01	8.90	8.44	310	354.1	12501	735.4	11.72
	—	—	—	1.000	32.000	804.2	103.67	8.90	8.38	353	348.6	14114	830.2	11.67
	—	—	—	1.125	31.750	791.3	116.13	8.90	8.31	395	343.2	15719	924.7	11.63
36 36.000	—	—	—	0.250	35.500	989.7	28.11	9.42	9.29	96	429.1	4491	249.5	12.64
	10	—	—	0.312	35.376	982.9	34.95	9.42	9.26	119	426.1	5565	309.1	12.62
	—	Std	—	0.375	35.250	975.8	42.01	9.42	9.23	143	423.1	6664	370.2	12.59
	20	XS	—	0.500	35.000	962.1	55.76	9.42	9.16	190	417.1	8785	488.1	12.55
	30	—	—	0.625	34.750	948.3	69.50	9.42	9.10	236	411.1	10872	604.0	12.51
	40	—	—	0.750	34.500	934.7	83.01	9.42	9.03	282	405.3	12898	716.5	12.46
	—	—	—	0.875	34.250	920.6	96.50	9.42	8.97	328	399.4	14903	827.9	12.42
	—	—	—	1.000	34.000	907.9	109.96	9.42	8.90	374	393.6	16851	936.2	12.38
	—	—	—	1.125	33.750	894.2	123.19	9.42	8.89	419	387.9	18763	1042.4	12.34
42 42.000	—	—	—	0.250	41.500	1352.6	32.82	10.99	10.86	112	586.4	7126	339.3	14.73
	—	Std	—	0.375	41.250	1336.3	49.08	10.99	10.80	167	579.3	10627	506.1	14.71
	20	XS	—	0.500	41.000	1320.2	65.18	10.99	10.73	222	572.3	14037	668.4	14.67
	30	—	—	0.625	40.750	1304.1	81.28	10.99	10.67	276	565.4	17373	827.3	14.62
	40	—	—	0.750	40.500	1288.2	97.23	10.99	10.60	330	558.4	20689	985.2	14.59
	—	—	—	1.000	40.000	1256.6	128.81	10.99	10.47	438	544.8	27080	1289.5	14.50
	—	—	—	1.250	39.500	1225.3	160.03	10.99	10.34	544	531.2	33233	1582.5	14.41
	—	—	—	1.500	39.000	1194.5	190.85	10.99	10.21	649	517.9	39181	1865.7	14.33

I-3   PRESSURE RATINGS OF STANDARD COMPONENTS

Pressure ratings provided in listed standards (those listed in Table 326.1 of ASME B31.3) are provided below. They are grouped by Iron (I-3.1), Steel (I-3.2), and Copper and Copper Alloys (I-3.3).

I-3.1   Iron

ASME B16.1 covers Cast Iron Pipe Flanges and Flanged Fittings, Classes 25, 125, and 250. It provides both dimensional information and pressure–temperature ratings. Sizes from NPS 1 to NPS 96 are included. Class A or B relates to class of iron. The pressure–temperature ratings are provided in the following table, Table 1 in ASME B16.1.

ASME B16.1 NONSHOCK GAGE PRESSURE–TEMPERATURE RATINGS

Temperature (°F)	Class 25 (1) ASTM A 126		Class 125 ASTM A 126				Class 250 (1) ASTM A 126			
	Class A		Class A	Class B			Class A	Class B		
	NPS 4–36	NPS 42–96	NPS 1–12	NPS 1–12	NPS 14–24	NPS 30–48	NPS 1–12	NPS 1–12	NPS 14–24	NPS 30–48
−20 to 150	45	25	175	200	150	150	400	500	300	300
200	40	25	165	190	135	115	370	460	280	250
225	35	25	155	180	130	100	355	440	270	225
250	30	25	150	175	125	85	340	415	260	200
275	25	25	145	170	120	65	325	395	250	175
300	–	–	140	165	110	50	310	375	240	150
325	–	–	130	155	105	–	295	355	230	125
353 (2)	–	–	125	150	100	–	280	335	220	100
375	–	–	–	145	–	–	265	315	210	–
406 (3)	–	–	–	140	–	–	250	290	200	–
425	–	–	–	130	–	–	–	270	–	–
450	–	–	–	125	–	–	–	250	–	–
Hydrostatic Shell Test Pressures (4)										
100	70	40	270	300	230	230	600	750	450	450

GENERAL NOTES:
(a) Pressure is in lb/sq in. gage.
(b) NPS is nominal pipe size.
(1) Limitations:
   a. Class 25. When Class 25 cast iron flanges and flanged fittings are used for gaseous service, the maximum pressure shall be limited to 25 psig. Tabulated pressure–temperature ratings above 25 psig for Class 25 cast iron flanges and flanged fittings are applicable for nonshock hydraulic service only.
   b. Class 250. When used for liquid service, the tabulated pressure–temperature ratings in NPS 14 and larger are applicable to Class 250 flanges only and not to Class 250 fittings.
(2) 353°F (max.) to reflect the temperature of saturated steam at 125 psig.
(3) 406°F (max.) to reflect the temperature of saturated steam at 250 psig.
(4) Hydrostatic tests are not required unless specified by the user.

ASME B16.3 covers Malleable Iron Threaded Fittings, Classes 150 and 300. It provides both dimensional information and pressure–temperature ratings. It includes sizes from NPS 1/8 to NPS 6. The pressure–temperature ratings are provided in the following table, Table 1 in ASME B16.3.

ASME B16.3 PRESSURE–TEMPERATURE RATINGS

Temperature (°F)	Class 150 (psig)	Class 300 (psig)		
		Sizes (1/4–1)	Sizes (1 1/4–2)	Sizes (2 1/2–3)
−20 to 150	300	2000	1500	1000
200	265	1785	1350	910
250	225	1575	1200	825
300	185	1360	1050	735
350	150 (1)	1150	900	650
400	–	935	750	560
450	–	725	600	475
500	–	510	450	385
550	–	300	300	300

NOTE:
(1) Permissible for service temperature up to 366°F, reflecting the temperature of saturated steam at 150 psig.

ASME **B16.4** covers **Gray Iron Threaded Fittings, Classes 125 and 250**. It provides both dimensional information and pressure–temperature ratings. It includes sizes from NPS 1/4 to NPS 12. The pressure–temperature ratings are provided in the following table, Table 1 in ASME B16.4

ASME B16.4   PRESSURE–TEMPERATURE RATINGS

Temperature (°F)	Class 125 (psi)	Class 250 (psi)
−20 to 150	175	400
200	165	370
250	150	340
300	140	310
350	125 (1)	300
400	—	250 (2)

NOTES:
(1) Permissible for service temperature up to 353°F reflecting the temperature of saturated steam at 125 psig.
(2) Permissible for service temperature up to 406°F reflecting the temperature of saturated steam at 250 psig.

ASME **B16.39** covers **Malleable Iron Threaded Pipe Unions, Classes 150, 250, and 300**. It provides both dimensional information and pressure–temperature ratings. It covers sizes from NPS 1/8 to NPS 4. The pressure–temperature ratings are provided in the following table, Table 1 in ASME B16.39. Unions with copper or copper alloy seats are not to be used for temperatures in excess of 232°C (450°F).

ASME B16.39   PRESSURE–TEMPERATURE RATINGS

Temperature (°F)	Pressure (psig)		
	Class 150	Class 250	Class 300
−20 to 150	300	500	600
200	265	455	550
250	225	405	505
300	185	360	460
350	150	315	415
400	110	270	370
450	75	225	325
500	—	180	280
550	—	130	230

ASME **B16.42** covers **Ductile Iron Pipe Flanges and Flanged Fittings, Classes 150 and 300**. It provides both dimensional information and pressure–temperature ratings. It covers sizes from NPS 1 to NPS 24. The pressure–temperature ratings are provided in the following table, Table 1 in ASME B16.42.

ASME B16.42   PRESSURE–TEMPERATURE RATINGS

Temperature (°F)	Working Pressure (psi gage)	
	Class 150	Class 300
−20 to 100	250	640
200	235	600
300	215	565
400	200	525
500	170	495
600	140	465
650	125	450

I-3.2   Steel

ASME B16.5 covers Pipe Flanges and Flanged Fittings, NPS ½ through NPS 24. It provides both dimensional information and pressure–temperature ratings. The pressure temperature ratings provided in ASME B16.5 are the commonly used pressure classes. These include Classes 150, 300, 400, 600, 900, 1500, and 2500. Tables of pressure–temperature ratings are provided for groups of materials. Examples for commonly used carbon steel and austenitic stainless steel materials are provided in the following tables, Tables 2–1.1 and 2–2.1 in ASME B16.5.

ASME B16.5   PRESSURE–TEMPERATURE RATINGS FOR GROUPS 1.1 THROUGH 3.17 MATERIALS

RATINGS FOR GROUP 1.1 MATERIALS

Nominal Designation	Forgings	Castings	Plates
C–Si	A 105 (1)	A 216 Gr. WCB (1)	A515 Gr. 70 (1)
C–Mn–Si	A 350 Gr. LF2 (1)		A 516 Gr. 70 (1)(2) A 537 Cl. 1 (3)
C–Mn–Si–V	A 350 Gr. LF6 Cl. 1 (4)		

NOTES:
(1) Upon prolonged exposure to temperatures above 800°F, the carbide phase of steel may be converted to graphite. Permissible, but not recommended for prolonged use above 800°F.
(2) Not to be used over 850°F.
(3) Not to be used over 700°F.
(4) Not to be used over 500°F.

Class Temperature (°F)	Working Pressures by Classes (psig)						
	150	300	400	600	900	1500	2500
−20 to 100	285	740	990	1480	2220	3705	6170
200	260	675	900	1350	2025	3375	5625
300	230	655	875	1315	1970	3280	5470
400	200	635	845	1270	1900	3170	5280
500	170	600	800	1200	1795	2995	4990
600	140	550	730	1095	1640	2735	4560
650	125	535	715	1075	1610	2685	4475
700	110	535	710	1065	1600	2665	4440
750	95	505	670	1010	1510	2520	4200
800	80	410	550	825	1235	2060	3430
850	65	270	355	535	805	1340	2230
900	50	170	230	345	515	860	1430
950	35	105	140	205	310	515	860
1000	20	50	70	105	155	260	430

RATINGS FOR GROUP 2.1 MATERIALS

Nominal Designation	Forgings	Castings	Plates
18Cr–8Ni	A 182 Gr. F304 (1)	A 351 Gr. CF3 (2)	A 240 Gr. 304 (1)
	A 182 Gr. F304H	A 351 Gr. CF8 (1)	A 240 Gr. 304H

NOTES:
(1) At temperatures over 1000°F, use only when the carbon content is 0.04% or higher.
(2) Not to be used over 800°F.

Class Temperature (°F)	Working Pressures by Classes (psig)						
	150	300	400	600	900	1500	2500
−20 to 100	275	720	960	1440	2160	3600	6000
200	230	600	800	1200	1800	3000	5000
300	205	540	720	1080	1620	2700	4500
400	190	495	660	995	1490	2485	4140
500	170	465	620	930	1395	2330	3880
600	140	435	580	875	1310	2185	3640
650	125	430	575	860	1290	2150	3580
700	110	425	565	850	1275	2125	3540
750	95	415	555	830	1245	2075	3460
800	80	405	540	805	1210	2015	3360
850	65	395	530	790	1190	1980	3300
900	50	390	520	780	1165	1945	3240
950	35	380	510	765	1145	1910	3180
1000	20	320	430	640	965	1605	2675
1050	—	310	410	615	925	1545	2570
1100	—	255	345	515	770	1285	2145
1150	—	200	265	400	595	995	1655
1200	—	155	205	310	465	770	1285
1250	—	115	150	225	340	565	945
1300	—	85	115	170	255	430	715
1350	—	60	80	125	185	310	515
1400	—	50	65	90	145	240	400
1450	—	35	45	70	105	170	285
1500	—	25	35	55	80	135	230

ASME B16.9 covers Factory-Made Wrought Buttwelding Fittings. It includes sizes up to NPS 24 for most fittings and up to NPS 48 for some. Pressure–temperature ratings are not provided; rather, the allowable pressure ratings for the fittings are stated to be the same as for straight seamless pipe of equivalent material (as shown by comparison of composition and mechanical properties in the respective material specifications) in accordance with the rules established in the applicable section of ASME B31. ASME B31.3 presently requires that the calculation be based on 87.5% of the wall thickness of matching straight seamless pipe.

ASME B16.11 covers Forged Fittings, Socket-Welding and Threaded. It includes sizes from NPS 1/8 to NPS 4. Pressure–temperature ratings are not provided; rather, the allowable pressure ratings for the fittings are stated to be the same as for straight seamless pipe of equivalent material (as shown by comparison of composition and mechanical properties in the respective material specifications) in accordance with the rules established in the applicable section of ASME B31. The correlation between fitting class and matching pipe is provided in the following table, Table 2 in ASME B16.11.

ASME B16.34 covers Valves—Flanged, Threaded and Welding End. It provides both dimensional information and pressure–temperature ratings. Ratings for both Standard Class and Special Class

ASME B16.11 CORRELATION OF FITTINGS CLASS WITH SCHEDULE NUMBER OR WALL DESIGNATION OF PIPE FOR CALCULATION OF RATINGS

Class Designation of Fitting	Type of Fitting	Pipe Used For Rating Basis (1)	
		Schedule No.	Wall Designation
2000	Threaded	80	XS
3000	Threaded	160	—
6000	Threaded	—	XXS
3000	Socket-Welding	80	XS
6000	Socket-Welding	160	—
9000	Socket-Welding	—	XXS

NOTE:

(1) This table is not intended to restrict the use of pipe of thinner or thicker wall with fittings. Pipe actually used may be thinner or thicker in nominal wall than that shown in this table. When thinner pipe is used, its strength may govern the rating. When thicker pipe is used (e.g., for mechanical strength), the strength of the fitting governs the rating.

valves are provided. Standard Class valves have the same pressure–temperature ratings as provided in ASME B16.5 for flanges, with the addition of Class 4500. Special Class valves have a higher pressure–temperature rating. Since the flanges are essentially governed by ASME B16.5, flanged valves are limited to Standard Class, and to Class 2500 and less. Special Class valves are subjected to additional examination requirements. The pressure–temperature ratings provided in ASME B16.34 are the commonly used pressure classes plus Class 4500. Tables of pressure–temperature ratings are provided for groups of materials. An example for commonly used carbon steel material is provided in the following table, Table 2-1.1 in ASME B16.34.

ASME B16.47 covers **Large Diameter Steel Flanges, NPS 26 through NPS 60.** It provides both dimensional information and pressure–temperature ratings. The pressure–temperature ratings are the same as in ASME B16.5, with the addition of Class 75 for Series B flanges. Two flange dimensions are provided; Series A flange dimensions were based on MSS SP 44 and Series B were based on API 605.

MSS SP-42 covers **Class 150 Corrosion-Resistant Gate, Globe, Angle and Check Valves With Flanged and Buttweld Ends.** The pressure–temperature ratings are provided in the following table, Table 4 in MSS SP-42.

MSS SP-42 PRESSURE–TEMPERATURE RATINGS

Service Temperature (°F)	Maximum Working Pressure (psig)				
	CF8, CF3, 304	CF8M, CF3M, 316	304L, 316L	CF8C, 347	CN7M
−20 to 100	275	275	230	275	230
150	253	255	213	265	215
200	230	235	195	255	200
250	218	225	185	243	195
300	205	215	175	230	190
350	198	205	168	215	190
400	190	195	160	200	190
450	180	183	153	185	180
500	170	170	145	170	170

GENERAL NOTE:

This table gives data in U.S. customary units, the temperature being indicated in °F and the actual values being represented in metric units.

MSS SP-42   PRESSURE–TEMPERATURE RATINGS

Service Temperature (°C)	Maximum Working Pressure (bar)				
	CF8, CF3, 304	CF8M, CF3M, 316	304L, 316L	CF8C, 347	CN7M
−29 to 38	19.0	19.0	15.9	19.0	15.9
50	18.3	18.4	15.3	18.7	15.4
100	15.7	16.0	13.3	17.4	13.7
150	14.1	14.8	12.0	15.8	13.1
200	13.2	13.6	11.1	14.0	13.1
250	12.0	12.0	10.2	12.1	12.0
300	10.2	10.2	9.8	10.3	10.2
260	11.7	11.7	10.0	11.7	11.7

MSS SP-43 covers Wrought and Fabricated Butt Welding Fittings for Low Pressure Corrosion Resistant Applications. It is limited to fittings for use with Schedule 5S and 10S pipe plus short pattern stub ends for Schedule 40S pipe. Seamless and ERW welded fittings are included. MSS SP-43 requires that they be suitable for a pressure equal to 30% of the allowable pressure for the matching pipe, as well as for a hydrostatic test of 1.5 times that pressure.

MSS SP-51 covers Class 150LW Corrosion-Resistant Cast Flanges and Flanged Fittings. The pressure–temperature ratings are provided in the following table, Table 1 in MSS SP-51.

MSS SP-51   PRESSURE–TEMPERATURE RATINGS

Temperature (°C)	Pressure Rating (Bar)	Temperature (°F)	Pressure Rating (psig)
−29 to 38	6.9	−20 to 100	100
50	6.8	200	90
100	6.1	300	80
150	5.5	400	70
200	4.9	500	60
260	4.1		

MSS SP-75 is a Specification for High Test Wrought Butt Welding Fittings. It includes seamless and ERW carbon and low alloy butt welding fittings for use in high pressure gas and oil transmission and distribution systems. The pressure rating for the fittings is the same as for matching straight seamless pipe.

MSS SP-79 covers Socket Welding Reducer Inserts for use with ASME B16.11 Class 3000 and 6000 socket welding fittings. The allowable pressure ratings for the fittings are the same as for ASME B16.11 socket weld fittings.

MSS SP-95 covers Swaged Nipples and Bull Plugs. They are required to be ordered for pipe with a specific wall thickness and material. The allowable pressure rating is the same as for matching straight seamless pipe.

MSS SP-97 covers Integrally Reinforced Forged Branch Outlet Fittings – Socket Welding, Threaded and Buttwelding Ends. The pressure ratings of the branch outlet fittings are the same as the pressure rating of matching straight seamless pipe.

I-3.3   Copper and Copper Alloys

ASME B16.15 covers Cast Bronze Threaded Fittings, Classes 125 and 250. It includes sizes from NPS 1/8 to NPS 4. The pressure–temperature ratings for fittings per this standard are provided in the following table, Table 1 in ASME B16.15.

API 602 PRESSURE–TEMPERATURE RATINGS FOR CLASS 800 GATE VALVES

Material Group Number

Service Temperature (1) °F	°C	1.1 A 105 (2) A 350-LF2 (3) A 216-VCB (2) psig	MPa	1.2 A 350-LF3 (3) A 352-LC2 (3) A 352-LC3 (3) psig	MPa	1.3 A -352-LCB (3) psig	MPa	1.9 A 182-F11 (4) A 217-WC6 (3) psig	MPa	1.10 A 182-F22 (4) A 217-WC9 (5) psig	MPa	1.13 A 182-F5 A 182-F5 (1) A 217-C5 psig	MPa	1.14 A 182-F9 A 217-C12 psig	MPa
−20 to 100	−29 to 38	1975	13.62	2000	13.79	1855	12.79	2000	13.79	2000	13.79	2000	13.79	2000	13.79
200	93.5	1800	12.41	2000	13.79	1750	12.06	1900	13.10	1910	13.17	2000	13.79	2000	13.79
300	149	1750	12.07	1940	13.38	1700	11.72	1795	12.38	1805	12.45	1940	13.38	1940	13.38
400	204.5	1690	11.65	1880	12.96	1645	11.34	1755	12.10	1730	11.93	1880	12.96	1880	12.96
500	260	1595	11.00	1775	12.24	1550	10.69	1710	11.79	1705	11.76	1775	12.24	1775	12.24
600	315.5	1460	10.07	1615	11.14	1420	9.79	1615	11.14	1615	11.14	1615	11.14	1615	11.14
650	343.5	1430	9.86	1570	10.82	1395	9.62	1570	10.82	1570	10.82	1570	10.82	1570	10.82
700	371	1420	9.79	—	—	—	—	1515	10.45	1515	10.45	1515	10.45	1515	10.45
750	399	1345	9.27	—	—	—	—	1420	9.79	1420	9.79	1420	9.79	1420	9.79
800	426.5	1100	7.58	—	—	—	—	1355	9.34	1355	9.34	1325	9.14	1355	9.34
850	454.5	715	4.93	—	—	—	—	1300	8.96	1300	8.96	1170	8.07	1300	8.96
900	482	460	3.17	—	—	—	—	1200	8.27	1200	8.27	940	6.48	1200	8.27
950	510	275	1.90	—	—	—	—	1005	6.93	1005	6.93	695	4.79	985	6.79
1000	538	140	0.96	—	—	—	—	595	4.10	715	4.93	510	3.52	780	5.38
1050	565.5	—	—	—	—	—	—	365	2.52	530	3.65	375	2.59	505	3.48
1100	593.5	—	—	—	—	—	—	255	1.76	275	2.07	275	1.90	300	2.07
1150	621	—	—	—	—	—	—	140	0.96	275	1.90	185	1.28	200	1.38
1200	649	—	—	—	—	—	—	95	0.65	145	1.00	120	0.83	140	0.97

GENERAL NOTES:

psig = pounds per square inch gauge; MPa = megapascals.

(1) For a material shown in this table that is acceptable for low temperature service the pressure rating for a service at any temperature below −20°F (−29°C) shall be no greater than the rating shown in this table for −20°F–100°F (−29°C–8°C).

(2) Permitted but not recommended for prolonged use above about 800°F (425°C).

(3) Not to be used over 650°F (345°C).

(4) Permitted but not recommended for prolonged use above about 1100°F (595°C).

(5) Not to be used over 1100°F (595°C).

API 602 PRESSURE–TEMPERATURE RATINGS FOR CLASS 800 GATE VALVES (CONTINUED)

Service Temperature		2.1 A 182-F304 A 351-CF3 (6) A 351-CF8		2.2 A 182-F316 A 351-CF3M (7) A 351-CF8M		2.3 A 182-F304L A 182-F316L		2.5 A 182 F347H A 351-CF8C	
°F	°C	psig	MPa	psig	MPa	psig	MPa	psig	MPa
-20 to 100	-29 to 38	1920	13.24	1920	13.24	1600	11.03	1920	13.24
200	93.5	1600	11.03	1655	11.41	1350	9.31	1695	11.69
300	149	1410	9.72	1495	10.31	1210	8.34	1570	10.82
400	204.5	1255	8.65	1370	9.45	1100	7.58	1480	10.20
500	260	1165	8.03	1275	8.79	1020	7.03	1380	9.51
600	315.5	1105	7.62	1205	8.31	960	6.62	1310	9.03
650	343.5	1090	7.52	1185	8.17	935	6.45	1280	8.83
700	371	1075	7.41	1150	7.93	915	6.31	1250	8.62
750	399	1060	7.31	1130	7.79	895	6.17	1230	8.48
800	426.5	1050	7.24	1105	7.62	875	6.03	1215	8.38
850	454.5	1035	7.14	1080	7.45	860	5.93	1185	8.17
900	482	1025	7.07	1050	7.24	—	—	1150	7.93
950	510	1000	6.89	1030	7.10	—	—	1030	7.10
1000	538	860	5.93	970	6.69	—	—	970	6.69
1050	565.5	825	5.69	960	6.62	—	—	960	6.62
1100	593.5	685	4.72	860	5.93	—	—	860	5.93
1150	621	520	3.58	735	5.07	—	—	735	5.07
1200	649	415	2.86	550	3.79	—	—	460	3.17
1250	676.5	295	2.03	485	3.34	—	—	330	2.28
1300	704.5	220	1.51	365	2.52	—	—	250	1.72
1350	732	165	1.14	275	1.90	—	—	180	1.24
1400	760	130	0.90	200	1.38	—	—	140	0.97
1450	788	95	0.66	155	1.07	—	—	110	0.76
1500	815.5	65	0.45	110	0.76	—	—	95	0.66

(6) Not to be used over 800 °F (425°C).
(7) Not to be used over 850 °F (455°C).

ASME B16.15   PRESSURE–TEMPERATURE RATINGS

Temperature (°F)	Class 125 (psi)	Class 250 (psi)
−20 to 150	200	400
200	190	385
250	180	365
300	165	335
350	150	300
400	125	250

ASME B16.18 covers **Cast Copper Alloy Solder Joint Pressure Fittings.** It also includes solder × threaded fittings. It provides both dimensional information and information on the permissible pressure–temperature. Sizes that are included range from Standard Water Tube 1/4 to 12. The pressure ratings of these fittings are required to be the same as ASTM B88 Type L Copper Water Tube, annealed. However, the pressure rating cannot exceed the rating of the solder joint, provided in the following table, Table A1 in ASME B16.18, Annex A.

ASME B16.22 covers **Wrought Copper and Copper Alloy Solder Joint Pressure Fittings.** It also includes solder × threaded fittings. It provides both dimensional information and pressure–temperature ratings. Sizes included range from Standard Water Tube 1/4 to 8. The pressure rating of the fittings is the lower of the rating shown in the following table, Table 1 in B16.22, and the rating of the solder joint, shown in the table on the following page from ASME B16.18.

ASME B16.22   RATED INTERNAL WORKING PRESSURE (1) FOR COPPER FITTINGS (PSI)

Standard Water Tube Size (2)	−20 to 100°F	150°F	200°F	250°F	300°F	350°F	400°F
1/4	912	775	729	729	714	608	456
3/8	779	662	623	623	610	519	389
1/2	722	613	577	577	565	481	361
5/8	631	537	505	505	495	421	316
3/4	582	495	466	466	456	388	291
1	494	420	395	395	387	330	247
1 1/4	439	373	351	351	344	293	219
1 1/2	408	347	327	327	320	272	204
2	364	309	291	291	285	242	182
2 1/2	336	285	269	269	263	224	168
3	317	270	254	254	248	211	159
3 1/2	304	258	243	243	238	202	152
4	293	249	235	235	230	196	147
5	269	229	215	215	211	179	135
6	251	213	201	201	196	167	125
8	270	230	216	216	212	180	135

NOTES:
(1) The fitting pressure rating applies to the largest opening of the fitting.
(2) For size designation of fittings, see Section 4 of ASME B16.22.

ASME B16.24 covers **Cast Copper Alloy Pipe Flanges and Flanged Fittings, Classes 150, 300, 400, 600, 900, 1500, and 2500.** It provides both dimensional information and pressure–temperature ratings. Sizes included range from NPS 1/2 to NPS 12. The pressure–temperature rating depends upon the alloy, and is provided in the following two tables, Tables 1 and 2 in ASME B16.24.

ASME B16.18 PRESSURE–TEMPERATURE RATINGS

Joining Material	Working Temperature		Maximum Working Gage Pressure									
			Size 1/8 thru 1 [1]		Size 1 1/4 thru 2 [1]		Size 2 1/2 thru 4 [1]		Size 5 thru 8 [1]		Size 10 and 12 [1]	
	°F	°C	psi	bar	psi	bar	psi	bar	psi	bar	psi	bar
50–50 Tin–Lead Solder[2]	100	38	200	14	175	12	150	10	135	9	100	7
	150	66	150	10	125	9	100	7	90	6	70	5
	200	93	100	7	90	6	75	5	70	5	50	3
	250	120	85	6	75	5	50	3	45	3	40	3
95–5 Tin–Antimony Solder[3]	100	38	500	35	400	28	300	20	270	19	150	10
	150	66	400	28	350	24	275	19	250	17	150	10
	200	93	300	20	250	17	200	14	180	13	140	10
	250	120	200	14	175	12	150	10	135	9	110	8
Joining[4] Materials Melting at or above 1000° (540°C)			Pressure–temperature ratings consistent with the materials and procedures employed									

GENERAL NOTES:

1 bar = 10^5 Pa.

(1) Standard water tube sizes.

(2) ANSI/ASTM B32 Alloy Grade 50A.

(3) ANSI/ASTM B32 Alloy Grade 95TA.

(4) These joining materials are defined as brazing alloys by the American Welding Society.

ASME B16.26 coversCast Copper Alloy Fittings for Flared Copper Tubes. They are rated for a maximum cold water service pressure of 175 psig.

ASME B16.24 PRESSURE–TEMPERATURE RATINGS FOR ALLOYS C83600 AND C92200

| Service Temperature (°F) | Working Pressure (psig) | | | |
| | Class 150 | | Class 300 | |
	ASTM B 62 C83600	ASTM B 61 C92200	ASTM B 62 C83600	ASTM B 61 C92200
−20 to 150	225	225	500	500
175	220	220	480	490
200	210	215	465	475
225	205	210	445	465
250	195	205	425	450
275	190	200	410	440
300	180	195	390	425
350	165	180	350	400
400	—	170	—	375
406	150	—	—	—
450	135 (1)	160	280 (1)	350
500	—	150	—	325
550	—	140	—	300
Test Pressure	350	350	750	750

NOTE:

(1) Some codes (e.g., ASME Boiler and Pressure Vessel Code, Section I; ASME B31.1; ASME B31.5) limit the rating temperature of the indicated material to 406°F.

ASME B16.24 PRESSURE–TEMPERATURE RATINGS FOR ASTM B 148, ALLOY C95200

| Service Temperature (°F) | Working Pressure (psig) | | | | | | |
	Class 150	Class 300	Class 400	Class 600	Class 900	Class 1500	Class 2500
−20 to 100	195	515	685	1030	1545	2575	4290
150	165	430	570	855	1285	2140	3570
200	155	400	535	800	1205	2005	3340
250	145	385	510	770	1150	1920	3200
300	140	370	495	740	1110	1850	3085
350	140	365	490	735	1100	1835	3060
400	140	365	485	725	1090	1820	3030
450	140	360	480	725	1085	1805	3010
500	140	360	480	720	1080	1800	3000
Test Pressure	300	776	1050	1550	2325	3875	6450

APPENDIX II

GUIDELINES FOR COMPUTER FLEXIBILITY ANALYSIS

Prior to initiating a computer piping flexibility analysis, the analyst should gather as much information as they can to perform the job, in an organized manner. This information includes the following.

1. Piping system operating and design conditions. The analyst should have an understanding of the various potential combinations of operating conditions that may occur. There may be various combinations of lines and vessels at different temperature conditions. For example, with parallel heat exchangers, operating with one heat exchanger in operation and hot, and the other blocked off and at an ambient temperature condition, may be one of the potential modes of operation. Note that while it is not a Code requirement that the flexibility analysis be conducted using the design temperature, it may be the requirement of the owner that it be done in that manner.

2. Determine if there are steam out or other conditions that may cause different temperatures (e.g. steam tracing temperature).

3. Standard piping classes or other information indicating the materials used to construct the piping system.

4. Dimensional information on the piping system, in the form of isometric, orthographic, or other format.

5. Information on attached equipment. This typically includes outline drawings of vessels, rotating equipment, etc. Load limits for the equipment. Operating conditions/temperatures for attached equipment to consider thermal growth. Thermal movements of fired equipment are typically provided on the equipment drawings.

6. If supports and restraints have been designed, drawings or other information to determine the types of supports, to appropriately model them.

7. If the piping system is existing, photographs, walkdown, or other means to confirm the supports used in construction. It is not unusual for the supports found in the field to differ from those shown on the original design drawings.

8. Allowable stresses. Code of construction.

9. Weights of piping components.

10. Density of fluid.

11. Insulation thickness and type (including properties such as density, if standard insulation type densities are not provided by the flexibility analysis program).

12. Information on expansion joints, if present in the system, including pressure thrust area and stiffnesses.

13. Information on refractory lining, if present in the system, to determine the effect of the refractory on the stiffness of the piping. The increase in stiffness can effect equipment loads and load sensitive components in a system (e.g. a hot walled slide valve in a refractory lined system).

14. Wind, seismic (seismic may be governed by a building code, e.g. UBC, BOCA, IBC) and other occasional loads, such as waterhammer, that require consideration.

15. Any special requirements, such as natural frequency limitations.

16. For vapor lines, information on how the hydrotest is to be conducted. Must the piping be designed to carry the weight of water during a hydrotest?

The first step in conducting a flexibility analysis is to gather this information. It is then used to construct a pipe stress isometric. The stress isometric is used to organize the information for input into the computer program. The following information should generally be on the piping stress isometric. Note that while it is certainly possible to input a pipe stress model without preparing a piping stress isometric, the chance of making errors is greatly increased.

1. A global coordinate system with the positive direction indicated for the x, y and z reference axes used in the stress analysis. The positive y axis is typically assigned to the vertical up direction. The plant north direction shown with respect to this global reference frame. It may be helpful on a project to consistently model plant north as the positive x axis and east as the positive z axis.

2. Node points used in the analysis, labeled and located. Dimensions between node points, resolved into components parallel to each of the three global axes.

3. Locations, function, and lines of action of all supports and restraints. Spring hanger stiffnesses, hot and cold load settings, and design travel. When known, spring hanger model number and manufacturer.

4. Piping design parameters such as pipe size, thickness or schedule, piping material, corrosion/erosion allowance, mechanical allowances, insulation, refractory density, valve and flange (and other piping component) weights and locations.

5. The general arrangement of equipment to which the piping is attached.

The final stress report should also contain the following information. Having this information gathered together makes it possible to check the analysis with relative ease, and provides complete documentation of the basis for the work performed.

1. Equipment outline drawings for all vessels, heat exchangers, fired heaters, rotating equipment, tankage, etc. to which the piping is attached. These drawings should include, as a minimum, basic overall dimensions, material, nozzle locations and details where piping is attached, and equipment support locations. As a minimum, references to the appropriate drawings should be provided.

2. Pressures, temperatures and content specific gravities for all load cases considered. Data on any pipe sections with high temperature gradients, thermal bowing, or transients should be included as well.

3. Expansion joint stiffness and movements (axial, transverse, and rotational), bellows dimensional information and overall expansion joint assembly weight.

4. Any dynamic load data used in the analysis.

5. Tabulations showing compliance of loads on rotating and other load sensitive equipment with their allowables.

6. Program input and output showing all analysis results for the specified load cases as well as a record of the software used, its version number, and record of its setup or configuration. This may be an electronic record.

7. Program generated plots showing principal nodes (annotated as appropriate).

8. Assumptions/supporting calculations for stress intensification factors when values other than program defaults are used.

9. Support stiffness calculations, when appropriate.

10. If the analysis is also a pressure design check, appropriate calculations for wall thickness, area replacement, etc.

The following is a checklist of considerations.

1. Is thermal movement of attached equipment included?

2. If design temperature is being used for the flexibility analysis, is the thermal movement of the attached equipment consistent?

3. Have design temperatures of the equipment been improperly used in lieu of operating temperatures? This can occur when pressure vessels are given higher design temperatures when the allowable stress does not change up to that temperature (e.g. 650F) even though the expected metal temperature may be substantially less. In that case, the design temperature has nothing to do with the expected metal temperature for determining thermal movement.

4. Is the pressure thrust in the piping properly considered in loads on attached equipment (e.g. pressure thrust typically must be added to the reported force in the piping, for conducting a nozzle analysis)?

5. Are there conditions that may cause reversing moments in the system, such that the range between these conditions must be considered in determining the stress range?

6. Has friction been included, when significant?

7. Has the effect of friction on sliding support loads been considered?

8. If friction has been included in the piping stress analysis, has an analysis been run without friction? In general, friction is not something that is relied on. The harmful effects of friction are considered, but the benefit is not. Thus, in some cases analyses should be run with and without friction, to determine which is worse.

9. Are the loads on attached equipment within the allowable limits?

10. May the flexibility of the structural steel or other supporting elements significantly effect the results?

11. Has thermal expansion of the pipe support, or equipment support, been considered?

12. Has corrosion allowance been included?

13. Is the geometry correct? Check the plots.

14. Are longitudinal stresses due to internal pressure included in the S_L calculation?

15. Are the correct valve and other component weights included?

16. Has the weight of valve actuators been included?

17. Does the flange weight include the weight of the bolting?

18. Has the effect of settlement (e.g. around tankage) been considered?

19. Have all combinations of operating conditions been considered?

20. Have environmental loads (e.g. wind, seismic, ice) been considered?

21. Has the correct minimum temperature for the location been considered in the analysis, in determining the range of temperatures the piping may be exposed to, or has a default value been used?

22. Are springs properly modeled?

23. If gaps at supports are included in the model, are adequate controls/documentation in place to assure that they will be installed in that manner in the field, and checked?

24. Has the effect of welding trunions on elbows (on their flexibility) been considered?

25. Is there a potential for elastic followup or other strain concentration condition?

26. Is radial thermal expansion of the line potentially significant (e.g. bottom supported large diameter hot lines)?

27. If the piping lifts off of supports modeled as nonlinear supports, has the lack of that support been considered in calculating the sustained stress for that operating condition?

28. Has pressure thrust of expansion joints been considered?

29. Has thermal expansion and the action of expansion joint hardware been considered?

30. Have axial load due to thermal expansion been considered in the thermal expansion stress range, if significant?

31. Have the effect of any flanges (or heavy walled components) welded adjacent to elbows been considered with respect to elbow stress intensification factors and flexibility?

32. Have the stresses in elbows and bends been considered at more than one location on the elbow or bend? These should typically be at least checked at the ends and the midpoint.

33. Has deflection between supports been considered, when necessary? Note that deflection is only calculated, reported, and plotted based on values at node points. Midspan nodes may be necessary.

34. On large diameter lines, has the effect of pipe radius on geometry been considered? Default modeling typically assumes the pipe spans between centerlines; however with short branch runs between large diameter lines, or between a large diameter line and some point of restraint, the assumption can be significantly unconservative.

35. Have sufficient means to direct the movements of a piping system been incorporated. Remember, ideal conditions do not typically exist, so floating systems without any intermediate points of fixity generally are not desirable.

36. Have any unlisted components been considered for pressure and fatigue?

37. On large D/t piping, have limitations to the SIF's in Appendix D been considered?

38. Should pressure stiffening of bends be considered in the analysis?

39. Are loads on the flanges high? Has the impact of loads on flange leakage been considered?

40. Should the change in pipe length due to internal pressure be considered? This depends upon the hoop stress, axial stress (which may be effected by the presence of expansion joints) and the material of construction (e.g. reinforced thermosetting resin piping may extend or contract due to internal pressure, depending upon the fiber wind angle).

APPENDIX III
USEFUL INFORMATION FOR FLEXIBILITY ANALYSIS

Weights of Piping Materials
Weights of Flanged Valves
Weights of Flange Bolting
Dimensions of Fittings (ASME B16.9)
Dimensions of Welding Neck Flanges (ASME B16.5)
Dimensions of Valves (ASME B16.10)
Coefficient of Thermal Expansion (ASME B31.3)

WEIGHT OF PIPING MATERIALS
(Courtesy of Anvil International)

The tabulation of weights of standard piping materials has been arranged for convenience of selection of data that formerly consumed considerable time to develop. For special materials, the three formulae listed below for weights of tubes, weights of contents of tubes, and weights of piping insulation will be helpful.

$$\text{Weight of tube} = F \times 10.68 \times T \times (D - T) \text{ lb/ft}$$

T = wall thickness in inches

D = outside diameter in inches

F = relative weight factor

The weight of tube furnished in this piping data is based on low carbon steel weighing 0.2833 pounds per cubic inch.

Relative Weight Factor F

Aluminum	0.34
Brass	1.09
Cast Iron	0.92
Copper	1.14
Ferritic stainless steel	0.95
Austenitic stainless steel	1.02
Steel	1.00
Wrought iron	0.99

Weight of contents of a tube –

$$G \times .3405 \times (D - 2T)^2 \text{ lb/ft}$$

G = specific gravity of contents

T = tube wall thickness in inches

D = tube outside diameter in inches

The weight per foot of steel pipe is subject to the following tolerances:

SPECIFICATION		TOLERANCE	
ASTM A-53 ASTM A-120	STD WT	+5%	-5%
	XS WT	+5%	-5%
	XXS WT	+10%	-10%
ASTM A-106	SCH 10-120	+6.5%	-3.5%
	SCH 140-160	+10%	-3.5%
ASTM A-335	12" and under	+6.5%	-3.5%
	over 12"	+10%	-5%
ASTM A-312 ASTM A-376	12" and under	+6.5%	-3.5%
API 5L	All sizes	+6.5%	-3.5%

The weight of welding tees and laterals are for full size fittings. The weights of reducing fittings are approximately the same as for full size fittings.

The weights of welding reducers are for one size reduction, and are approximately correct for other reductions.

Weights of valves of the same type may vary because of individual manufacturer's designs. Listed valve weights are approximate only. Specific valve weights should be used when available.

Where specific insulation thicknesses and densities differ from those shown, refer to "Weight of Piping Insulation" formula below.

Weight of piping insulation -

$$I \times .0218 \times T \times (D + T) \text{ lb/ft}$$

I = insulation density in pounds per cubic foot

T = insulation thickness in inches

D = outside diameter of pipe in inches

WEIGHT OF PIPING MATERIALS
(Courtesy of Anvil International)

1.313" O.D. 1" PIPE

PIPE													
	Schedule No.	5S	10S	40	80	160							
	Wall Designation			Std.	XS		XXS						
	Thickness — In.	0.065	0.109	.133	.179	.250	.358						
	Pipe — Lbs/Ft	**0.868**	**1.404**	**1.68**	**2.17**	**2.84**	**3.66**						
	Water — Lbs/Ft	0.478	0.409	.37	.31	.23	.12						

WELDING FITTINGS

Fitting	5S	10S	40	80	160	XXS							
L.R. 90° Elbow	.2 / .3	.4 / .3	.4 / .3	.4 / .3	.6 / .3	1.0 / .3							
S.R. 90° Elbow			.3 / .2										
L.R. 45° Elbow	.1 / .2	.3 / .2	.3 / .2	.3 / .2	.4 / .2	.5 / .2							
Tee	.4 / .4	.6 / .4	.8 / .4	.9 / .4	1.1 / .4	1.3 / .4							
Lateral	.7 / 1.1	1.2 / 1.1	1.7 / 1.1	2.5 / 1.1									
Reducer	.2 / .2	.4 / .2	.3 / .2	.4 / .2	.5 / .2	.5 / .2							
Cap	.1 / .3	.1 / .3	.3 / .3	.3 / .3	.4 / .3	.5 / .3							

INSULATION

	Temperature Range °F	100-199	200-299	300-399	400-499	500-599	600-699	700-799	800-899	900-999	1000-1099	1100-1200
85% Magnesia Calcium Silicate	Nom. Thick., In.	1	1	1½	2	2						
	Lbs./Ft	.57	.57	.97	1.54	1.54						
Combination	Nom. Thick., In.						2½	2½	2½	3	3	3
	Lbs/Ft						3.30	3.30	3.30	4.70	4.70	4.70

FLANGES

	Pressure Rating psi	Cast Iron		Steel						
		125	250	150	300	400	600	900	1500	2500
Screwed or Slip-On		2.3 / 1.5	4 / 1.5	2.5 / 1.5	4 / 1.5	5 / 1.5	5 / 1.5	12 / 1.5	12 / 1.5	15 / 1.5
Welding Neck				3 / 1.5	5 / 1.5	7 / 1.5	7 / 1.5	12 / 1.5	12 / 1.5	16 / 1.5
Lap Joint				2.5 / 1.5	4 / 1.5	5 / 1.5	5 / 1.5	12 / 1.5	12 / 1.5	15 / 1.5
Blind		2.5 / 1.5	5 / 1.5	2.5 / 1.5	5 / 1.5	5 / 1.5	5 / 1.5	12 / 1.5	12 / 1.5	15 / 1.5

FLANGED FITTINGS

	125	250	150	300	400	600	900	1500	2500
S.R. 90° Elbow						15 / 3.7		28 / 3.8	
L.R. 90° Elbow									
45° Elbow						14 / 3.4		26 / 3.6	
Tee						20 / 5.6		39 / 5.7	

VALVES

	125	250	150	300	400	600	900	1500	2500
Flanged Bonnet Gate				20 / 1.2		25 / 1.5		80 / 4.3	
Flanged Bonnet Globe or Angle								84 / 3.5	
Flanged Bonnet Check									
Pressure Seal Bonnet — Gate						31 / 1.7	31 / 1.7		
Pressure Seal Bonnet — Globe									

Boldface type is weight in pounds. Lightface type beneath weight is weight factor for insulation.

Insulation thicknesses and weights are based on average conditions and do not constitute a recommendation for specific thicknesses of materials. Insulation weights are based on 85% magnesia and hydrous calcium silicate at 11 lbs/cubic foot. The listed thicknesses and weights of combination covering are the sums of the inner layer of diatomaceous earth at 21 lbs/cubic foot and the outer layer at 11 lbs/cubic foot.

Insulation weights include allowances for wire, cement, canvas, bands and paint, but not special surface finishes.

To find the weight of covering on flanges, valves or fittings, multiply the weight factor by the weight per foot of covering used on straight pipe.

Valve weights are approximate. When possible, obtain weights from the manufacturer.

Cast iron valve weights are for flanged end valves; steel weights for welding end valves.

All flanged fitting, flanged valve and flange weights include the proportional weight of bolts or studs to make up all joints.

WEIGHT OF PIPING MATERIALS
(Courtesy of Anvil International)

1.660" O.D. 1¼" PIPE

PIPE								
	Schedule No.	5S	10S	40	80	160		
	Wall Designation			Std.	XS		XXS	
	Thickness — In.	.065	.109	.140	.191	.250	.382	
	Pipe — Lbs/Ft	1.11	1.81	2.27	3.00	3.77	5.22	
	Water — Lbs/Ft	.80	.71	.65	.56	.46	.27	

WELDING FITTINGS

	5S	10S	40	80	160	XXS
L.R. 90° Elbow	.3 / .3	.5 / .3	.6 / .3	.8 / .3	1.0 / .3	1.3 / .3
S.R. 90° Elbow			.4 / .2			
L.R. 45° Elbow	.2 / .2	.3 / .2	.3 / .2	.5 / .2	.6 / .2	.7 / .2
Tee	.7 / .5	1.1 / .5	1.6 / .5	1.6 / .5	1.9 / .5	2.4 / .5
Lateral	1.1 / 1.2	1.9 / 1.2	2.4 / 1.2	3.8 / 1.2		
Reducer	.3 / .2	.4 / .2	.5 / .2	.6 / .2	.7 / .2	.8 / .2
Cap	.1 / .3	.1 / .3	.4 / .3	.4 / .3	.6 / .3	.6 / .3

INSULATION

Temperature Range °F	100-199	200-299	300-399	400-499	500-599	600-699	700-799	800-899	900-999	1000-1099	1100-1200
85% Magnesia Calcium Silicate — Nom. Thick., In.	1	1	1½	2	2	2½	2½	2½	3	3	3
Lbs./Ft	.65	.65	1.47	1.83	1.83	2.65	2.65	2.65	3.58	3.58	3.58
Combination — Nom. Thick., In.						2½	2½	2½	3	3	3
Lbs/Ft						3.17	3.17	3.17	5.76	5.76	5.76

FLANGES

Pressure Rating psi	Cast Iron		Steel						
	125	250	150	300	400	600	900	1500	2500
Screwed or Slip-On	2.5 / 1.5	4.8 / 1.5	3.5 / 1.5	5 / 1.5	7 / 1.5	7 / 1.5	13 / 1.5	13 / 1.5	23 / 1.5
Welding Neck			3 / 1.5	7 / 1.5	8 / 1.5	8 / 1.5	13 / 1.5	13 / 1.5	25 / 1.5
Lap Joint			3.5 / 1.5	5 / 1.5	7 / 1.5	7 / 1.5	13 / 1.5	13 / 1.5	22 / 1.5
Blind	2.8 / 1.5	5.5 / 1.5	3.5 / 1.5	4 / 1.5	7 / 1.5	7 / 1.5	13 / 1.5	13 / 1.5	23 / 1.5

FLANGED FITTINGS

	150	300	400	600	900	1500	2500
S.R. 90° Elbow		17 / 3.7		18 / 3.8		33 / 3.9	
L.R. 90° Elbow		18 / 3.9					
45° Elbow		15 / 3.4		16 / 3.5		31 / 3.7	
Tee		23 / 5.6		28 / 5.6		49 / 5.9	

VALVES

	150	300	400	600	900	1500	2500
Flanged Bonnet Gate		40 / 4		60 / 4.2		97 / 4.6	
Flanged Bonnet Globe or Angle							
Flanged Bonnet Check		21 / 4					
Pressure Seal Bonnet — Gate					38 / 1.8	38 / 1.1	
Pressure Seal Bonnet — Globe							

Boldface type is weight in pounds. Lightface type beneath weight is weight factor for insulation.

Insulation thicknesses and weights are based on average conditions and do not constitute a recommendation for specific thicknesses of materials. Insulation weights are based on 85% magnesia and hydrous calcium silicate at 11 lbs/cubic foot. The listed thicknesses and weights of combination covering are the sums of the inner layer of diatomaceous earth at 21 lbs/cubic foot and the outer layer at 11 lbs/cubic foot.

Insulation weights include allowances for wire, cement, canvas, bands and paint, but not special surface finishes.

To find the weight of covering on flanges, valves or fittings, multiply the weight factor by the weight per foot of covering used on straight pipe.

Valve weights are approximate. When possible, obtain weights from the manufacturer.

Cast iron valve weights are for flanged end valves; steel weights for welding end valves.

All flanged fitting, flanged valve and flange weights include the proportional weight of bolts or studs to make up all joints.

WEIGHT OF PIPING MATERIALS
(Courtesy of Anvil International)

1.900″ O.D. 1½″ PIPE

PIPE

Schedule No.	5S	10S	40	80	160			
Wall Designation			Std.	XS		XXS		
Thickness — In.	.065	.109	.145	.200	.281	.400	.525	.650
Pipe — Lbs/Ft	1.27	2.09	2.72	3.63	4.86	6.41	7.71	8.68
Water — Lbs/Ft	1.07	.96	.88	.77	.61	.41	.25	.12

WELDING FITTINGS (boldface = weight; lightface = weight factor for insulation)

	5S	10S	40	80	160	XXS		
L.R. 90° Elbow	.4 / .4	.8 / .4	.9 / .4	1.2 / .4	1.5 / .4	2.0 / .4		
S.R. 90° Elbow			.6 / .3	.8 / .3				
L.R. 45° Elbow	.3 / .2	.5 / .2	.5 / .2	.7 / .2	.8 / .2	.1 / .2		
Tee	.9 / .6	1.5 / .6	2 / .6	2.4 / .6	3.0 / .6	3.7 / .6		
Lateral	1.3 / 1.3	2.1 / 1.3	3.3 / 1.3	5.5 / 1.3				
Reducer	.3 / .2	.6 / .2	.6 / .2	.8 / .2	1.0 / .2	1.2 / .2		
Cap	.1 / .3	.2 / .3	.4 / .3	.5 / .3	.7 / .3	.8 / .3		

INSULATION

Temperature Range °F	100-199	200-299	300-399	400-499	500-599	600-699	700-799	800-899	900-999	1000-1099	1100-1200
85% Magnesia Calcium Silicate — Nom. Thick., In.	1	1	1½	2	2	2½	2½	2½	3	3	3
85% Magnesia Calcium Silicate — Lbs./Ft	.84	.84	1.35	2.52	2.52	3.47	3.47	3.47	4.52	4.52	4.52
Combination — Nom. Thick., In.						2½	2½	2½	3	3	3
Combination — Lbs/Ft						4.20	4.20	4.20	5.62	5.62	5.62

FLANGES (boldface = weight; lightface = weight factor for insulation)

Pressure Rating psi	Cast Iron		Steel						
	125	250	150	300	400	600	900	1500	2500
Screwed or Slip-On	3 / 1.5	6 / 1.5	3.5 / 1.5	6 / 1.5	9 / 1.5	9 / 1.5	19 / 1.5	19 / 1.5	30 / 1.5
Welding Neck			4.5 / 1.5	8 / 1.5	12 / 1.5	12 / 1.5	19 / 1.5	19 / 1.5	34 / 1.5
Lap Joint			3.5 / 1.5	6 / 1.5	9 / 1.5	9 / 1.5	19 / 1.5	19 / 1.5	30 / 1.5
Blind	4 / 1.5	6 / 1.5	3.5 / 1.5	8 / 1.5	10 / 1.5	10 / 1.5	19 / 1.5	19 / 1.5	31 / 1.5

FLANGED FITTINGS

	125	250	150	300	400	600	900	1500	2500
S.R. 90° Elbow	9 / 3.7		12 / 3.7	23 / 3.8		26 / 3.9		46 / 4	
L.R. 90° Elbow	12 / 4		13 / 4	24 / 4					
45° Elbow	8 / 3.4		11 / 3.4	21 / 3.5		23 / 3.5		39 / 3.7	
Tee	15 / 5.6		20 / 5.6	30 / 5.7		37 / 5.8		70 / 6	

VALVES

	125	250	150	300	400	600	900	1500	2500
Flanged Bonnet Gate	27 / 6.8			55 / 4.2		70 / 4.5		125 / 5	
Flanged Bonnet Globe or Angle				40 / 4.2		45 / 4.2		170 / 5	
Flanged Bonnet Check			30 / 4.1	35 / 4.1		40 / 4.2		110 / 4.5	
Pressure Seal Bonnet — Gate							42 / 1.9	42 / 1.2	
Pressure Seal Bonnet — Globe									

Boldface type is weight in pounds. Lightface type beneath weight is weight factor for insulation.

Insulation thicknesses and weights are based on average conditions and do not constitute a recommendation for specific thicknesses of materials. Insulation weights are based on 85% magnesia and hydrous calcium silicate at 11 lbs/cubic foot. The listed thicknesses and weights of combination covering are the sums of the inner layer of diatomaceous earth at 21 lbs/cubic foot and the outer layer at 11 lbs/cubic foot.

Insulation weights include allowances for wire, cement, canvas, bands and paint, but not special surface finishes.

To find the weight of covering on flanges, valves or fittings, multiply the weight factor by the weight per foot of covering used on straight pipe.

Valve weights are approximate. When possible, obtain weights from the manufacturer.

Cast iron valve weights are for flanged end valves; steel weights for welding end valves.

All flanged fitting, flanged valve and flange weights include the proportional weight of bolts or studs to make up all joints.

WEIGHT OF PIPING MATERIALS
(Courtesy of Anvil International)

2.375" O.D. 2" PIPE

PIPE														
Schedule No.	5S	10S	40	80	160								
Wall Designation			Std.	XS		XXS							
Thickness — In.	.065	.109	.154	.218	.343	.436	.562	.687					
Pipe — Lbs/Ft	1.60	2.64	3.65	5.02	7.44	9.03	10.88	12.39					
Water — Lbs/Ft	1.72	1.58	1.46	1.28	.97	.77	.53	.34					

WELDING FITTINGS

	5S	10S	40	80	160	XXS		
L.R. 90° Elbow	.6 / .5	1.1 / .5	1.5 / .5	2.1 / .5	3.0 / .5	4.0 / .5		
S.R. 90° Elbow			1 / .3	1.4 / .3				
L.R. 45° Elbow	.4 / .2	.6 / .2	.9 / .2	1.1 / .2	1.6 / .2	2.0 / .2		
Tee	1.1 / .6	1.8 / .6	2.9 / .6	3.7 / .6	4.9 / .6	5.7 / .6		
Lateral	1.9 / 1.4	3.2 / 1.4	5 / 1.4	7.7 / 1.4				
Reducer	.4 / .3	.9 / .3	.9 / .3	1.2 / .3	1.6 / .3	1.9 / .3		
Cap	.2 / .4	.3 / .4	.6 / .4	.7 / .4	1.1 / .4	1.2 / .4		

INSULATION

Temperature Range °F	100-199	200-299	300-399	400-499	500-599	600-699	700-799	800-899	900-999	1000-1099	1100-1200
85% Magnesia Calcium Silicate — Nom. Thick., In.	1	1	1½	2	2	2½	2½	2½	3	3	3
85% Magnesia Calcium Silicate — Lbs./Ft	1.01	1.01	1.71	2.53	2.53	3.48	3.48	4.42	4.42	4.42	5.59
Combination — Nom. Thick., In.						2½	2½	2½	3	3	3
Combination — Lbs/Ft						4.28	4.28	5.93	5.93	5.93	7.80

FLANGES

Pressure Rating psi	Cast Iron		Steel						
	125	250	150	300	400	600	900	1500	2500
Screwed or Slip-On	5 / 1.5	7 / 1.5	6 / 1.5	9 / 1.5	11 / 1.5	11 / 1.5	32 / 1.5	32 / 1.5	49 / 1.5
Welding Neck			7 / 1.5	11 / 1.5	14 / 1.5	14 / 1.5	32 / 1.5	32 / 1.5	53 / 1.5
Lap Joint			6 / 1.5	9 / 1.5	11 / 1.5	11 / 1.5	32 / 1.5	32 / 1.5	48 / 1.5
Blind	5 / 1.5	8 / 1.5	5 / 1.5	10 / 1.5	12 / 1.5	12 / 1.5	32 / 1.5	32 / 1.5	50 / 1.5

FLANGED FITTINGS

	125	250	150	300	400	600	900	1500	2500
S.R. 90° Elbow	14 / 3.8	20 / 3.8	19 / 3.8	29 / 3.8		35 / 4		83 / 4.2	
L.R. 90° Elbow	16 / 4.1	27 / 4.1	22 / 4.1	31 / 4.1					
45° Elbow	12 / 3.4	18 / 3.5	16 / 3.4	24 / 3.5		33 / 3.7		73 / 3.9	
Tee	21 / 5.7	32 / 5.7	27 / 5.7	41 / 5.7		52 / 6		129 / 6.3	

VALVES

	125	250	150	300	400	600	900	1500	2500
Flanged Bonnet Gate	37 / 6.9	52 / 7.1	40 / 4	65 / 4.2		80 / 4.5		190 / 5	
Flanged Bonnet Globe or Angle	30 / 7	64 / 7.3	30 / 3.8	45 / 4		85 / 4.5		235 / 5.5	
Flanged Bonnet Check	26 / 7	51 / 7.3	35 / 3.8	40 / 4		60 / 4.2		300 / 5.8	
Pressure Seal Bonnet — Gate								150 / 2.5	
Pressure Seal Bonnet — Globe								165 / 3	

Boldface type is weight in pounds. Lightface type beneath weight is weight factor for insulation.

Insulation thicknesses and weights are based on average conditions and do not constitute a recommendation for specific thicknesses of materials. Insulation weights are based on 85% magnesia and hydrous calcium silicate at 11 lbs/cubic foot. The listed thicknesses and weights of combination covering are the sums of the inner layer of diatomaceous earth at 21 lbs/cubic foot and the outer layer at 11 lbs/cubic foot.

Insulation weights include allowances for wire, cement, canvas, bands and paint, but not special surface finishes.

To find the weight of covering on flanges, valves or fittings, multiply the weight factor by the weight per foot of covering used on straight pipe.

Valve weights are approximate. When possible, obtain weights from the manufacturer.

Cast iron valve weights are for flanged end valves; steel weights for welding end valves.

All flanged fitting, flanged valve and flange weights include the proportional weight of bolts or studs to make up all joints.

WEIGHT OF PIPING MATERIALS
(Courtesy of Anvil International)

2.875" O.D. 2½" PIPE

PIPE

Schedule No.	5S	10S	40	80	160			
Wall Designation			Std.	XS		XXS		
Thickness — In.	.083	.120	.203	.276	.375	.553	.675	.800
Pipe — Lbs/Ft	2.48	3.53	5.79	7.66	10.01	13.70	15.86	17.73
Water — Lbs/Ft	2.50	2.36	2.08	1.84	1.54	1.07	.79	.55

WELDING FITTINGS

	5S	10S	40	80	160			
L.R. 90° Elbow	1.2 / .6	1.8 / .6	3.0 / .6	3.8 / .6	5.0 / .6	7.0 / .6		
S.R. 90° Elbow			2.2 / .4	2.5 / .4				
L.R. 45° Elbow	.7 / .3	1.0 / .3	1.6 / .3	2.1 / .3	3.0 / .3	3.5 / .3		
Tee	2.1 / .8	3.0 / .8	5.2 / .8	6.4 / .8	7.8 / .8	9.8 / .8		
Lateral	3.5 / 1.5	4.9 / 1.5	9.0 / 1.5	13 / 1.5				
Reducer	.6 / .3	1.2 / .3	1.6 / .3	2.0 / .3	2.7 / .3	3.3 / .3		
Cap	.3 / .4	.4 / .4	.9 / .4	.1 / .4	1.9 / .4	2.0 / .4		

INSULATION

Temperature Range °F	100-199	200-299	300-399	400-499	500-599	600-699	700-799	800-899	900-999	1000-1099	1100-1200
85% Magnesia Calcium Silicate — Nom. Thick., In.	1	1	1½	2	2	2½	2½	3	3	3½	3½
Lbs./Ft	1.14	1.14	2.29	3.23	3.23	4.28	4.28	5.46	5.46	6.86	6.86
Combination — Nom. Thick., In.						2½	2½	3	3	3½	3½
Lbs./Ft						5.20	5.20	7.36	7.36	9.58	9.58

FLANGES

Pressure Rating psi	Cast Iron		Steel						
	125	250	150	300	400	600	900	1500	2500
Screwed or Slip-On	7 / 1.5	12.5 / 1.5	8 / 1.5	14 / 1.5	17 / 1.5	17 / 1.5	46 / 1.5	46 / 1.5	69 / 1.5
Welding Neck			11 / 1.5	16 / 1.5	22 / 1.5	22 / 1.5	46 / 1.5	46 / 1.5	66 / 1.5
Lap Joint			8 / 1.5	14 / 1.5	16 / 1.5	16 / 1.5	45 / 1.5	45 / 1.5	67 / 1.5
Blind	7.8 / 1.5	10 / 1.5	8 / 1.5	16 / 1.5	19 / 1.5	19 / 1.5	45 / 1.5	45 / 1.5	70 / 1.5

FLANGED FITTINGS

	125	250	150	300	400	600	900	1500	2500
S.R. 90° Elbow	20 / 3.8	33 / 3.9	27 / 3.8	42 / 3.9		50 / 4.1		114 / 4.4	
L.R. 90° Elbow	24 / 4.2		30 / 4.2	47 / 4.2					
45° Elbow	18 / 3.5	31 / 3.6	22 / 3.5	35 / 3.6		46 / 3.8		99 / 3.9	
Tee	31 / 5.7	49 / 5.8	42 / 5.7	61 / 5.9		77 / 6.2		169 / 6.6	

VALVES

	125	250	150	300	400	600	900	1500	2500
Flanged Bonnet Gate	50 / 7	82 / 7.1	60 / 4	100 / 4.2		105 / 4.6		275 / 5.2	
Flanged Bonnet Globe or Angle	43 / 7.1	87 / 7.4	50 / 4	70 / 4.1		120 / 4.6		325 / 5.5	
Flanged Bonnet Check	36 / 7.1	71 / 7.4	40 / 4	50 / 4		105 / 4.6		320 / 5.5	
Pressure Seal Bonnet — Gate								215 / 2.5	
Pressure Seal Bonnet — Globe								230 / 2.8	

Boldface type is weight in pounds. Lightface type beneath weight is weight factor for insulation.

Insulation thicknesses and weights are based on average conditions and do not constitute a recommendation for specific thicknesses of materials. Insulation weights are based on 85% magnesia and hydrous calcium silicate at 11 lbs/cubic foot. The listed thicknesses and weights of combination covering are the sums of the inner layer of diatomaceous earth at 21 lbs/cubic foot and the outer layer at 11 lbs/cubic foot.

Insulation weights include allowances for wire, cement, canvas, bands and paint, but not special surface finishes.

To find the weight of covering on flanges, valves or fittings, multiply the weight factor by the weight per foot of covering used on straight pipe.

Valve weights are approximate. When possible, obtain weights from the manufacturer.

Cast iron valve weights are for flanged end valves; steel weights for welding end valves.

All flanged fitting, flanged valve and flange weights include the proportional weight of bolts or studs to make up all joints.

WEIGHT OF PIPING MATERIALS
(Courtesy of Anvil International)

3.500″ O.D. 3″ PIPE

PIPE

Schedule No.	5S	10S	40	80	160								
Wall Designation			Std.	XS		XXS							
Thickness — In.	.083	.120	.216	.300	.438	.600	.725	.850					
Pipe — Lbs/Ft	3.03	4.33	7.58	10.25	14.32	18.58	21.49	24.06					
Water — Lbs/Ft	3.78	3.61	3.20	2.86	2.35	1.80	1.43	1.10					

WELDING FITTINGS

	5S	10S	40	80	160	XXS			
L.R. 90° Elbow	1.7 / .8	2.5 / .8	4.7 / .8	6.0 / .8	8.5 / .8	11 / .8			
S.R. 90° Elbow			3.3 / .5	4.1 / .5					
L.R. 45° Elbow	.9 / .3	1.3 / .3	2.5 / .3	3.3 / .3	4.5 / .3	5.5 / .3			
Tee	2.7 / .8	3.9 / .8	7.0 / .8	10 / .8	12.2 / .8	14.8 / .8			
Lateral	4.5 / 1.8	6.4 / 1.8	12.5 / 1.8	18 / 1.8					
Reducer	.8 / .3	1.5 / .3	2.1 / .3	2.8 / .3	3.7 / .3	4.6 / .3			
Cap	.5 / .5	.7 / .5	1.4 / .5	1.8 / .5	3.5 / .5	3.6 / .5			

INSULATION

Temperature Range °F	100-199	200-299	300-399	400-499	500-599	600-699	700-799	800-899	900-999	1000-1099	1100-1200
85% Magnesia Calcium Silicate — Nom. Thick., In.	1	1	1½	2	2	2½	3	3	3	3½	3½
Lbs./Ft	1.25	1.25	2.08	3.01	3.01	4.07	5.24	5.24	5.24	6.65	6.65
Combination — Nom. Thick., In.						2½	3	3	3	3½	3½
Lbs/Ft						5.07	6.94	6.94	6.94	9.17	9.17

FLANGES

Pressure Rating psi	Cast Iron		Steel						
	125	250	150	300	400	600	900	1500	2500
Screwed or Slip-On	8.6 / 1.5	15.8 / 1.5	9 / 1.5	17 / 1.5	20 / 1.5	20 / 1.5	37 / 1.5	61 / 1.5	102 / 1.5
Welding Neck			12 / 1.5	19 / 1.5	27 / 1.5	27 / 1.5	38 / 1.5	61 / 1.5	113 / 1.5
Lap Joint			9 / 1.5	17 / 1.5	19 / 1.5	19 / 1.5	36 / 1.5	60 / 1.5	99 / 1.5
Blind	9 / 1.5	17.5 / 1.5	10 / 1.5	20 / 1.5	24 / 1.5	24 / 1.5	38 / 1.5	61 / 1.5	105 / 1.5

FLANGED FITTINGS

	125	250	150	300	400	600	900	1500	2500
S.R. 90° Elbow	25 / 3.9	44 / 4	32 / 3.9	53 / 4		67 / 4.1	98 / 4.3	150 / 4.6	
L.R. 90° Elbow	29 / 4.3		40 / 4.3	63 / 4.3					
45° Elbow	21 / 3.5	39 / 3.6	28 / 3.5	46 / 3.6		60 / 3.8	93 / 3.9	135 / 4	
Tee	38 / 5.9	62 / 6	52 / 5.9	81 / 6		102 / 6.2	151 / 6.5	238 / 6.9	

VALVES

	125	250	150	300	400	600	900	1500	2500
Flanged Bonnet Gate	66 / 7	112 / 7.4	70 / 4	125 / 4.4		155 / 4.8	260 / 5	410 / 5.5	
Flanged Bonnet Globe or Angle	56 / 7.2	87 / 7.6	60 / 4.3	95 / 4.5		155 / 4.8	225 / 5	495 / 5.5	
Flanged Bonnet Check	46 / 7.2	100 / 7.6	60 / 4.3	70 / 4.4		120 / 4.8	150 / 4.9	440 / 5.8	
Pressure Seal Bonnet — Gate							208 / 3	235 / 3.2	
Pressure Seal Bonnet — Globe							135 / 2.5	180 / 3	

Boldface type is weight in pounds. Lightface type beneath weight is weight factor for insulation.

Insulation thicknesses and weights are based on average conditions and do not constitute a recommendation for specific thicknesses of materials. Insulation weights are based on 85% magnesia and hydrous calcium silicate at 11 lbs/cubic foot. The listed thicknesses and weights of combination covering are the sums of the inner layer of diatomaceous earth at 21 lbs/cubic foot and the outer layer at 11 lbs/cubic foot.

Insulation weights include allowances for wire, cement, canvas, bands and paint, but not special surface finishes.

To find the weight of covering on flanges, valves or fittings, multiply the weight factor by the weight per foot of covering used on straight pipe.

Valve weights are approximate. When possible, obtain weights from the manufacturer.

Cast iron valve weights are for flanged end valves; steel weights for welding end valves.

All flanged fitting, flanged valve and flange weights include the proportional weight of bolts or studs to make up all joints.

WEIGHT OF PIPING MATERIALS
(Courtesy of Anvil International)

4.000" O.D. 3½" PIPE

		Schedule No.	5S	10S	40	80	160								
PIPE		Wall Designation			Std.	XS		XXS							
		Thickness — In.	.083	.120	.226	.318	.636								
		Pipe — Lbs/Ft	3.47	4.97	9.11	12.51	22.85								
		Water — Lbs/Ft	5.01	4.81	4.28	3.85	2.53								
WELDING FITTINGS		L.R. 90° Elbow	2.4 / .9	3.4 / .9	6.7 / .9	8.7 / .9	15 / .9								
		S.R. 90° Elbow			4.2 / .6	5.75 / .6									
		L.R. 45° Elbow	1.2 / .4	1.7 / .4	3.3 / .4	4.4 / .4	8 / .4								
		Tee	3.4 / .9	4.9 / .9	10.3 / .9	13.8 / .9	20.2 / .9								
		Lateral	6.2 / 1.8	8.9 / 1.8	17.2 / 1.8	25 / 1.8									
		Reducer	1.2 / .3	2.1 / .3	3.0 / .3	4.0 / .3	6.8 / .3								
		Cap	.6 / .6	.8 / .6	2.1 / .6	2.8 / .6	5.5 / .6								

	Temperature Range °F	100-199	200-299	300-399	400-499	500-599	600-699	700-799	800-899	900-999	1000-1099	1100-1200
INSULATION 85% Magnesia Calcium Silicate	Nom. Thick., In.	1	1	1½	2	2½	2½	3	3	3½	3½	3½
	Lbs./Ft	1.83	1.83	2.77	3.71	4.88	4.88	6.39	6.39	7.80	7.80	7.80
Combination	Nom. Thick., In.						2½	3	3	3½	3½	3½
	Lbs/Ft						6.49	8.71	8.71	10.8	10.8	10.8

		Pressure Rating psi	Cast Iron		Steel						
			125	250	150	300	400	600	900	1500	2500
FLANGES		Screwed or Slip-On	11 / 1.5	20 / 1.5	13 / 1.5	21 / 1.5	27 / 1.5	27 / 1.5			
		Welding Neck			14 / 1.5	22 / 1.5	32 / 1.5	32 / 1.5			
		Lap Joint			13 / 1.5	21 / 1.5	26 / 1.5	26 / 1.5			
		Blind	13 / 1.5	23 / 1.5	15 / 1.5	25 / 1.5	35 / 1.5	35 / 1.5			
FLANGED FITTINGS		S.R. 90° Elbow	33 / 4		49 / 4			82 / 4.3			
		L.R. 90° Elbow			54 / 4.4						
		45° Elbow	29 / 3.6		39 / 3.6			75 / 3.6			
		Tee	51 / 6	103 / 6.2	70 / 6			133 / 6.4			
VALVES		Flanged Bonnet Gate	82 / 7.1	143 / 7.5	90 / 4.1	155 / 4.5		180 / 4.8	360 / 5	510 / 5.5	
		Flanged Bonnet Globe or Angle	74 / 7.3	137 / 7.7				160 / 4.7			
		Flanged Bonnet Check	71 / 7.3	125 / 7.7				125 / 4.7			
		Pressure Seal Bonnet — Gate						140 / 2.5	295 / 2.8	380 / 3	
		Pressure Seal Bonnet — Globe									

Boldface type is weight in pounds. Lightface type beneath weight is weight factor for insulation.

Insulation thicknesses and weights are based on average conditions and do not constitute a recommendation for specific thicknesses of materials. Insulation weights are based on 85% magnesia and hydrous calcium silicate at 11 lbs/cubic foot. The listed thicknesses and weights of combination covering are the sums of the inner layer of diatomaceous earth at 21 lbs/cubic foot and the outer layer at 11 lbs/cubic foot.

Insulation weights include allowances for wire, cement, canvas, bands and paint, but not special surface finishes.

To find the weight of covering on flanges, valves or fittings, multiply the weight factor by the weight per foot of covering used on straight pipe.

Valve weights are approximate. When possible, obtain weights from the manufacturer.

Cast iron valve weights are for flanged end valves; steel weights for welding end valves.

All flanged fitting, flanged valve and flange weights include the proportional weight of bolts or studs to make up all joints.

WEIGHT OF PIPING MATERIALS
(Courtesy of Anvil International)

4.500" O.D. 4" PIPE

PIPE

	5S	10S		40	80	120		160			
Schedule No.	5S	10S		40	80	120		160			
Wall Designation				Std.	XS				XXS		
Thickness — In.	.083	.120	.188	.237	.337	.438	.500	.531	.674	.800	.925
Pipe — Lbs/Ft	3.92	5.61	8.56	10.79	14.98	18.96	21.36	22.51	27.54	31.61	35.32
Water — Lbs/Ft	6.40	6.17	5.80	5.51	4.98	4.48	4.16	4.02	3.38	2.86	2.39

WELDING FITTINGS (boldface = weight in lbs; lightface = weight factor)

	5S	10S		40	80	120		160			
L.R. 90° Elbow	3.0 / 1	4.3 / 1		8.7 / 1	12.0 / 1			18.0 / 1	20.5 / 1		
S.R. 90° Elbow				6.7 / .7	8.3 / .7						
L.R. 45° Elbow	1.5 / .4	2.2 / .4		4.3 / .4	5.9 / .4			8.5 / .4	10 / .4		
Tee	3.9 / 1	5.7 / 1		13.5 / 1	16.4 / 1			22.8 / 1	26.6 / 1		
Lateral	6.6 / 2.1	10.0 / 2.1		20.5 / 2.1	32 / 2.1						
Reducer	1.2 / .3	2.4 / .3		3.6 / .3	4.8 / .3			6.6 / .3	8.2 / .3		
Cap	.8 / .3	1.2 / .3		2.5 / .5	3.4 / .5			6.5 / 6.5	6.6 / 6.6		

INSULATION

Temperature Range °F	100-199	200-299	300-399	400-499	500-599	600-699	700-799	800-899	900-999	1000-1099	1100-1200
85% Magnesia Calcium Silicate — Nom. Thick., In.	1	1	1½	2	2½	2½	3	3	3½	3½	4
85% Magnesia Calcium Silicate — Lbs./Ft	1.62	1.62	2.55	3.61	4.66	4.66	6.07	6.07	7.48	7.48	9.10
Combination — Nom. Thick., In.						2½	3	3	3½	3½	3½
Combination — Lbs/Ft						6.07	8.30	8.30	10.6	10.6	10.6

FLANGES (psi pressure rating; boldface = weight, lightface = weight factor)

	Cast Iron		Steel						
Pressure Rating psi	125	250	150	300	400	600	900	1500	2500
Screwed or Slip-On	14 / 1.5	24 / 1.5	15 / 1.5	26 / 1.5	32 / 1.5	43 / 1.5	66 / 1.5	90 / 1.5	158 / 1.5
Welding Neck			17 / 1.5	29 / 1.5	41 / 1.5	48 / 1.5	648 / 1.5	90 / 1.5	177 / 1.5
Lap Joint			15 / 1.5	26 / 1.5	31 / 1.5	42 / 1.5	64 / 1.5	92 / 1.5	153 / 1.5
Blind	16 / 1.5	27 / 1.5	19 / 1.5	31 / 1.5	39 / 1.5	47 / 1.5	67 / 1.5	90 / 1.5	164 / 1.5

FLANGED FITTINGS

	125	250	150	300	400	600	900	1500	2500
S.R. 90° Elbow	43 / 4.1	69 / 4.2	59 / 4.1	85 / 4.2	99 / 4.3	128 / 4.4	185 / 4.5	254 / 4.8	
L.R. 90° Elbow	50 / 4.5		72 / 4.5	98 / 4.5					
45° Elbow	38 / 3.7	62 / 3.8	51 / 3.7	78 / 3.8	82 / 3.9	119 / 4	170 / 4.1	214 / 4.2	
Tee	66 / 6.1	103 / 6.3	86 / 6.1	121 / 6.3	153 / 6.4	187 / 6.6	262 / 6.8	386 / 7.2	

VALVES

	125	250	150	300	400	600	900	1500	2500
Flanged Bonnet Gate	109 / 7.2	188 / 7.5	100 / 4.2	175 / 4.5	195 / 5	255 / 5.1	455 / 5.4	735 / 6	
Flanged Bonnet Globe or Angle	97 / 7.4	177 / 7.8	95 / 4.3	145 / 4.8	215 / 5	230 / 5.1	415 / 5.5	800 / 6	
Flanged Bonnet Check	80 / 7.4	146 / 7.8	80 / 4.3	105 / 4.5	160 / 4.8	195 / 5	320 / 5.6	780 / 6	
Pressure Seal Bonnet — Gate						215 / 2.8	380 / 3	520 / 4	
Pressure Seal Bonnet — Globe							240 / 2.7	290 / 3	

Boldface type is weight in pounds. Lightface type beneath weight is weight factor for insulation.

Insulation thicknesses and weights are based on average conditions and do not constitute a recommendation for specific thicknesses of materials. Insulation weights are based on 85% magnesia and hydrous calcium silicate at 11 lbs/cubic foot. The listed thicknesses and weights of combination covering are the sums of the inner layer of diatomaceous earth at 21 lbs/cubic foot and the outer layer at 11 lbs/cubic foot.

Insulation weights include allowances for wire, cement, canvas, bands and paint, but not special surface finishes.

To find the weight of covering on flanges, valves or fittings, multiply the weight factor by the weight per foot of covering used on straight pipe.

Valve weights are approximate. When possible, obtain weights from the manufacturer.

Cast iron valve weights are for flanged end valves; steel weights for welding end valves.

All flanged fitting, flanged valve and flange weights include the proportional weight of bolts or studs to make up all joints.

WEIGHT OF PIPING MATERIALS
(Courtesy of Anvil International)

5.563" O.D. **5"** PIPE

		5S	10S	40	80	120	160				
PIPE	Schedule No.	5S	10S	40	80	120	160				
	Wall Designation			Std.	XS			XXS			
	Thickness — In.	.109	.134	.258	.375	.500	.625	.750	.875	1.000	
	Pipe — Lbs/Ft	6.35	7.77	14.62	20.78	27.04	32.96	38.55	43.81	47.73	
	Water — Lbs/Ft	9.73	9.53	8.66	7.89	7.09	6.33	5.62	4.95	4.23	

WELDING FITTINGS

	5S	10S	40	80	120	160	XXS		
L.R. 90° Elbow	6.0 / 1.3	7.4 / 1.3	16.0 / 1.3	21.4 / 1.3		33 / 1.3	34 / 1.3		
S.R. 90° Elbow	4.2 / .8	5.2 / .8	10.4 / .8	14.5 / .8					
L.R. 45° Elbow	3.1 / .5	3.8 / .5	8.3 / .5	10.5 / .5		14 / .5	18 / .5		
Tee	9.8 / 1.2	12.0 / 1.2	19.8 / 1.2	26.9 / 1.2		38.5 / 1.2	43.4 / 1.2		
Lateral	15.3 / 2.5	18.4 / 2.5	31 / 2.5	49 / 2.5					
Reducer	2.5 / .4	4.3 / .4	5.9 / .4	8.3 / .4		12.4 / .4	14.2 / .4		
Cap	1.3 / .7	1.6 / .7	4.2 / .7	5.7 / .7		11 / .7	11 / .7		

INSULATION

Temperature Range °F	100-199	200-299	300-399	400-499	500-599	600-699	700-799	800-899	900-999	1000-1099	1100-1200
85% Magnesia Calcium Silicate — Nom. Thick., In.	1	1½	1½	2	2½	2½	3	3½	3½	4	4
Lbs./Ft	1.86	2.92	2.92	4.08	5.38	5.38	6.90	8.41	8.41	10.4	10.4
Combination — Nom. Thick., In.						2½	3	3½	3½	4	4
Lbs/Ft						7.01	9.30	11.8	11.8	14.9	14.9

FLANGES

Pressure Rating psi	Cast Iron		Steel						
	125	250	150	300	400	600	900	1500	2500
Screwed or Slip-On	17 / 1.5	28 / 1.5	18 / 1.5	32 / 1.5	37 / 1.5	73 / 1.5	100 / 1.5	162 / 1.5	259 / 1.5
Welding Neck			22 / 1.5	36 / 1.5	49 / 1.5	78 / 1.5	103 / 1.5	162 / 1.5	293 / 1.5
Lap Joint			18 / 1.5	32 / 1.5	35 / 1.5	71 / 1.5	98 / 1.5	179 / 1.5	253 / 1.5
Blind	21 / 1.5	35 / 1.5	23 / 1.5	39 / 1.5	50 / 1.5	78 / 1.5	104 / 1.5	172 / 1.5	272 / 1.5

FLANGED FITTINGS

	125	250	150	300	400	600	900	1500	2500
S.R. 90° Elbow	55 / 4.3	91 / 4.3	80 / 4.3	113 / 4.3	123 / 4.5	205 / 4.7	268 / 4.8	435 / 5.2	
L.R. 90° Elbow	65 / 4.7		91 / 4.7	128 / 4.7					
45° Elbow	48 / 3.8	80 / 3.8	66 / 3.8	98 / 3.8	123 / 4	180 / 4.2	239 / 4.3	350 / 4.5	
Tee	84 / 6.4	139 / 6.5	119 / 6.4	172 / 6.4	179 / 6.8	304 / 7	415 / 7.2	665 / 7.8	

VALVES

	125	250	150	300	400	600	900	1500	2500
Flanged Bonnet Gate	138 / 7.3	264 / 7.9	150 / 4.3	265 / 4.9	310 / 5.3	455 / 5.5	615 / 6	1340 / 7	
Flanged Bonnet Globe or Angle	138 / 7.6	247 / 8	155 / 4.3	215 / 5	355 / 5.2	515 / 5.8	555 / 5.8	950 / 6	
Flanged Bonnet Check	118 / 7.6	210 / 8	110 / 4.3	165 / 5	185 / 5	350 / 5.8	570 / 6	1150 / 7	
Pressure Seal Bonnet — Gate						350 / 3.1	520 / 3.8	865 / 4.5	
Pressure Seal Bonnet — Globe						280 / 4	450 / 4.5		

Boldface type is weight in pounds. Lightface type beneath weight is weight factor for insulation.

Insulation thicknesses and weights are based on average conditions and do not constitute a recommendation for specific thicknesses of materials. Insulation weights are based on 85% magnesia and hydrous calcium silicate at 11 lbs/cubic foot. The listed thicknesses and weights of combination covering are the sums of the inner layer of diatomaceous earth at 21 lbs/cubic foot and the outer layer at 11 lbs/cubic foot.

Insulation weights include allowances for wire, cement, canvas, bands and paint, but not special surface finishes.

To find the weight of covering on flanges, valves or fittings, multiply the weight factor by the weight per foot of covering used on straight pipe.

Valve weights are approximate. When possible, obtain weights from the manufacturer.

Cast iron valve weights are for flanged end valves; steel weights for welding end valves.

All flanged fitting, flanged valve and flange weights include the proportional weight of bolts or studs to make up all joints.

WEIGHT OF PIPING MATERIALS
(Courtesy of Anvil International)

6.625″ O.D. **6″** PIPE

PIPE

Schedule No.	5S	10S		40	80	120	160			
Wall Designation				Std.	XS			XXS		
Thickness — In.	.109	.134	.219	.280	.432	.562	.718	.864	1.000	1.125
Pipe — Lbs/Ft	**5.37**	**9.29**	**15.02**	**18.97**	**28.57**	**36.39**	**45.3**	**53.2**	**60.01**	**66.08**
Water — Lbs/Ft	**13.98**	**13.74**	**13.10**	**12.51**	**11.29**	**10.30**	**9.2**	**8.2**	**7.28**	**6.52**

WELDING FITTINGS (boldface = weight in lbs; lightface = weight factor)

Fitting	5S	10S		40	80	120	160	XXS		
L.R. 90° Elbow	**8.9** 1.5	**11.0** 1.5		**22.8** 1.5	**32.2** 1.5	**43** 1.5	**55** 1.6	**62** 1.5		
S.R. 90° Elbow	**6.1** 1	**7.5** 1		**16.6** 1	**22.9** 1	**30** 1				
L.R. 45° Elbow	**4.5** .6	**5.5** .6		**11.3** .6	**16.4** .6	**21** .6	**26** .6	**30** .6		
Tee	**13.8** 1.4	**17.0** 1.4		**31.3** 1.4	**39.5** 1.4		**59** 1.4	**68** 1.4		
Lateral	**16.7** 2.9	**20.5** 2.9		**42** 2.9	**78** 2.9					
Reducer	**3.3** .5	**5.8** .5		**8.6** .6	**12.6** .5		**18.8** .5	**21.4** .5		
Cap	**1.6** .9	**1.9** .9		**6.4** .9	**9.2** .9	**13.3** .9	**17.5** .9	**17.5** .9		

INSULATION

Temperature Range °F	100-199	200-299	300-399	400-499	500-599	600-699	700-799	800-899	900-999	1000-1099	1100-1200
85% Magnesia Calcium Silicate — Nom. Thick., In.	1	1½	2	2	2½	3	3	3½	3½	4	4
Lbs./Ft	2.11	3.28	4.57	4.57	6.09	7.60	7.60	9.82	9.82	11.5	11.4
Combination — Nom. Thick., In.						3	3	3½	3½	4	4
Lbs/Ft						10.3	10.3	13.4	13.4	16.6	16.6

FLANGES (boldface = weight in lbs; lightface = weight factor)

Pressure Rating psi	Cast Iron		Steel						
	125	250	150	300	400	600	900	1500	2500
Screwed or Slip-On	**20** 1.5	**38** 1.5	**22** 1.5	**45** 1.5	**54** 1.5	**95** 1.5	**128** 1.5	**202** 1.5	**396** 1.5
Welding Neck			**27** 1.5	**48** 1.5	**67** 1.5	**96** 1.5	**130** 1.5	**202** 1.5	**451** 1.5
Lap Joint			**22** 1.5	**45** 1.5	**52** 1.5	**93** 1.5	**125** 1.5	**208** 1.5	**387** 1.5
Blind	**26** 1.5	**48** 1.5	**29** 1.5	**56** 1.5	**71** 1.5	**101** 1.5	**133** 1.5	**197** 1.5	**418** 1.5

FLANGED FITTINGS

Fitting	125	250	150	300	400	600	900	1500	2500
S.R. 90° Elbow	**71** 4.3	**121** 4.4	**90** 4.3	**147** 4.4	**184** 4.6	**275** 4.8	**375** 5	**566** 5.3	
L.R. 90° Elbow	**88** 4.9		**126** 4.9	**182** 4.9					
45° Elbow	**63** 3.8	**111** 3.9	**82** 3.8	**132** 3.9	**149** 4.1	**240** 4.3	**320** 4.3	**487** 4.6	
Tee	**108** 6.5	**186** 6.6	**149** 6.5	**218** 6.6	**280** 6.9	**400** 7.2	**565** 7.5	**839** 8	

VALVES

Valve	125	250	150	300	400	600	900	1500	2500
Flanged Bonnet Gate	**172** 7.3	**359** 8	**190** 4.3	**360** 5	**435** 5.5	**620** 5.8	**835** 6	**1595** 7	
Flanged Bonnet Globe or Angle	**184** 7.8	**345** 8.2	**185** 4.4	**275** 5	**415** 5.3	**645** 5.8	**765** 6	**1800** 7	
Flanged Bonnet Check	**154** 7.8	**286** 8.2	**150** 4.8	**200** 5	**360** 5.4	**445** 6	**800** 6.4	**1730** 7	
Pressure Seal Bonnet — Gate						**580** 3.5	**750** 4	**1215** 5	
Pressure Seal Bonnet — Globe							**730** 4	**780** 5	

Boldface type is weight in pounds. Lightface type beneath weight is weight factor for insulation.

Insulation thicknesses and weights are based on average conditions and do not constitute a recommendation for specific thicknesses of materials. Insulation weights are based on 85% magnesia and hydrous calcium silicate at 11 lbs/cubic foot. The listed thicknesses and weights of combination covering are the sums of the inner layer of diatomaceous earth at 21 lbs/cubic foot and the outer layer at 11 lbs/cubic foot.

Insulation weights include allowances for wire, cement, canvas, bands and paint, but not special surface finishes.

To find the weight of covering on flanges, valves or fittings, multiply the weight factor by the weight per foot of covering used on straight pipe.

Valve weights are approximate. When possible, obtain weights from the manufacturer.

Cast iron valve weights are for flanged end valves; steel weights for welding end valves.

All flanged fitting, flanged valve and flange weights include the proportional weight of bolts or studs to make up all joints.

WEIGHT OF PIPING MATERIALS
(Courtesy of Anvil International)

8.625" O.D. 8" PIPE

PIPE

Schedule No.	5S	10S		20	30	40	60	80	100	120	140	160
Wall Designation						Std.		XS				
Thickness — In.	.109	.148	.219	.250	.277	.322	.406	.500	.593	.718	.812	.906
Pipe — Lbs/Ft	9.91	13.40	19.64	22.36	24.70	28.55	35.64	43.4	50.9	60.6	67.8	74.7
Water — Lbs/Ft	24.07	23.59	22.90	22.48	22.18	21.69	20.79	19.8	18.8	17.6	16.7	15.8

WELDING FITTINGS

	5S	10S		20	30	40	60	80	100	120	140	160
L.R. 90° Elbow	15.4 / 2	21.0 / 2				44.9 / 2		70.3 / 2				120 / 2
S.R. 90° Elbow	6.6 / 1.3	14.3 / 1.3				34.5 / 1.3		50.2 / 1.3				
L.R. 45° Elbow	8.1 / .8	11.0 / .8				22.8 / .8		32.8 / .8				56 / .8
Tee	18.4 / 1.8	25.0 / 1.8				60.2 / 1.8		78 / 1.8				120 / 1.8
Lateral	25.3 / 3.8	41.1 / 3.8				76 / 3.8		140 / 3.8				
Reducer	4.5 / .5	7.8 / .5				13.9 / .5		20.4 / .5				32.1 / .5
Cap	2.1 / 1	2.8 / 1				11.3 / 1		16.3 / 1				32 / 1

INSULATION

Temperature Range °F	100-199	200-299	300-399	400-499	500-599	600-699	700-799	800-899	900-999	1000-1099	1100-1200
85% Magnesia Calcium Silicate — Nom. Thick., In.	1½	1½	2	2	2½	3	3½	3½	4	4	4½
85% Magnesia Calcium Silicate — Lbs./Ft	4.13	4.13	5.64	5.64	7.85	9.48	11.5	11.5	13.8	13.8	16.0
Combination — Nom. Thick., In.						3	3½	3½	4	4	4½
Combination — Lbs/Ft						12.9	16.2	16.2	20.4	20.4	23.8

FLANGES

Pressure Rating	Cast Iron		Steel						
psi	125	250	150	300	400	600	900	1500	2500
Screwed or Slip-On	29 / 1.5	60 / 1.5	33 / 1.5	67 / 1.5	82 / 1.5	135 / 1.5	207 / 1.5	319 / 1.5	601 / 1.5
Welding Neck			42 / 1.5	76 / 1.5	104 / 1.5	137 / 1.5	222 / 1.5	334 / 1.5	692 / 1.5
Lap Joint			33 / 1.5	77 / 1.5	79 / 1.5	132 / 1.5	223 / 1.5	347 / 1.5	587 / 1.5
Blind	43 / 1.5	79 / 1.5	48 / 1.5	90 / 1.5	115 / 1.5	159 / 1.5	232 / 1.5	363 / 1.5	649 / 1.5

FLANGED FITTINGS

	125	250	150	300	400	600	900	1500	2500
S.R. 90° Elbow	113 / 4.5	194 / 4.7	157 / 4.5	238 / 4.7	310 / 5	435 / 5.2	639 / 5.4	995 / 5.7	
L.R. 90° Elbow	148 / 5.3		202 / 5.3	283 / 5.3					
45° Elbow	97 / 3.9	164 / 4	127 / 3.9	203 / 4	215 / 4.1	360 / 4.4	507 / 4.5	870 / 4.8	
Tee	168 / 6.8	289 / 7.1	230 / 6.8	337 / 7.1	445 / 7.5	610 / 7.8	978 / 8.1	1465 / 8.6	

VALVES

	125	250	150	300	400	600	900	1500	2500
Flanged Bonnet Gate	251 / 7.5	583 / 8.1	305 / 4.5	505 / 5.1	703 / 6	960 / 6.3	1180 / 6.6	2740 / 7	
Flanged Bonnet Globe or Angle	317 / 8.4	554 / 8.6	475 / 5.4	505 / 5.5	610 / 5.9	1130 / 6.3	1160 / 6.3	2865 / 7	
Flanged Bonnet Check	302 / 8.4	454 / 8.6	235 / 5.2	310 / 5.3	475 / 5.6	725 / 6	1140 / 6.4	2075 / 7	
Pressure Seal Bonnet — Gate							925 / 4.5	1185 / 4.7	2345 / 5.5
Pressure Seal Bonnet — Globe								1550 / 4	1680 / 5

Boldface type is weight in pounds. Lightface type beneath weight is weight factor for insulation.

Insulation thicknesses and weights are based on average conditions and do not constitute a recommendation for specific thicknesses of materials. Insulation weights are based on 85% magnesia and hydrous calcium silicate at 11 lbs/cubic foot. The listed thicknesses and weights of combination covering are the sums of the inner layer of diatomaceous earth at 21 lbs/cubic foot and the outer layer at 11 lbs/cubic foot.

Insulation weights include allowances for wire, cement, canvas, bands and paint, but not special surface finishes.

To find the weight of covering on flanges, valves or fittings, multiply the weight factor by the weight per foot of covering used on straight pipe.

Valve weights are approximate. When possible, obtain weights from the manufacturer.

Cast iron valve weights are for flanged end valves; steel weights for welding end valves.

All flanged fitting, flanged valve and flange weights include the proportional weight of bolts or studs to make up all joints.

WEIGHT OF PIPING MATERIALS
(Courtesy of Anvil International)

10.750" O.D. 10" PIPE

PIPE

Schedule No.	5S	10S		20	30	40	60	80	100	120	140	160
Wall Designation						Std.		XS				
Thickness — In.	.134	.165	.219	.250	.307	.365	.500	.593	.718	.843	1.000	1.125
Pipe — Lbs/Ft	15.15	18.70	24.63	28.04	34.24	40.5	54.7	64.3	76.9	89.2	104.1	115.7
Water — Lbs/Ft	37.4	36.9	36.2	35.77	34.98	34.1	32.3	31.1	29.5	28.0	26.1	24.6

WELDING FITTINGS

	5S	10S		20	30	40	60	80	100	120	140	160
L.R. 90° Elbow	29.2	36.0				84	112					230
	2.5	2.5				2.5	2.5					2.5
S.R. 90° Elbow	20.3	24.9				62.2	74					
	1.7	1.7				1.7	1.7					
L.R. 45° Elbow	14.6	18.0				42.4	53.8					109
	1	1				1	1					1
Tee	30.0	37.0				104	132					222
	2.1	2.1				2.1	2.1					2.1
Lateral	47.5	70.0				124	200					
	4.4	4.4				4.4	4.4					
Reducer	8.1	14.0				23.2	31.4					58
	.6	.6				.6	.6					.6
Cap	3.8	4.7				20	26.3					59
	1.3	1.3				1.3	1.3					1.3

INSULATION

Temperature Range °F	100-199	200-299	300-399	400-499	500-599	600-699	700-799	800-899	900-999	1000-1099	1100-1200
85% Magnesia Calcium Silicate — Nom. Thick., In.	1½	1½	2	2½	2½	3	3½	3½	4	4	4½
Lbs./Ft	5.20	5.20	7.07	8.93	8.93	11.0	13.2	13.2	15.5	15.5	18.1
Combination — Nom. Thick., In.						3	3½	3½	4	4	4½
Lbs/Ft						15.4	19.3	19.3	23	23	27.2

FLANGES

Pressure Rating psi	Cast Iron		Steel						
	125	250	150	300	400	600	900	1500	2500
Screwed or Slip-On	45	93	50	100	117	213	293	528	1148
	1.5	1.5	1.5	1.5	1.5	1.5	1.5	1.5	1.5
Welding Neck			59	110	152	225	316	546	1291
			1.5	1.5	1.5	1.5	1.5	1.5	1.5
Lap Joint			50	110	138	231	325	577	1120
			1.5	1.5	1.5	1.5	1.5	1.5	1.5
Blind	66	120	77	146	181	267	338	599	1248
	1.5	1.5	1.5	1.5	1.5	1.5	1.5	1.5	1.5

FLANGED FITTINGS

	125	250	150	300	400	600	900	1500	2500
S.R. 90° Elbow	182	306	240	343	462	747	995		
	4.8	4.9	4.8	4.9	5.2	5.6	5.8		
L.R. 90° Elbow	237		290	438					
	5.8		5.8	5.8					
45° Elbow	152	256	185	288	332	572	732		
	4.1	4.2	4.1	4.2	4.3	4.6	4.7		
Tee	277	446	353	527	578	1007	1417		
	7.2	7.4	7.2	7.4	7.8	8.4	8.7		

VALVES

	125	250	150	300	400	600	900	1500	2500
Flanged Bonnet Gate	471	899	455	750	1035	1575	2140	3690	
	7.7	8.3	4.5	5	6	6.9	7.1	8	
Flanged Bonnet Globe or Angle	541	943	485	855	1070	1500	2500	4160	
	9.1	9.1	4.5	5.5	6	6.3	6.8	8	
Flanged Bonnet Check	453	751	370	485	605	1030	1350	2280	
	9.1	9.1	6	6.1	6.3	6.8	7	7.5	
Pressure Seal Bonnet — Gate						1450	1860	3150	
						4.9	5.5	6	
Pressure Seal Bonnet — Globe							1800	1910	
							5	6	

Boldface type is weight in pounds. Lightface type beneath weight is weight factor for insulation.

Insulation thicknesses and weights are based on average conditions and do not constitute a recommendation for specific thicknesses of materials. Insulation weights are based on 85% magnesia and hydrous calcium silicate at 11 lbs/cubic foot. The listed thicknesses and weights of combination covering are the sums of the inner layer of diatomaceous earth at 21 lbs/cubic foot and the outer layer at 11 lbs/cubic foot.

Insulation weights include allowances for wire, cement, canvas, bands and paint, but not special surface finishes.

To find the weight of covering on flanges, valves or fittings, multiply the weight factor by the weight per foot of covering used on straight pipe.

Valve weights are approximate. When possible, obtain weights from the manufacturer.

Cast iron valve weights are for flanged end valves; steel weights for welding end valves.

All flanged fitting, flanged valve and flange weights include the proportional weight of bolts or studs to make up all joints.

WEIGHT OF PIPING MATERIALS
(Courtesy of Anvil International)

12.750" O.D. 12" PIPE

PIPE

Schedule No.	5S	10S	20	30		40		60	80	120	140	160
Wall Designation					Std.		XS					
Thickness — In.	.156	.180	.250	.330	.375	.406	.500	.562	.687	1.000	1.125	1.312
Pipe — Lbs/Ft	20.99	24.20	33.38	43.8	49.6	53.5	65.4	73.2	88.5	125.5	139.7	160.3
Water — Lbs/Ft	52.7	52.2	51.10	49.7	49.0	48.5	47.0	46.0	44.0	39.3	37.5	34.9

WELDING FITTINGS

(Weight in pounds; lightface figure beneath is weight factor for insulation)

Fitting	5S	10S	40	XS	160
L.R. 90° Elbow	51.2	57.0	122	156	375
(factor)	3	3	3	3	3
S.R. 90° Elbow	33.6	38.1	82	104	
(factor)	2	2	2	2	
L.R. 45° Elbow	25.5	29.0	60.3	78	182
(factor)	1.3	1.3	1.3	1.3	1.3
Tee	46.7	54.0	162	180	360
(factor)	2.5	2.5	2.5	2.5	2.5
Lateral	74.7	86.2	180	273	
(factor)	5.4	5.4	5.4	5.4	
Reducer	14.1	20.9	33.4	43.6	94
(factor)	.7	.7	.7	.7	.7
Cap	6.2	7.1	29.5	38.1	95
(factor)	1.5	1.5	1.5	1.5	1.5

INSULATION

Temperature Range °F	100-199	200-299	300-399	400-499	500-599	600-699	700-799	800-899	900-999	1000-1099	1100-1200
85% Magnesia Calcium Silicate Nom. Thick., In.	1½	1½	2	2½	3	3	3½	4	4	4½	4½
Lbs./Ft	6.04	6.04	8.13	10.5	12.7	12.7	15.1	17.9	17.9	20.4	20.4
Combination Nom. Thick., In.						3	3½	4	4	4½	4½
Lbs/Ft						17.7	21.9	26.7	26.7	31.1	31.1

FLANGES

Pressure Rating psi	Cast Iron 125	250	Steel 150	300	400	600	900	1500	2500
Screwed or Slip-On	58	123	71	140	164	261	388	820	1611
(factor)	1.5	1.5	1.5	1.5	1.5	1.5	1.5	1.5	1.5
Welding Neck			87	163	212	272	434	843	1919
(factor)			1.5	1.5	1.5	1.5	1.5	1.5	1.5
Lap Joint			71	164	187	286	433	902	1583
(factor)			1.5	1.5	1.5	1.5	1.5	1.5	1.5
Blind	95	165	117	209	261	341	475	928	1775
(factor)	1.5	1.5	1.5	1.5	1.5	1.5	1.5	1.5	1.5

FLANGED FITTINGS

	Cast Iron 125	250	Steel 150	300	400	600	900	1500	2500
S.R. 90° Elbow	257	430	345	509	669	815	1474		
(factor)	5	5.2	5	5.2	5.5	5.8	6.2		
L.R. 90° Elbow	357		485	624			1598		
(factor)	6.2		6.2	6.2			6.2		
45° Elbow	227	360	282	414	469	705	1124		
(factor)	4.3	4.3	4.3	4.3	4.5	4.7	4.8		
Tee	387	640	513	754	943	1361	1928		
(factor)	7.5	7.8	7.5	7.8	8.3	8.7	9.3		

VALVES

	Cast Iron 125	250	Steel 150	300	400	600	900	1500	2500
Flanged Bonnet Gate	687	1298	635	1015	1420	2155	2770	4650	
(factor)	7.8	8.5	4	5	5.5	7	7.2	8	
Flanged Bonnet Globe or Angle	808	1200	710	1410					
(factor)	9.4	9.5	5	5.5					
Flanged Bonnet Check	674	1160	560	720			1410	2600	3370
(factor)	9.4	9.5	6	6.5			7.2	8	8
Pressure Seal Bonnet — Gate							1975	2560	4515
(factor)							5.5	6	7
Pressure Seal Bonnet — Globe									

Boldface type is weight in pounds. Lightface type beneath weight is weight factor for insulation.

Insulation thicknesses and weights are based on average conditions and do not constitute a recommendation for specific thicknesses of materials. Insulation weights are based on 85% magnesia and hydrous calcium silicate at 11 lbs/cubic foot. The listed thicknesses and weights of combination covering are the sums of the inner layer of diatomaceous earth at 21 lbs/cubic foot and the outer layer at 11 lbs/cubic foot.

Insulation weights include allowances for wire, cement, canvas, bands and paint, but not special surface finishes.

To find the weight of covering on flanges, valves or fittings, multiply the weight factor by the weight per foot of covering used on straight pipe.

Valve weights are approximate. When possible, obtain weights from the manufacturer.

Cast iron valve weights are for flanged end valves; steel weights for welding end valves.

All flanged fitting, flanged valve and flange weights include the proportional weight of bolts or studs to make up all joints.

WEIGHT OF PIPING MATERIALS
(Courtesy of Anvil International)

14″ O.D. 14″ PIPE

PIPE

Schedule No.	5S	10S	10	20	30	40		60	80	120	140	160
Wall Designation					Std.		XS					
Thickness — In.	.156	.188	.250	.312	.375	.438	.500	.593	.750	1.093	1.250	1.406
Pipe — Lbs/Ft	23.0	27.7	36.71	45.7	54.6	63.4	72.1	84.9	106.1	150.7	170.2	189.1
Water — Lbs/Ft	63.7	63.1	62.06	60.92	59.7	58.7	57.5	55.9	53.2	47.5	45.0	42.6

WELDING FITTINGS (boldface = weight; lightface = weight factor)

	5S	10S	Std.	XS
L.R. 90° Elbow	**65.6** 3.5	**78.0** 3.5	**157** 3.5	**200** 3.5
S.R. 90° Elbow	**43.1** 2.3	**51.7** 2.3	**108** 2.3	**135** 2.3
L.R. 45° Elbow	**32.5** 1.5	**39.4** 1.5	**80** 1.5	**98** 1.5
Tee	**49.4** 2.8	**59.6** 2.8	**196** 2.8	**220** 2.8
Lateral	**94.4** 5.8	**113** 5.8	**218** 5.8	**340** 5.8
Reducer	**25.0** 1.1	**31.2** 1.1	**63** 1.1	**83** 1.1
Cap	**7.6** 1.7	**9.2** 1.7	**35.3** 1.7	**45.9** 1.7

INSULATION

Temperature Range °F	100-199	200-299	300-399	400-499	500-599	600-699	700-799	800-899	900-999	1000-1099	1100-1200
85% Magnesia Calcium Silicate — Nom. Thick., In.	1½	1½	2	2½	3	3	3½	4	4	4½	4½
85% Magnesia Calcium Silicate — Lbs./Ft	6.16	6.16	8.38	10.7	13.1	13.1	15.8	18.5	18.5	21.3	21.3
Combination — Nom. Thick., In.						3	3½	4	4	4½	4½
Combination — Lbs/Ft						18.2	22.8	27.5	27.5	32.4	32.4

FLANGES (boldface = weight; lightface = weight factor)

Pressure Rating psi	Cast Iron		Steel						
	125	250	150	300	400	600	900	1500	2500
Screwed or Slip-On	**90** 1.5	**184** 1.5	**95** 1.5	**195** 1.5	**235** 1.5	**318** 1.5	**460** 1.5	**1016** 1.5	
Welding Neck			**130** 1.5	**217** 1.5	**277** 1.5	**406** 1.5	**642** 1.5	**1241** 1.5	
Lap Joint			**119** 1.5	**220** 1.5	**254** 1.5	**349** 1.5	**477** 1.5	**1076** 1.5	
Blind	**125** 1.5	**239** 1.5	**141** 1.5	**267** 1.5	**354** 1.5	**437** 1.5	**574** 1.5		

FLANGED FITTINGS

	125	250	150	300	400	600	900	1500	2500
S.R. 90° Elbow	**360** 5.3	**617** 5.5	**497** 5.3	**632** 5.5	**664** 5.7	**918** 5.9	**1549** 6.4		
L.R. 90° Elbow	**480** 6.6	**767** 6.6	**622** 6.6	**772** 6.6					
45° Elbow	**280** 4.3	**497** 4.4	**377** 4.3	**587** 4.4	**638** 4.6	**883** 4.8	**1246** 4.9		
Tee	**540** 8	**956** 8.3	**683** 8	**968** 8.3	**1131** 8.6	**1652** 8.9	**2318** 9.6		

VALVES

	125	250	150	300	400	600	900	1500	2500
Flanged Bonnet Gate	**921** 7.9	**1762** 8.8	**905** 4.9	**1525** 6	**1920** 6.3	**2960** 7	**4170** 8	**6425** 8.8	
Flanged Bonnet Globe or Angle	**1171** 9.9								
Flanged Bonnet Check	**885** 9.9		**1010** 5	**1155** 5.2					
Pressure Seal Bonnet — Gate						**2620** 6	**3475** 6.5	**6380** 7.5	
Pressure Seal Bonnet — Globe									

Boldface type is weight in pounds. Lightface type beneath weight is weight factor for insulation.

Insulation thicknesses and weights are based on average conditions and do not constitute a recommendation for specific thicknesses of materials. Insulation weights are based on 85% magnesia and hydrous calcium silicate at 11 lbs/cubic foot. The listed thicknesses and weights of combination covering are the sums of the inner layer of diatomaceous earth at 21 lbs/cubic foot and the outer layer at 11 lbs/cubic foot.

Insulation weights include allowances for wire, cement, canvas, bands and paint, but not special surface finishes.

To find the weight of covering on flanges, valves or fittings, multiply the weight factor by the weight per foot of covering used on straight pipe.

Valve weights are approximate. When possible, obtain weights from the manufacturer.

Cast iron valve weights are for flanged end valves; steel weights for welding end valves.

All flanged fitting, flanged valve and flange weights include the proportional weight of bolts or studs to make up all joints.

WEIGHT OF PIPING MATERIALS
(Courtesy of Anvil International)

16" O.D. **16"** PIPE

		Schedule No.	5S	10S	10	20	30	40	60	80	100	120	140	160
PIPE		Wall Designation					Std.		XS					
		Thickness — In.	.165	.188	.250	.312	.375	.500	.656	.843	1.031	1.218	1.438	1.593
		Pipe — Lbs/Ft	28.0	32.0	42.1	52.4	62.6	82.8	107.5	136.5	164.8	192.3	223.6	245.1
		Water — Lbs/Ft	83.5	83.0	81.8	80.5	79.1	76.5	73.4	69.7	66.1	62.6	58.6	55.9

WELDING FITTINGS

Fitting	5S	10S			30	40
L.R. 90° Elbow	89.8 / 4	102.0 / 4			208 / 4	270 / 4
S.R. 90° Elbow	59.7 / 2.5	67.7 / 2.5			135 / 2.5	177 / 2.5
L.R. 45° Elbow	44.9 / 1.7	51.0 / 1.7			104 / 1.7	136 / 1.7
Tee	66.8 / 3.2	75.9 / 3.2			250 / 3.2	278 / 3.2
Lateral	127.0 / 6.7	144.0 / 6.7			275 / 6.7	431 / 6.7
Reducer	31.3 / 1.2	35.7 / 1.2			77 / 1.2	102 / 1.2
Cap	10.1 / 1.8	11.5 / 1.8			44.3 / 1.8	57 / 1.8

INSULATION

	Temperature Range °F	100-199	200-299	300-399	400-499	500-599	600-699	700-799	800-899	900-999	1000-1099	1100-1200
85% Magnesia Calcium Silicate	Nom. Thick., In.	1½	1½	2	2½	3	3	3½	4	4	4½	4½
	Lbs./Ft	6.90	6.90	9.33	12.0	14.6	14.6	17.5	20.5	20.5	23.6	23.6
Combination	Nom. Thick., In.							3	3½	4	4	4½	4½
	Lbs/Ft							20.3	25.2	30.7	30.7	36.0	36.0

FLANGES

	Pressure Rating psi	Cast Iron		Steel						
		125	250	150	300	400	600	900	1500	2500
	Screwed or Slip-On	114 / 1.5	233 / 1.5	107 / 1.5	262 / 1.5	310 / 1.5	442 / 1.5	559 / 1.5	1297 / 1.5	
	Welding Neck			141 / 1.5	288 / 1.5	351 / 1.5	577 / 1.5	785 / 1.5	1597 / 1.5	
	Lap Joint			142 / 1.5	282 / 1.5	337 / 1.5	476 / 1.5	588 / 1.5	1372 / 1.5	
	Blind	174 / 1.5	308 / 1.5	184 / 1.5	349 / 1.5	455 / 1.5	603 / 1.5	719 / 1.5		

FLANGED FITTINGS

	125	250	150	300	400	600	900	1500	2500
S.R. 90° Elbow	484 / 5.5	826 / 5.8	656 / 5.5	958 / 5.8	1014 / 6	1402 / 6.3	1886 / 6.7		
L.R. 90° Elbow	684 / 7	1036 / 7	781 / 7	1058 / 7					
45° Elbow	374 / 4.3	696 / 4.6	481 / 4.3	708 / 4.6	839 / 4.7	1212 / 5	1586 / 5		
Tee	714 / 8.3	1263 / 8.7	961 / 8.3	1404 / 8.6	1671 / 9	2128 / 9.4	3054 / 10		

VALVES

	125	250	150	300	400	600	900	1500	2500
Flanged Bonnet Gate	1254 / 8	2321 / 9	1190 / 5	2015 / 7	2300 / 7.2	3675 / 7.9	4950 / 8.2	7875 / 9	
Flanged Bonnet Globe or Angle									
Flanged Bonnet Check	1166 / 10.5			1225 / 6					
Pressure Seal Bonnet — Gate							3230 / 7	8130 / 8	
Pressure Seal Bonnet — Globe									

Boldface type is weight in pounds. Lightface type beneath weight is weight factor for insulation.

Insulation thicknesses and weights are based on average conditions and do not constitute a recommendation for specific thicknesses of materials. Insulation weights are based on 85% magnesia and hydrous calcium silicate at 11 lbs/cubic foot. The listed thicknesses and weights of combination covering are the sums of the inner layer of diatomaceous earth at 21 lbs/cubic foot and the outer layer at 11 lbs/cubic foot.

Insulation weights include allowances for wire, cement, canvas, bands and paint, but not special surface finishes.

To find the weight of covering on flanges, valves or fittings, multiply the weight factor by the weight per foot of covering used on straight pipe.

Valve weights are approximate. When possible, obtain weights from the manufacturer.

Cast iron valve weights are for flanged end valves; steel weights for welding end valves.

All flanged fitting, flanged valve and flange weights include the proportional weight of bolts or studs to make up all joints.

WEIGHT OF PIPING MATERIALS
(Courtesy of Anvil International)

18″ O.D. **18″** PIPE

		5S	10S	10	20		30		40	60	80	120	160
PIPE	Schedule No.	5S	10S	10	20		30		40	60	80	120	160
	Wall Designation					Std.		XS					
	Thickness — In.	.165	.188	.250	.312	.375	.438	.500	.562	750	937	1.375	1.781
	Pipe — Lbs/Ft	31.0	36.0	47.4	59.0	70.6	82.1	93.5	104.8	138.2	170.8	244.1	308.5
	Water — Lbs/Ft	106.2	105.7	104.3	102.8	101.2	99.9	98.4	97	92.7	88.5	79.2	71.0

WELDING FITTINGS

	5S	10S				30		XS	
L.R. 90° Elbow	114.0 / 4.5	129.0 / 4.5				256 / 4.5		332 / 4.5	
S.R. 90° Elbow	75.7 / 2.8	85.7 / 2.8				176 / 2.8		225 / 2.8	
L.R. 45° Elbow	57.2 / 1.9	64.5 / 1.9				132 / 1.9		168 / 1.9	
Tee	83.2 / 3.6	94.7 / 3.6				282 / 3.6		351 / 3.6	
Lateral	157.0 / 7.5	179.0 / 7.5				326 / 7.5		525 / 7.5	
Reducer	42.6 / 1.3	48.5 / 1.3				94 / 1.3		123 / 1.3	
Cap	12.7 / 2.1	14.5 / 2.1				57 / 2.1		75 / 2.1	

INSULATION

Temperature Range °F	100-199	200-299	300-399	400-499	500-599	600-699	700-799	800-899	900-999	1000-1099	1100-1200
85% Magnesia Calcium Silicate — Nom. Thick., In.	1½	1½	2	2½	3	3	3½	4	4	4½	4½
Lbs./Ft	7.73	7.73	10.4	13.3	16.3	16.3	19.3	22.6	22.6	25.9	25.9
Combination — Nom. Thick., In.						3	3½	4	4	4½	4½
Lbs/Ft						22.7	28.0	33.8	33.8	39.5	39.5

FLANGES

	Cast Iron		Steel						
Pressure Rating psi	125	250	150	300	400	600	900	1500	2500
Screwed or Slip-On	125 / 1.5		139 / 1.5	331 / 1.5	380 / 1.5	573 / 1.5	797 / 1.5	1694 / 1.5	
Welding Neck			159 / 1.5	355 / 1.5	430 / 1.5	652 / 1.5	1074 / 1.5	2069 / 1.5	
Lap Joint			165 / 1.5	355 / 1.5	415 / 1.5	566 / 1.5	820 / 1.5	1769 / 1.5	
Blind	209 / 1.5	396 / 1.5	228 / 1.5	440 / 1.5	572 / 1.5	762 / 1.5	1030 / 1.5		

FLANGED FITTINGS

	125	250	150	300	400	600	900	1500	2500
S.R. 90° Elbow	599 / 5.8	1060 / 6	711 / 5.8	1126 / 6	1340 / 6.2	1793 / 6.6	2817 / 7		
L.R. 90° Elbow		1350 / 7.4	941 / 7.4	1426 / 7.4					
45° Elbow	439 / 4.4	870 / 4.7	521 / 4.4	901 / 4.7	1040 / 4.8	1543 / 5	2252 / 5.2		
Tee	879 / 8.6	1625 / 9	1010 / 8.6	1602 / 9	1909 / 9.3	2690 / 9.9	4327 / 10.5		

VALVES

	125	250	150	300	400	600	900	1500	2500
Flanged Bonnet Gate	1629 / 8.2	2578 / 9.3	1510 / 6	2505 / 6.5	3765 / 7	4460 / 7.8	6675 / 8.5		
Flanged Bonnet Globe or Angle									
Flanged Bonnet Check	1371 / 10.5								
Pressure Seal Bonnet — Gate						3100 / 5.5	3400 / 5.6	4200 / 6	
Pressure Seal Bonnet — Globe									

Boldface type is weight in pounds. Lightface type beneath weight is weight factor for insulation.

Insulation thicknesses and weights are based on average conditions and do not constitute a recommendation for specific thicknesses of materials. Insulation weights are based on 85% magnesia and hydrous calcium silicate at 11 lbs/cubic foot. The listed thicknesses and weights of combination covering are the sums of the inner layer of diatomaceous earth at 21 lbs/cubic foot and the outer layer at 11 lbs/cubic foot.

Insulation weights include allowances for wire, cement, canvas, bands and paint, but not special surface finishes.

To find the weight of covering on flanges, valves or fittings, multiply the weight factor by the weight per foot of covering used on straight pipe.

Valve weights are approximate. When possible, obtain weights from the manufacturer.

Cast iron valve weights are for flanged end valves; steel weights for welding end valves.

All flanged fitting, flanged valve and flange weights include the proportional weight of bolts or studs to make up all joints.

WEIGHT OF PIPING MATERIALS
(Courtesy of Anvil International)

20" O.D. **20"** PIPE

PIPE

Schedule No.	5S	10S	10	20	30	40	60	80	100	120	140	160
Wall Designation				Std.	XS							
Thickness — In.	.188	.218	.250	.375	.500	.593	.812	1.031	1.281	1.500	1.750	1.968
Pipe — Lbs/Ft	40.0	46.0	52.7	78.6	104.1	122.9	166.4	208.9	256.1	296.4	341.1	379.0
Water — Lbs/Ft	131.0	130.2	129.5	126.0	122.8	120.4	115.0	109.4	103.4	98.3	92.6	87.9

WELDING FITTINGS

Fitting	5S	10S	10	20	30
L.R. 90° Elbow	160.0 / 5	185.0 / 5		322 / 5	438 / 5
S.R. 90° Elbow	106.0 / 3.4	122.0 / 3.4		238 / 3.4	278 / 3.4
L.R. 45° Elbow	80.3 / 2.1	92.5 / 2.1		160 / 2.1	228 / 2.1
Tee	112.0 / 4	130.0 / 4		378 / 4	490 / 4
Lateral	228.0 / 8.3	265.0 / 8.3		396 / 8.3	625 / 8.3
Reducer	71.6 / 1.7	87.6 / 1.7		142 / 1.7	186 / 1.7
Cap	17.7 / 2.3	20.5 / 2.3		71 / 2.3	93 / 2.3

INSULATION

Temperature Range °F	100-199	200-299	300-399	400-499	500-599	600-699	700-799	800-899	900-999	1000-1099	1100-1200
85% Magnesia Calcium Silicate — Nom. Thick., In.	1½	1½	2	2½	3	3	3½	4	4	4½	4½
85% Magnesia Calcium Silicate — Lbs./Ft	8.45	8.45	11.6	14.6	17.7	17.7	21.1	24.6	24.6	28.1	28.1
Combination — Nom. Thick., In.						3	3½	4	4	4½	4½
Combination — Lbs/Ft						24.7	30.7	37.0	37.0	43.1	43.1

FLANGES

Pressure Rating psi	Cast Iron 125	Cast Iron 250	Steel 150	Steel 300	Steel 400	Steel 600	Steel 900	Steel 1500	Steel 2500
Screwed or Slip-On	153 / 1.5		180 / 1.5	378 / 1.5	468 / 1.5	733 / 1.5	972 / 1.5	2114 / 1.5	
Welding Neck			195 / 1.5	431 / 1.5	535 / 1.5	811 / 1.5	1344 / 1.5	2614 / 1.5	
Lap Joint			210 / 1.5	428 / 1.5	510 / 1.5	725 / 1.5	1048 / 1.5	2189 / 1.5	
Blind	275 / 1.5	487 / 1.5	297 / 1.5	545 / 1.5	711 / 1.5	976 / 1.5	1287 / 1.5		

FLANGED FITTINGS

Fitting	Cast Iron 125	Cast Iron 250	Steel 150	Steel 300	Steel 400	Steel 600	Steel 900	Steel 1500	Steel 2500
S.R. 90° Elbow	792 / 6	1315 / 6.3	922 / 6	1375 / 6.3	1680 / 6.5	2314 / 6.9	3610 / 7.3		
L.R. 90° Elbow	1132 / 7.8	1725 / 7.8	1352 / 7.8	1705 / 7.8					
45° Elbow	592 / 4.6	1055 / 4.8	652 / 4.6	1105 / 4.8	1330 / 4.9	1917 / 5.2	2848 / 5.4		
Tee	1178 / 9	2022 / 9.5	1378 / 9	1908 / 9.5	2370 / 9.7	3463 / 10.1	5520 / 11		

VALVES

Valve	Cast Iron 125	Cast Iron 250	Steel 150	Steel 300	Steel 400	Steel 600	Steel 900	Steel 1500	Steel 2500
Flanged Bonnet Gate	1934 / 8.3	3823 / 9.5	1855 / 6	3370 / 7	5700 / 8	5755 / 8			
Flanged Bonnet Globe or Angle									
Flanged Bonnet Check	1772 / 11								
Pressure Seal Bonnet — Gate									
Pressure Seal Bonnet — Globe									

Boldface type is weight in pounds. Lightface type beneath weight is weight factor for insulation.

Insulation thicknesses and weights are based on average conditions and do not constitute a recommendation for specific thicknesses of materials. Insulation weights are based on 85% magnesia and hydrous calcium silicate at 11 lbs/cubic foot. The listed thicknesses and weights of combination covering are the sums of the inner layer of diatomaceous earth at 21 lbs/cubic foot and the outer layer at 11 lbs/cubic foot.

Insulation weights include allowances for wire, cement, canvas, bands and paint, but not special surface finishes.

To find the weight of covering on flanges, valves or fittings, multiply the weight factor by the weight per foot of covering used on straight pipe.

Valve weights are approximate. When possible, obtain weights from the manufacturer.

Cast iron valve weights are for flanged end valves; steel weights for welding end valves.

All flanged fitting, flanged valve and flange weights include the proportional weight of bolts or studs to make up all joints.

WEIGHT OF PIPING MATERIALS
(Courtesy of Anvil International)

24" O.D. **24"** PIPE

PIPE

Schedule No.	5S	10	20		30	40	60	80	120	140	160
Wall Designation			Std.	XS							
Thickness — In.	.218	.250	.375	.500	.562	.687	.968	1.218	1.812	2.062	2.343
Pipe — Lbs/Ft	55.0	63.4	94.6	125.5	140.8	171.2	238.1	296.4	429.4	483.1	541.9
Water — Lbs/Ft	188.9	188.0	183.8	180.1	178.1	174.3	165.8	158.3	141.4	134.5	127.0

WELDING FITTINGS

	5S	20	XS
L.R. 90° Elbow	260.0 / 6	500 / 6	578 / 6
S.R. 90° Elbow	178.0 / 3.7	305 / 3.7	404 / 3.7
L.R. 45° Elbow	130.0 / 2.5	252 / 2.5	292 / 2.5
Tee	174.0 / 4.9	544 / 4.9	607 / 4.9
Lateral	361.0 / 10	544 / 10	875 / 10
Reducer	107.0 / 1.7	167 / 1.7	220 / 1.7
Cap	28.6 / 2.8	102 / 2.8	134 / 2.8

INSULATION

Temperature Range °F	100-199	200-299	300-399	400-499	500-599	600-699	700-799	800-899	900-999	1000-1099	1100-1200
85% Magnesia Calcium Silicate — Nom. Thick., In.	1½	1½	2	2½	3	3	3½	4	4	4½	4½
85% Magnesia Calcium Silicate — Lbs./Ft	10.0	10.0	13.4	17.0	21.0	21.0	24.8	28.7	28.7	32.9	32.9
Combination — Nom. Thick., In.						3	3½	4	4	4½	4½
Combination — Lbs/Ft						29.2	36.0	43.1	43.1	50.6	50.6

FLANGES

Pressure Rating psi	Cast Iron		Steel						
	125	250	150	300	400	600	900	1500	2500
Screwed or Slip-On	236 / 1.5		245 / 1.5	577 / 1.5	676 / 1.5	1056 / 1.5	1823 / 1.5	3378 / 1.5	
Welding Neck			295 / 1.5	632 / 1.5	777 / 1.5	1157 / 1.5	2450 / 1.5	4153 / 1.5	
Lap Joint			295 / 1.5	617 / 1.5	752 / 1.5	1046 / 1.5	2002 / 1.5	3478 / 1.5	
Blind	404 / 1.5	757 / 1.5	446 / 1.5	841 / 1.5	1073 / 1.5	1355 / 1.5	2442 / 1.5		

FLANGED FITTINGS

	125	250	150	300	400	600	900	1500	2500
S.R. 90° Elbow	1231 / 6.7	2014 / 6.8	1671 / 6.7	2174 / 6.8	2474 / 7.1	3506 / 7.6	6155 / 8.1		
L.R. 90° Elbow	1711 / 8.7	2644 / 8.7	1821 / 8.7	2874 / 8.7					
45° Elbow	871 / 4.8	1604 / 5	1121 / 4.8	1634 / 5	1974 / 5.1	2831 / 5.5	5124 / 6		
Tee	1836 / 10	3061 / 10.2	2276 / 10	3161 / 10.2	3811 / 10.6	5184 / 11.4	9387 / 12.1		

VALVES

	125	250	150	300	400	600	900	1500	2500
Flanged Bonnet Gate	3062 / 8.5	6484 / 9.8	2500 / 5	4675 / 7	6995 / 8.7	8020 / 9.5			
Flanged Bonnet Globe or Angle									
Flanged Bonnet Check	2956 / 12								
Pressure Seal Bonnet — Gate									
Pressure Seal Bonnet — Globe									

Boldface type is weight in pounds. Lightface type beneath weight is weight factor for insulation.

Insulation thicknesses and weights are based on average conditions and do not constitute a recommendation for specific thicknesses of materials. Insulation weights are based on 85% magnesia and hydrous calcium silicate at 11 lbs/cubic foot. The listed thicknesses and weights of combination covering are the sums of the inner layer of diatomaceous earth at 21 lbs/cubic foot and the outer layer at 11 lbs/cubic foot.

Insulation weights include allowances for wire, cement, canvas, bands and paint, but not special surface finishes.

To find the weight of covering on flanges, valves or fittings, multiply the weight factor by the weight per foot of covering used on straight pipe.

Valve weights are approximate. When possible, obtain weights from the manufacturer.

Cast iron valve weights are for flanged end valves; steel weights for welding end valves.

All flanged fitting, flanged valve and flange weights include the proportional weight of bolts or studs to make up all joints.

WEIGHT OF PIPING MATERIALS
(Courtesy of Anvil International)

26″ O.D. **26″** PIPE

PIPE	Schedule No.		10		20									
	Wall Designation			Std.	XS									
	Thickness — In.	.250	.312	.375	.500	.625	.750	.875	1.000	1.125				
	Pipe — Lbs/Ft	67.0	85.7	102.6	136.2	169.0	202.0	235.0	267.0	299.0				
	Water — Lbs/Ft	221.4	219.2	216.8	212.5	208.6	204.4	200.2	196.1	192.1				

WELDING FITTINGS

L.R. 90° Elbow			502 / 8.5	713 / 8.5						
S.R. 90° Elbow			359 / 5	474 / 5						
L.R. 45° Elbow			269 / 3.5	355 / 3.5						
Tee			634 / 6.8	794 / 6.8						
Lateral										
Reducer			200 / 4.3	272 / 4.3						
Cap			110 / 4.3	145 / 4.3						

INSULATION

		100-199	200-299	300-399	400-499	500-599	600-699	700-799	800-899	900-999	1000-1099	1100-1200
Temperature Range °F		100-199	200-299	300-399	400-499	500-599	600-699	700-799	800-899	900-999	1000-1099	1100-1200
85% Magnesia Calcium Silicate	Nom. Thick., In.	1½	1½	2	2½	3	3½	4	4½	5	5	6
	Lbs./Ft	10.4	10.4	14.1	18.0	21.9	26.0	30.2	34.6	39.1	39.1	48.4
Combination	Nom. Thick., In.						3½	4½	5½	6	6½	7
	Lbs/Ft						37.0	51.9	67.8	76.0	84.5	93.2

FLANGES

		Cast Iron		Steel						
Pressure Rating psi		125	250	150	300	400	600	900	1500	2500
Screwed or Slip-On				292 / 1.5	699 / 1.5	650 / 1.5	950 / 1.5	1525 / 1.5		
Welding Neck				342 / 1.5	799 / 1.5	750 / 1.5	1025 / 1.5	1575 / 1.5		
Lap Joint										
Blind				567 / 1.5	1179 / 1.5	1125 / 1.5	1525 / 1.5	2200 / 1.5		

FLANGED FITTINGS

S.R. 90° Elbow	
L.R. 90° Elbow	
45° Elbow	
Tee	

VALVES

Flanged Bonnet Gate	
Flanged Bonnet Globe or Angle	
Flanged Bonnet Check	
Pressure Seal Bonnet — Gate	
Pressure Seal Bonnet — Globe	

Boldface type is weight in pounds. Lightface type beneath weight is weight factor for insulation.

Insulation thicknesses and weights are based on average conditions and do not constitute a recommendation for specific thicknesses of materials. Insulation weights are based on 85% magnesia and hydrous calcium silicate at 11 lbs/cubic foot. The listed thicknesses and weights of combination covering are the sums of the inner layer of diatomaceous earth at 21 lbs/cubic foot and the outer layer at 11 lbs/cubic foot.

Insulation weights include allowances for wire, cement, canvas, bands and paint, but not special surface finishes.

To find the weight of covering on flanges, valves or fittings, multiply the weight factor by the weight per foot of covering used on straight pipe.

Valve weights are approximate. When possible, obtain weights from the manufacturer.

Cast iron valve weights are for flanged end valves; steel weights for welding end valves.

All flanged fitting, flanged valve and flange weights include the proportional weight of bolts or studs to make up all joints.

WEIGHT OF PIPING MATERIALS
(Courtesy of Anvil International)

28" O.D. 28" PIPE

PIPE

Schedule No.		10		20	30								
Wall Designation			Std.	XS									
Thickness — In.	.250	.312	.375	.500	.625	.750	.875	1.000	1.125				
Pipe — Lbs/Ft	74.0	92.4	110.6	146.9	182.7	218.0	253.0	288.0	323.0				
Water — Lbs/Ft	257.3	255.0	252.7	248.1	243.6	238.9	234.4	230.0	225.6				

WELDING FITTINGS

L.R. 90° Elbow			626 / 9	829 / 9					
S.R. 90° Elbow			415 / 5.4	551 / 5.4					
L.R. 45° Elbow			312 / 3.6	413 / 3.6					
Tee			729 / 7	910 / 7					
Lateral									
Reducer			210 / 2.7	290 / 2.7					
Cap			120 / 4.5	160 / 4.5					

INSULATION

Temperature Range °F	100-199	200-299	300-399	400-499	500-599	600-699	700-799	800-899	900-999	1000-1099	1100-1200
85% Magnesia Calcium Silicate Nom. Thick., In.	1½	1½	2	2½	3	3½	4	4½	5	5	6
Lbs./Ft	11.2	11.2	15.1	19.2	23.4	27.8	32.3	36.9	41.6	41.6	51.4
Combination Nom. Thick., In.						3½	4½	5½	6	6½	7
Lbs/Ft						39.5	55.4	72.2	80.9	89.8	99.0

FLANGES

Pressure Rating psi	Cast Iron		Steel						
	125	250	150	300	400	600	900	1500	2500
Screwed or Slip-On			334 / 1.5	853 / 1.5	780 / 1.5	1075 / 1.5	1800 / 1.5		
Welding Neck			364 / 1.5	943 / 1.5	880 / 1.5	1175 / 1.5	1850 / 1.5		
Lap Joint									
Blind			669 / 1.5	1408 / 1.5	1425 / 1.5	1750 / 1.5	2575 / 1.5		

FLANGED FITTINGS

S.R. 90° Elbow									
L.R. 90° Elbow									
45° Elbow									
Tee									

VALVES

Flanged Bonnet Gate									
Flanged Bonnet Globe or Angle									
Flanged Bonnet Check									
Pressure Seal Bonnet — Gate									
Pressure Seal Bonnet — Globe									

Boldface type is weight in pounds. Lightface type beneath weight is weight factor for insulation.

Insulation thicknesses and weights are based on average conditions and do not constitute a recommendation for specific thicknesses of materials. Insulation weights are based on 85% magnesia and hydrous calcium silicate at 11 lbs/cubic foot. The listed thicknesses and weights of combination covering are the sums of the inner layer of diatomaceous earth at 21 lbs/cubic foot and the outer layer at 11 lbs/cubic foot.

Insulation weights include allowances for wire, cement, canvas, bands and paint, but not special surface finishes.

To find the weight of covering on flanges, valves or fittings, multiply the weight factor by the weight per foot of covering used on straight pipe.

Valve weights are approximate. When possible, obtain weights from the manufacturer.

Cast iron valve weights are for flanged end valves; steel weights for welding end valves.

All flanged fitting, flanged valve and flange weights include the proportional weight of bolts or studs to make up all joints.

WEIGHT OF PIPING MATERIALS
(Courtesy of Anvil International)

30" O.D. 30" PIPE

		5S	10 & 10S		20	30								
PIPE	Schedule No.	5S	10 & 10S		20	30								
	Wall Designation			Std.	XS									
	Thickness — In.	.250	.312	.375	.500	.625	.750	.875	1.000	1.125				
	Pipe — Lbs/Ft	79.0	98.9	118.7	157.6	196.1	234.0	272.0	310.0	347.0				
	Water — Lbs/Ft	296.3	293.5	291.0	286.0	281.1	276.6	271.8	267.0	262.2				

WELDING FITTINGS

Fitting	5S	10 & 10S	Std.	XS					
L.R. 90° Elbow	478.0 / 10			775 / 10	953 / 10		596.0 / 10		
S.R. 90° Elbow	319.0 / 5.9			470 / 5.9	644 / 5.9		388.0 / 5.9		
L.R. 45° Elbow	239.0 / 3.9			358 / 3.9	475 / 3.9		298.0 / 3.9		
Tee				855 / 7.8	1065 / 7.8				
Lateral									
Reducer				220 / 3.9	315 / 3.9				
Cap				125 / 4.8	175 / 4.8				

INSULATION

Temperature Range °F	100-199	200-299	300-399	400-499	500-599	600-699	700-799	800-899	900-999	1000-1099	1100-1200
85% Magnesia Calcium Silicate Nom. Thick., In.	1½	1½	2	2½	3	3½	4	4½	5	5	6
Lbs./Ft	11.9	11.9	16.1	20.5	25.0	29.5	34.3	39.1	44.1	44.1	54.4
Combination Nom. Thick., In.						3½	4½	5½	6	6½	7
Lbs/Ft						42.1	58.9	76.5	85.7	95.1	104.7

FLANGES

Pressure Rating psi	Cast Iron		Steel						
	125	250	150	300	400	600	900	1500	2500
Screwed or Slip-On			365 / 1.5	975 / 1.5	900 / 1.5	1175 / 1.5	2075 / 1.5		
Welding Neck			410 / 1.5	1095 / 1.5	1000 / 1.5	1300 / 1.5	2150 / 1.5		
Lap Joint									
Blind			770 / 1.5	1665 / 1.5	1675 / 1.5	2000 / 1.5	3025 / 1.5		

FLANGED FITTINGS

Fitting	125	250	150	300	400	600	900	1500	2500
S.R. 90° Elbow									
L.R. 90° Elbow									
45° Elbow									
Tee									

VALVES

Valve	125	250	150	300	400	600	900	1500	2500
Flanged Bonnet Gate									
Flanged Bonnet Globe or Angle									
Flanged Bonnet Check									
Pressure Seal Bonnet — Gate									
Pressure Seal Bonnet — Globe									

Boldface type is weight in pounds. Lightface type beneath weight is weight factor for insulation.

Insulation thicknesses and weights are based on average conditions and do not constitute a recommendation for specific thicknesses of materials. Insulation weights are based on 85% magnesia and hydrous calcium silicate at 11 lbs/cubic foot. The listed thicknesses and weights of combination covering are the sums of the inner layer of diatomaceous earth at 21 lbs/cubic foot and the outer layer at 11 lbs/cubic foot.

Insulation weights include allowances for wire, cement, canvas, bands and paint, but not special surface finishes.

To find the weight of covering on flanges, valves or fittings, multiply the weight factor by the weight per foot of covering used on straight pipe.

Valve weights are approximate. When possible, obtain weights from the manufacturer.

Cast iron valve weights are for flanged end valves; steel weights for welding end valves.

All flanged fitting, flanged valve and flange weights include the proportional weight of bolts or studs to make up all joints.

WEIGHT OF PIPING MATERIALS
(Courtesy of Anvil International)

32″ O.D. **32″** PIPE

PIPE

Schedule No.		10		20	30	40					
Wall Designation			Std.	XS							
Thickness — In.	.250	.312	.375	.500	.625	.688	.750	.875	1.000	1.125	
Pipe — Lbs/Ft	85.0	105.8	126.7	168.2	209.4	229.9	250.0	291.0	331.0	371.0	
Water — Lbs/Ft	337.8	335.0	323.3	327.0	321.8	319.2	316.7	311.6	306.4	301.3	

WELDING FITTINGS

L.R. 90° Elbow			818 / 10.5	1090 / 10.5							
S.R. 90° Elbow			546 / 6.3	722 / 6.3							
L.R. 45° Elbow			408 / 4.2	541 / 4.2							
Tee			991 / 8.4	1230 / 8.4							
Lateral											
Reducer			255 / 3.1	335 / 3.1							
Cap			145 / 5.2	190 / 5.2							

INSULATION

Temperature Range °F	100-199	200-299	300-399	400-499	500-599	600-699	700-799	800-899	900-999	1000-1099	1100-1200
85% Magnesia Calcium Silicate — Nom. Thick., In.	1½	1½	2	2½	3	3½	4	4½	5	5	6
85% Magnesia Calcium Silicate — Lbs./Ft	12.7	12.7	17.1	21.7	26.5	31.3	36.3	41.4	46.6	46.6	57.5
Combination — Nom. Thick., In.						3½	4½	5½	6	6½	7
Combination — Lbs/Ft						44.7	62.3	80.9	90.5	100.4	110.5

FLANGES

Pressure Rating psi	Cast Iron		Steel						
	125	250	150	300	400	600	900	1500	2500
Screwed or Slip-On			476 / 1.5	1093 / 1.5	1025 / 1.5	1375 / 1.5	2500 / 1.5		
Welding Neck			516 / 1.5	1228 / 1.5	1150 / 1.5	1500 / 1.5	2575 / 1.5		
Lap Joint									
Blind			951 / 1.5	1978 / 1.5	1975 / 1.5	2300 / 1.5	3650 / 1.5		

FLANGED FITTINGS

S.R. 90° Elbow	
L.R. 90° Elbow	
45° Elbow	
Tee	

VALVES

Flanged Bonnet Gate	
Flanged Bonnet Globe or Angle	
Flanged Bonnet Check	
Pressure Seal Bonnet — Gate	
Pressure Seal Bonnet — Globe	

Boldface type is weight in pounds. Lightface type beneath weight is weight factor for insulation.

Insulation thicknesses and weights are based on average conditions and do not constitute a recommendation for specific thicknesses of materials. Insulation weights are based on 85% magnesia and hydrous calcium silicate at 11 lbs/cubic foot. The listed thicknesses and weights of combination covering are the sums of the inner layer of diatomaceous earth at 21 lbs/cubic foot and the outer layer at 11 lbs/cubic foot.

Insulation weights include allowances for wire, cement, canvas, bands and paint, but not special surface finishes.

To find the weight of covering on flanges, valves or fittings, multiply the weight factor by the weight per foot of covering used on straight pipe.

Valve weights are approximate. When possible, obtain weights from the manufacturer.

Cast iron valve weights are for flanged end valves; steel weights for welding end valves.

All flanged fitting, flanged valve and flange weights include the proportional weight of bolts or studs to make up all joints.

WEIGHT OF PIPING MATERIALS
(Courtesy of Anvil International)

34" O.D. 34" PIPE

PIPE

Schedule No.		10		20	30	40						
Wall Designation			Std.	XS								
Thickness — In.	.250	.312	.375	.500	.625	.688	.750	.875	1.000	1.125		
Pipe — Lbs/Ft	90.0	112.4	134.7	178.9	222.8	244.6	266.0	310.0	353.0	395.0		
Water — Lbs/Ft	382.0	379.1	376.0	370.3	365.0	362.2	359.5	354.1	348.6	343.2		

WELDING FITTINGS

L.R. 90° Elbow			926 / 11	1230 / 11								
S.R. 90° Elbow			617 / 5.5	817 / 5.5								
L.R. 45° Elbow			463 / 4.4	615 / 4.4								
Tee			1136 / 8.9	1420 / 8.9								
Lateral												
Reducer			270 / 3.3	355 / 3.3								
Cap			160 / 5.6	210 / 5.6								

INSULATION

Temperature Range °F	100-199	200-299	300-399	400-499	500-599	600-699	700-799	800-899	900-999	1000-1099	1100-1200
85% Magnesia Calcium Silicate — Nom. Thick., In.	1½	1½	2	2½	3	3½	4	4½	5	5	6
85% Magnesia Calcium Silicate — Lbs./Ft	13.4	13.4	18.2	23.0	28.0	33.1	38.3	43.7	49.1	49.1	60.5
Combination — Nom. Thick., In.						3½	4½	5½	6	6½	7
Combination — Lbs/Ft						47.2	65.8	85.3	95.4	105.7	116.3

FLANGES

Pressure Rating psi	Cast Iron		Steel						
	125	250	150	300	400	600	900	1500	2500
Screwed or Slip-On			515 / 1.5	1281 / 1.5	1150 / 1.5	1500 / 1.5	2950 / 1.5		
Welding Neck			560 / 1.5	1406 / 1.5	1300 / 1.5	1650 / 1.5	3025 / 1.5		
Lap Joint									
Blind			1085 / 1.5	2231 / 1.5	2250 / 1.5	2575 / 1.5	4275 / 1.5		

Boldface type is weight in pounds. Lightface type beneath weight is weight factor for insulation.

Insulation thicknesses and weights are based on average conditions and do not constitute a recommendation for specific thicknesses of materials. Insulation weights are based on 85% magnesia and hydrous calcium silicate at 11 lbs/cubic foot. The listed thicknesses and weights of combination covering are the sums of the inner layer of diatomaceous earth at 21 lbs/cubic foot and the outer layer at 11 lbs/cubic foot.

Insulation weights include allowances for wire, cement, canvas, bands and paint, but not special surface finishes.

To find the weight of covering on flanges, valves or fittings, multiply the weight factor by the weight per foot of covering used on straight pipe.

Valve weights are approximate. When possible, obtain weights from the manufacturer.

Cast iron valve weights are for flanged end valves; steel weights for welding end valves.

All flanged fitting, flanged valve and flange weights include the proportional weight of bolts or studs to make up all joints.

FLANGED FITTINGS

S.R. 90° Elbow	
L.R. 90° Elbow	
45° Elbow	
Tee	

VALVES

Flanged Bonnet Gate	
Flanged Bonnet Globe or Angle	
Flanged Bonnet Check	
Pressure Seal Bonnet — Gate	
Pressure Seal Bonnet — Globe	

WEIGHT OF PIPING MATERIALS
(Courtesy of Anvil International)

36″ O.D. 36″ PIPE

PIPE

	Schedule No.		10		20	30	40						
	Wall Designation			Std.	XS								
	Thickness — In.	.250	.312	.375	.500	.625	.750	.875	1.000	1.125			
	Pipe — Lbs/Ft	96.0	119.1	142.7	189.6	236.1	282.4	328.0	374.0	419.0			
	Water — Lbs/Ft	429.1	425.9	422.6	416.6	411.0	405.1	399.4	393.6	387.9			

WELDING FITTINGS

L.R. 90° Elbow				1040 12	1380 12								
S.R. 90° Elbow				692 5	913 5								
L.R. 45° Elbow				518 4.8	686 4.8								
Tee				1294 9.5	1610 9.5								
Lateral													
Reducer				340 3.6	360 3.6								
Cap				175 6	235 6								

INSULATION

Temperature Range °F	100-199	200-299	300-399	400-499	500-599	600-699	700-799	800-899	900-999	1000-1099	1100-1200
85% Magnesia Calcium Silicate — Nom. Thick., In.	1½	1½	2	2½	3	3½	4	4½	5	5	6
Lbs./Ft	14.2	14.2	19.2	24.2	29.5	34.8	40.3	45.9	51.7	51.7	63.5
Combination — Nom. Thick., In.						3½	4½	5½	6	6½	7
Lbs/Ft						49.8	69.3	89.7	100.2	111.0	122.0

FLANGES

	Pressure Rating psi	Cast Iron		Steel						
		125	250	150	300	400	600	900	1500	2500
Screwed or Slip-On				588 1.5	1485 1.5	1325 1.5	1600 1.5	3350 1.5		
Welding Neck				628 1.5	1585 1.5	1475 1.5	1750 1.5	3450 1.5		
Lap Joint										
Blind				1233 1.5	2560 1.5	2525 1.5	2950 1.5	4900 1.5		

FLANGED FITTINGS

S.R. 90° Elbow	
L.R. 90° Elbow	
45° Elbow	
Tee	

VALVES

Flanged Bonnet Gate	
Flanged Bonnet Globe or Angle	
Flanged Bonnet Check	
Pressure Seal Bonnet — Gate	
Pressure Seal Bonnet — Globe	

Boldface type is weight in pounds. Lightface type beneath weight is weight factor for insulation.

Insulation thicknesses and weights are based on average conditions and do not constitute a recommendation for specific thicknesses of materials. Insulation weights are based on 85% magnesia and hydrous calcium silicate at 11 lbs/cubic foot. The listed thicknesses and weights of combination covering are the sums of the inner layer of diatomaceous earth at 21 lbs/cubic foot and the outer layer at 11 lbs/cubic foot.

Insulation weights include allowances for wire, cement, canvas, bands and paint, but not special surface finishes.

To find the weight of covering on flanges, valves or fittings, multiply the weight factor by the weight per foot of covering used on straight pipe.

Valve weights are approximate. When possible, obtain weights from the manufacturer.

Cast iron valve weights are for flanged end valves; steel weights for welding end valves.

All flanged fitting, flanged valve and flange weights include the proportional weight of bolts or studs to make up all joints.

WEIGHT OF PIPING MATERIALS
(Courtesy of Anvil International)

42" O.D. 42" PIPE

PIPE					20	30	40									
Schedule No.					20	30	40									
Wall Designation			Std.	XS												
Thickness — In.		.250	.375	.500	.625	.750	1.000	1.250	1.500							
Pipe — Lbs/Ft		112.0	166.7	221.6	276.0	330.0	438.0	544.0	649.0							
Water — Lbs/Ft		586.4	578.7	571.7	565.4	558.4	544.8	531.2	517.9							

WELDING FITTINGS

L.R. 90° Elbow		1420 / 15	1880 / 15	
S.R. 90° Elbow		1079 / 9	1430 / 9	
L.R. 45° Elbow		707 / 6	937 / 6	
Tee		1870	2415	
Lateral				
Reducer		310 / 4.5	410 / 4.5	
Cap		230 / 7.5	300 / 7.5	

INSULATION

Temperature Range °F	100-199	200-299	300-399	400-499	500-599	600-699	700-799	800-899	900-999	1000-1099	1100-1200
85% Magnesia Calcium Silicate — Nom. Thick., In.	1½	1½	2	2½	3	3½	4	4½	5	5	6
Lbs./Ft	16.5	16.5	22.2	28.0	34.0	40.1	46.4	52.7	59.2	59.2	72.6
Combination — Nom. Thick., In.						3½	4½	5½	6	6½	7
Lbs/Ft						57.4	79.7	102.8	114.8	126.9	139.3

FLANGES

	Cast Iron		Steel						
Pressure Rating psi	125	250	150	300	400	600	900	1500	2500
Screwed or Slip-On			792 / 1.5	1895 / 1.5	1759 / 1.5	2320 / 1.5			
Welding Neck			862 / 1.5	2024 / 1.5	1879 / 1.5	2414 / 1.5			
Lap Joint									
Blind			1733 / 1.5	3449 / 1.5	3576 / 1.5	4419 / 1.5			

FLANGED FITTINGS

S.R. 90° Elbow	
L.R. 90° Elbow	
45° Elbow	
Tee	

VALVES

Flanged Bonnet Gate	
Flanged Bonnet Globe or Angle	
Flanged Bonnet Check	
Pressure Seal Bonnet — Gate	
Pressure Seal Bonnet — Globe	

Boldface type is weight in pounds. Lightface type beneath weight is weight factor for insulation.

Insulation thicknesses and weights are based on average conditions and do not constitute a recommendation for specific thicknesses of materials. Insulation weights are based on 85% magnesia and hydrous calcium silicate at 11 lbs/cubic foot. The listed thicknesses and weights of combination covering are the sums of the inner layer of diatomaceous earth at 21 lbs/cubic foot and the outer layer at 11 lbs/cubic foot.

Insulation weights include allowances for wire, cement, canvas, bands and paint, but not special surface finishes.

To find the weight of covering on flanges, valves or fittings, multiply the weight factor by the weight per foot of covering used on straight pipe.

Valve weights are approximate. When possible, obtain weights from the manufacturer.

Cast iron valve weights are for flanged end valves; steel weights for welding end valves.

All flanged fitting, flanged valve and flange weights include the proportional weight of bolts or studs to make up all joints.

WEIGHT OF FLANGED GATE VALVES (lb)

Size (in.)	150#	300#	600#
2	46	74	84
3	76	108	160
4	110	165	300
5	155	235	
6	175	320	640
8	310	500	1080
10	455	760	1550
12	650	1020	2100
14	860	1380	
16	1120	1960	
18	1400	2450	
20	2125	3890	
24	3120	5955	

Courtesy of Crane Valves

WEIGHT OF FLANGED CHECK (SWING) VALVES (lb)

Size (in.)	150#	300#	600#
2	33	46	62
3	59	86	115
4	93	154	192
5	152	255	
6	165	276	495
8	275	420	780
10	440	640	1400
12	680	1000	1750
14	950	1550	
16	1225	1700	
18	1700	2200	
20	1850	2800	
24	2900	3650	

Courtesy of Crane Valves

WEIGHT OF FLANGED GLOBE VALVES (lb)

Size (in.)	150#	300#	600#
2	47	60	88
3	82	117	160
4	134	176	270
5	199	290	
6	240	340	550
8	370	530	1000
10	525	750	
12	900	1100	
14	1000		

Courtesy of Crane Valves

Bolt Weights (lb) [1]

Flange Size (in.)	Flange Schedule								
	Cast Iron		Steel						
	125	250	150	300	400	600	900	1500	2500
½	1.0		1.0	1.0		1.0		3.2	3.4
¾	1.0		1.0	2.0		2.0		3.3	3.6
1	1.0	2.0	1.0	2.0		2.0		6.0	6.0
1¼	1.0	2.0	1.0	2.0		2.0		6.0	9.0
1½	1.0	2.5	1.0	3.5		3.5		9.0	12.0
2	1.5	3.5	1.5	4.0		4.5		12.5	21.0
2½	1.5	6.0	1.5	7.0		8.0	19.0	19.0	27.0
3	1.5	6.0	1.5	7.5		8.0	12.5	25.0	37.0
3½	3.5	6.5	3.5	7.5		12.0			
4	4.0	6.5	4.0	7.5	12.0	12.5	25.0	34.0	61.0
5	6.0	6.5	6.0	8.0	12.5	19.5	33.0	60.0	98.0
6	6.0	10.0	6.0	11.5	19.0	30.0	40.0	76.0	145.0
8	6.5	16.0	6.5	18.0	30.0	40.0	69.0	121.0	232.0
10	15.0	33.0	15.0	38.0	52.0	72.0	95.0	184.0	445.0
12	15.0	44.0	15.0	49.0	69.0	91.0	124.0	306.0	622.0
14	22.0	57.0	22.0	62.0	88.0	118.0	159.0	425.0	
16	31.0	76.0	31.0	83.0	114.0	152.0	199.0	570.0	
18	41.0	93.0	41.0	101.0	139.0	193.0	299.0	770.0	
20	52.0	95.0	52.0	105.0	180.0	242.0	361.0	1010.0	
24	71.0	174.0	71.0	174.0	274.0	360.0	687.0	1560.0	

[1] Weights are for one complete flanged joint and include bolts and nuts.

Steel Butt-Welding Fitting Dimensions

These dimensions are per ASME B16.9 – 2001 edition.

(all dimensions in inches)

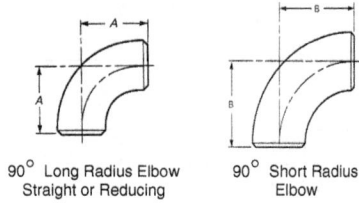

90° Long Radius Elbow Straight or Reducing

90° Short Radius Elbow

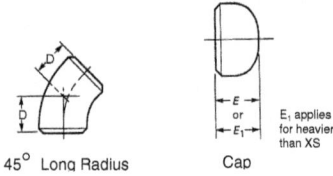

45° Long Radius

Cap   E₁ applies for heavier than XS

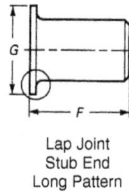

Lap Joint Stub End Long Pattern

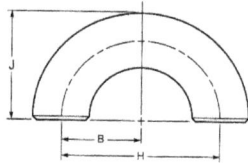

Short Radius Return Bend

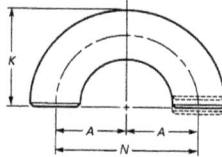

Long Radius Return Bend

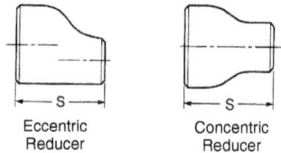

Eccentric Reducer

Concentric Reducer

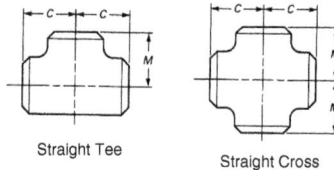

Straight Tee

Straight Cross

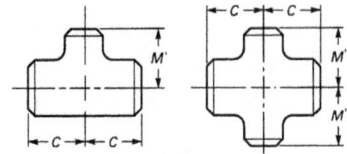

Reducing Tee

Reducing Cross

Dimension "M" is shown in table below; refer to large table for dimension "C"

Size a	b	M'	Size a	b	M'
1	x 3/4	1.50	6	x 3 1/2	5.00
1 1/4	x 1	1.88		x 3	4.88
	x 3/4	1.88		x 2 1/2	4.75
1 1/2	x 1 1/4	2.25	8	x 6	6.62
	x 1	2.25		x 5	6.38
	x 3/4	2.25		x 4	6.12
2	x 1 1/2	2.38		x 3 1/2	6.00
	x 1 1/4	2.25	10	x 8	8.00
	x 1	2.00		x 6	7.62
	x 3/4	1.75		x 5	7.50
2 1/2	x 2	2.75		x 4	7.25
	x 1 1/2	2.62	12	x 10	9.50
	x 1 1/4	2.50		x 8	9.00
	x 1	2.25		x 6	8.62
3	x 2 1/2	3.25		x 5	8.50
	x 2	3.00	14	x 12	10.62
	x 1 1/2	2.88		x 10	10.12
	x 1 1/4	2.75		x 8	9.75
3 1/2	x 3	3.62		x 6	9.38
	x 2 1/2	3.50	16	x 14	12.00
	x 2	3.25		x 12	11.62
	x 1 1/2	3.12		x 10	11.12
4	x 3 1/2	4.00		x 8	10.75
	x 3	3.88	18	x 16	13.00
	x 2 1/2	3.75		x 14	13.00
	x 2	3.50		x 12	12.62
	x 1 1/2	3.38		x 10	12.12
5	x 4	4.62	20	x 18	14.50
	x 3 1/2	4.50		x 16	14.00
	x 3	4.38		x 14	14.00
	x 2 1/2	4.25		x 12	13.62
	x 2	4.12	24	x 20	17.00
6	x 5	5.38		x 18	16.50
	x 4	5.12		x 16	16.00

Size	A	B	D	E	E¹	F	G	H	J	K	M	N	S*	Pipe Schedule Numbers for: Std. Ftgs.	Extra Strong
1/2	1.50		0.62	1.00	1.00	3.00	1.38			1.88	1.00	3.00		40	80
3/4	1.50		0.75	1.00	1.00	3.00	1.38			2.00	1.12	3.00	1.50	40	80
1	1.50	1.00	0.88	1.50	1.50	4.00	2.00	2.00	1.62	2.19	1.50	3.00	2.00	40	80
1 1/4	1.88	1.25	1.00	1.50	1.50	4.00	2.50	2.50	2.06	2.75	1.88	3.75	2.00	40	80
1 1/2	2.25	1.50	1.12	1.50	1.50	4.00	2.88	3.00	2.44	3.25	2.25	4.50	2.50	40	80
2	3.00	2.00	1.38	1.50	1.75	6.00	3.62	4.00	3.19	4.19	2.50	6.00	3.00	40	80
2 1/2	3.75	2.50	1.75	1.50	2.00	6.00	4.12	5.00	3.94	5.19	3.00	7.50	3.50	40	80
3	4.50	3.00	2.00	2.00	2.50	6.00	5.00	6.00	4.75	6.25	3.38	9.00	3.50	40	80
3 1/2	5.25	3.40	2.25	2.50	3.00	6.00	5.50	7.00	5.50	7.25	3.75	10.50	4.00	40	80
4	6.00	4.00	2.50	2.50	3.00	6.00	5.19	8.00	6.25	8.25	4.12	12.00	4.00	40	80
5	7.50	5.00	3.12	3.00	3.50	8.00	7.31	10.00	7.75	10.31	4.88	15.00	5.00	40	80
6	9.00	6.00	3.75	3.50	4.00	8.00	8.50	12.00	9.31	12.31	5.62	18.00	5.50	40	80
8	12.00	8.00	5.00	4.00	5.00	8.00	10.62	16.00	12.31	16.31	7.00	24.00	6.00	40	80
10	15.00	10.00	6.25	5.00	6.00	10.00	12.75	20.00	15.38	20.38	8.50	30.00	7.00	40	60
12	18.00	12.00	7.50	6.00	7.00	10.00	15.00	24.00	18.38	24.38	10.00	36.00	8.00		
14	21.00	14.00	8.75	6.50	7.50	12.00	16.25	28.00	21.00	28.00	11.00	42.00	13.00	30	
16	24.00	16.00	10.00	7.00	8.00	12.00	18.50	32.00	24.00	32.00	12.00	48.00	14.00	30	40
18	27.00	18.00	11.25	8.00	9.00	12.00	21.00	36.00	27.00	36.00	13.50	54.00	15.00		
20	30.00	20.00	12.50	9.00	10.00	12.00	23.00	40.00	30.00	40.00	15.00	60.00	20.00	20	30
22	33.00	22.00	13.50	10.00	10.00	12.00	25.25	44.00	33.00	44.00	16.50	66.00	20.00		
24	36.00	24.00	15.00	10.50	12.00	12.00	27.25	48.00	36.00	48.00	17.00	72.00	20.00	20	

¹ Size is that of large end.

Welding Neck Flange Overall Length, Y (in.)
(ASME B16.5, 1998 Edition)

NPS	Flange Class						
	150	300	400	600	900	1500	2500
½	1.88	2.06		2.06		2.38	2.88
¾	2.06	2.25		2.25		2.75	3.12
1	2.19	2.44		2.44		2.88	3.50
1 ¼	2.25	2.56		2.62		2.88	3.75
1 ½	2.44	2.69		2.75		3.25	4.38
2	2.50	2.75		2.88		4.00	5.00
2 ½	2.75	3.00		3.12		4.12	5.62
3	2.75	3.12		3.25	4.00	4.62	6.62
3 ½	2.81	3.19		3.38			
4	3.00	3.38	3.50	4.00	4.50	4.88	7.50
5	3.50	3.88	4.00	4.50	5.00	6.12	9.00
6	3.50	3.88	4.06	4.62	5.50	6.75	10.75
8	4.00	4.38	4.62	5.25	6.38	8.38	12.50
10	4.00	4.62	4.88	6.00	7.25	10.00	16.50
12	4.50	5.12	5.38	6.12	7.88	11.12	18.25
14	5.00	5.62	5.88	6.50	8.38	11.75	
16	5.00	5.75	6.00	7.00	8.50	12.25	
18	5.50	6.25	6.50	7.25	9.00	12.88	
20	5.69	6.38	6.62	7.50	9.75	14.00	
24	6.00	6.62	6.88	8.00	11.50	16.00	

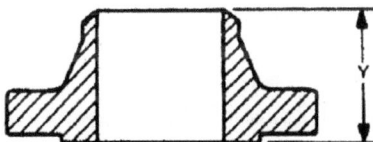

Class 150 and 300   Class 400 - 2500

Gate Valves [1]
Solid Wedge, Double Disc, and Conduit
End-to-End Dimensions "A" Per ASME B16.10 (1992 Rev)
Flanged and Buttwelding Ends

| Nominal Size | | 125/150 | | 250/300 [2] | 600 | | 900 | | 1500 | | 2500 | |
NPS	DN	Flanged End [2]	Butt Weld [5]		Long Pattern [3]	Short Pattern [4]	Long Pattern [3]	Short Pattern [4]	Long Pattern [3]	Short Pattern [4]	Long Pattern [3]	Short Pattern [4]
1/4	8	4.00	4.00									
3/8	10	4.00	4.00									
1/2	15	4.25	4.25	5.50	6.50						10.38	
3/4	20	4.62	4.62	6.00	7.50						10.75	
1	25	5.00	5.00	6.50	8.50	5.25	10.00	5.50	10.00	5.50	12.12	7.31
1 1/4	32	5.50	5.50	7.00	9.00	5.75	11.00	6.50	11.00	6.50	13.75	9.12
1 1/2	40	6.50	6.50	7.50	9.50	6.00	12.00	7.00	12.00	7.00	15.12	9.12
2	50	7.00	8.50	8.50	11.50	7.00	14.50	8.50	14.50	8.50	17.75	11.00
2 1/2	65	7.50	9.50	9.50	13.00	8.50	16.50	10.00	16.50	10.00	20.00	13.00
3	80	8.00	11.12	11.12	14.00	10.00	15.00	12.00	18.50	12.00	22.75	14.50
4	100	9.00	12.00	12.00	17.00	12.00	18.00	14.00	21.50	16.00	26.50	18.00
5	125	10.00	15.00	15.00	20.00	15.00	22.00	17.00	26.50	19.00	31.25	21.00
6	150	10.50	15.88	15.88	22.00	18.00	24.00	20.00	27.75	22.00	36.00	24.00
8	200	11.50	16.50	16.50	26.00	23.00	29.00	26.00	32.75	28.00	40.25	30.00
10	250	13.00	18.00	18.00	31.00	28.00	33.00	31.00	39.00	34.00	50.00	36.00
12	300	14.00	19.75	19.75	33.00	32.00	38.00	36.00	44.50	39.00	56.00	41.00
14	350	15.00	22.50	22.50 (30.00)	35.00	35.00	40.50	39.00	49.50	42.00		44.00
16	400	16.00	24.00	24.00 (33.00)	39.00	39.00	44.50	43.00	54.50	47.00		49.00
18	450	17.00	26.00	26.00 (36.00)	43.00	43.00	48.00		60.50	53.00		55.00
20	500	18.00	28.00	28.00 (39.00)	47.00	47.00	52.00		65.50	58.00		
22	550	[20.00]	30.00	43.00	51.00							
24	600	20.00	32.00	31.00 (45.00)	55.00	55.00	61.00		76.50			
26	650	22.00	34.00	49.00	57.00							
28	700	24.00	36.00	53.00	61.00							
30	750	24.00 [26.00]	36.00	55.00	65.00							
32	800	[28.00]	38.00	60.00	70.00							
34	850	[30.00]	40.00	64.00	76.00							
36	900	28.00 [32.00]	40.00	68.00	82.00							

[1] Not all valve types are available in all sizes, refer to valve manufacturer information and/or ASME B16.10.
[2] Steel Valves in parenthesis, when different. Conduit gate valves in brackets "[ ]" when different.
[3] Flanged and Buttwelding, Long Pattern.
[4] Buttwelding, Short Pattern.
[5] Steel, Class 150 only.

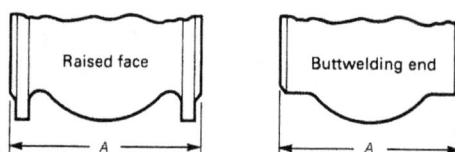

Raised face   Buttwelding end

A   A

Globe, Lift Check and Swing Check Valves [1]
End-to-End Dimensions "A" Per ASME B16.10 (1992 Rev)
Flanged and Buttwelding Ends

Nominal Size		125/150 [2]	250/300 [2]	600		900		1500		2500	
NPS	DN			Long Pattern [3]	Short Pattern [4]	Long Pattern [3]	Short Pattern [4]	Long Pattern [3]	Short Pattern [4]	Long Pattern [3]	Short Pattern [4]
1/4	8	4.00									
3/8	10	4.00									
1/2	15	4.25	6.00	6.50				8.50		10.38	
3/4	20	4.62	7.00	7.50		9.00		9.00		10.75	
1	25	5.00	8.00	8.50	5.25	10.00		10.00		12.12	
1 1/4	32	5.50	8.50	9.00	5.75	11.00		11.00		13.75	
1 1/2	40	6.50	9.00	9.50	6.00	12.00		12.00		15.12	
2	50	8.00	10.50	11.50	7.00	14.50		14.50	8.50	17.75	11.00
2 1/2	65	8.50	11.50	13.00	8.50	16.50	10.00	16.50	10.00	20.00	13.00
3	80	9.50	12.50	14.00	10.00	15.00	12.00	18.50	12.00	22.75	14.50
4	100	11.50	14.00	17.00	12.00	18.00	14.00	21.50	16.00	26.50	18.00
5	125	13.00 (14.00)	15.75	20.00	15.00	22.00	17.00	26.50	19.00	31.25	21.00
6	150	14.00 (16.00)	17.50	22.00	18.00	24.00	20.00	27.75	22.00	36.00	24.00
8	200	19.50	21.00 (22.00)	26.00	23.00	29.00	26.00	32.75	28.00	40.25	30.00
10	250	24.00	24.50	31.00	28.00	33.00	31.00	39.00	34.00	50.00	36.00
12	300	27.50	28.00	33.00	32.00	38.00	36.00	44.50	39.00	56.00	41.00
14	350	31.00 (21.00)		35.00		40.50	39.00	49.50	42.00		
16	400	36.00 [34.00]		39.00		44.50	43.00	54.50	47.00		
18	450	38.50		43.00		48.00		60.50			
20	500	38.50		47.00		52.00		65.50			
22	550	42.00		51.00							
24	600	51.00		55.00		61.00		76.50			
26	650	51.00		57.00							
28	700	57.00		63.00							
30	750	60.00		65.00							
36	900	77.00		82.00							

[1] Not all valve types are available in all sizes, refer to valve manufacturer information and/or ASME B16.10.
[2] Steel Valves in parenthesis, when different. Swing check steel valve in brackets "[ ]" when different.
[3] Flanged and Buttwelding, Long Pattern.
[4] Buttwelding, Short Pattern.

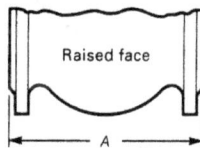

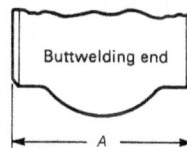

Raised face   Buttwelding end

A   A

Angle and Lift Check [1]
End-to-End Dimensions "D" Per ASME B16.10 (1992 Rev)
Flanged and Buttwelding Ends

Nominal Size		125/150 [2]	250/300 [2]	600		900		1500 Long Pattern [3]	2500 Long Pattern [3]
NPS	DN			Long Pattern [3]	Short Pattern [4]	Long Pattern [3]	Short Pattern [4]		
¼	8	2.00							
⅜	10	2.00							
½	15	2.25	3.00	3.25				4.25	5.19
¾	20	2.50	3.50	3.75		4.50		4.50	5.38
1	25	2.75	4.00	4.25		5.00		5.00	6.06
1¼	32	3.00	4.25	4.50		5.50		5.50	6.88
1½	40	3.25	4.50	4.75		6.00		6.00	7.56
2	50	4.00	5.25	5.75	4.25	7.25		7.25	8.88
2½	65	4.25	5.75	6.50	5.00	8.25		8.25	10.00
3	80	4.75	6.25	7.00	6.00	7.50	6.00	9.25	11.38
4	100	5.75	7.00	8.50	7.00	9.00	7.00	10.75	13.25
5	125	6.50 (7.00)	7.88	10.00	8.50	11.00	8.50	13.25	15.62
6	150	7.00 (8.00)	8.75	11.00	10.00	12.00	10.00	13.88	18.00
8	200	9.75	10.50 (11.00)	13.00		14.50	13.00	16.38	20.12
10	250	12.25	12.25	15.50		16.50	15.50	19.50	25.00
12	300	13.75	14.00	16.50		19.00	18.00	22.25	28.00
14	350	15.50				20.25	19.50	24.75	
16	400	18.00				26.00			
18	450					29.00			
20	500					32.50			
22	550								
24	600					39.00			

[1] Not all valve types are available in all sizes, refer to valve manufacturer information and/or ASME B16.10.

[2] Steel Valves in parenthesis, when different.

[3] Flanged and Buttwelding, Long Pattern.

[4] Buttwelding, Short Pattern. Dimensions apply to pressure seal or flangeless bonnet valves. They may be applied at the manufacturer's option to valves with flanged bonnets.

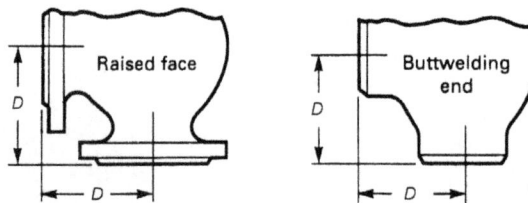

Total Thermal Expansion, U.S. Units, for Metals
Total Linear Thermal Expansion Between 70°F and Indicated Temperature, in./100 ft
Per ASME B31.3 – 1999 Edition

Temp., °F	Material					Temp., °F	Material				
	Carbon Steel Carbon-Moly- Low-Chrome (Through 3Cr-Mo)	5Cr-Mo Through 9Cr-Mo	Austenitic Stainless Steels 18 Cr-8Ni	UNS No8XXX Series Ni-Fe-Cr	UNS No6XXX Series Ni-Cr-Fe		Carbon Steel Carbon-Moly- Low-Chrome (Through 3Cr-Mo)	5Cr-Mo Through 9Cr-Mo	Austenitic Stainless Steels 18 Cr-8Ni	UNS No8XXX Series Ni-Fe-Cr	UNS No6XXX Series Ni-Cr-Fe
-325	-2.37	-2.22	-3.85			575	4.35	4.02	5.93	5.44	4.77
-300	-2.24	-2.10	-3.63			600	4.60	4.24	6.24	5.72	5.02
-275	-2.11	-1.98	-3.41			625	4.86	4.47	6.55	6.01	5.27
-250	-1.98	-1.86	-3.19			650	5.11	4.69	6.87	6.30	5.53
-225	-1.85	-1.74	-2.96			675	5.37	4.92	7.18	6.58	5.79
-200	-1.71	-1.62	-2.73			700	5.63	5.14	7.50	6.88	6.05
-175	-1.58	-1.50	-2.50			725	5.90	5.38	7.82	7.17	6.31
-150	-1.45	-1.37	-2.27			750	6.16	5.62	8.15	7.47	6.57
-125	-1.30	-1.23	-2.01			775	6.43	5.86	8.47	7.76	6.84
-100	-1.15	-1.08	-1.75			800	6.70	6.10	8.80	8.06	7.10
-75	-1.00	-0.94	-1.50			825	6.97	6.34	9.13	8.35	
-50	-0.84	-0.79	-1.24			850	7.25	6.59	9.46	8.66	
-25	-0.68	-0.63	-0.98			875	7.53	6.83	9.79	8.95	
0	-0.49	-0.46	-0.72			900	7.81	7.07	10.12	9.26	
25	-0.32	-0.30	-0.46			925	8.08	7.31	10.46	9.56	
50	-0.14	-0.13	-0.21			950	8.35	7.56	10.80	9.87	
70	0	0	0	0	0	975	8.62	7.81	11.14	10.18	
100	0.23	0.22	0.34	0.28	0.26	1000	8.89	8.06	11.48	10.49	
125	0.42	0.40	0.62	0.52	0.48	1025	9.17	8.30	11.82	10.80	
150	0.61	0.58	0.90	0.76	0.70	1050	9.46	8.55	12.16	11.11	
175	0.80	0.76	1.18	0.99	0.92	1075	9.75	8.80	12.50	11.42	
200	0.99	0.94	1.46	1.23	1.15	1100	10.04	9.05	12.84	11.74	
225	1.21	1.13	1.75	1.49	1.38	1125	10.31	9.28	13.18	12.05	
250	1.40	1.33	2.03	1.76	1.61	1150	10.57	9.52	13.52	12.38	
275	1.61	1.52	2.32	2.03	1.85	1175	10.83	9.76	13.86	12.69	
300	1.82	1.71	2.61	2.30	2.09	1200	11.10	10.00	14.20	13.02	
325	2.04	1.90	2.90	2.59	2.32	1225	11.38	10.26	14.54	13.36	
350	2.26	2.10	3.20	2.88	2.56	1250	11.66	10.53	14.88	13.71	
375	2.48	2.30	3.50	3.18	2.80	1275	11.94	10.79	15.22	14.04	
400	2.70	2.50	3.80	3.48	3.05	1300	12.22	11.06	15.56	14.39	
425	2.93	2.72	4.10	3.76	3.29	1325	12.50	11.30	15.90	14.74	
450	3.16	2.93	4.41	4.04	3.53	1350	12.78	11.55	16.24	15.10	
475	3.39	3.14	4.71	4.31	3.78	1375	13.06	11.80	16.58	15.44	
500	3.62	3.35	5.01	4.59	4.02	1400	13.34	12.05	16.92	15.80	
525	3.86	3.58	5.31	4.87	4.27	1425			17.30	16.16	
550	4.11	3.80	5.62	5.16	4.52	1450			17.69	16.53	
						1475			18.08	16.88	
						1500			18.47	17.25	

Appendix IV

A Practical Guide to Expansion Joints

This appendix is largely a copy of the publication, "A Practical Guide to Expansion Joints" by the Expansion Joint Manufacturers Association, Inc. 25 North Broadway, Tarrytown, NY 10591. It is reproduced, courtesy of EJMA.

IV-1 WHAT ARE EXPANSION JOINTS

IV-1.1 Definition of an Expansion Joint

For purposes of this publication, the definition of an expansion joint is "any device containing one or more metal bellows used to absorb dimensional changes such as those caused by thermal expansion or contraction of a pipe-line, duct, or vessel."

IV-1.2 Expansion Devices (Fig. IV-1.1)

Bellows expansion joints provide an alternative to pipe expansion loops and slip-type joints.

IV-1.3 Manufacturing Methods

There are several methods of manufacturing bellows, the two most common bellows types are:

(1) FORMED BELLOWS – (Fig. IV-1.2.1) Whether formed hydraulically or mechanically, this is by far the most common type available today. Formed from a thin-walled tube, a formed bellows contains only longitudinal welds, and exhibits significant flexibility. Formed bellows of varying shapes are made from single or multiple plies of suitable material (usually stainless steel) ranging in thickness from 0.004″ to 0.125″ and greater, and in diameters of 3/4″ to over 12 ft. These bellows are usually categorized according to convolution shape (Fig. IV-1.2.1 and IV-1.4).

(2) FABRICATED BELLOWS – Made by welding together a series of thin gage diaphragms or discs (Fig. IV-1.2.2 and IV-1.4). Fabricated bellows are usually made of heavier gage material than formed bellows and are therefore able to withstand higher pressures.

Due to the variations in manufacturing and design configurations, the number, height, and pitch of the convolutions should never be specified in the purchase of an expansion joint as a condition of design, but should be left to the decision of the manufacturer. These dimensions will vary from one manufacturer to another and in particular will depend on the specific design conditions and the manufacturer's own rating procedure which has been established on the basis of laboratory tests and field experience.

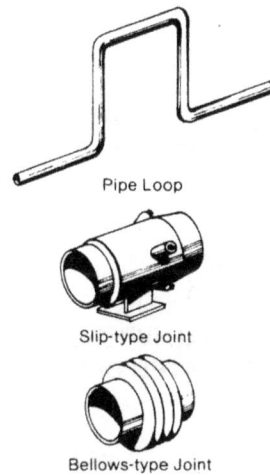

FIG. IV-1.1
EXPANSION DEVICES

IV-1.4   Design Variables

If an expansion joint is to fulfill its intended function safely and reliably, it must be kept in mind that it is a highly specialized product. Interchangeability is rare in expansion joints, and in a real sense, each unit is custom-made for an intended application. It becomes necessary then, to supply the expansion joint manufacturer with accurate information regarding the conditions of design that the expansion joint will be subjected to in service. Because many of these design conditions interact with each other, designing a bellows becomes much like assembling a jigsaw puzzle; one needs all of the pieces (design conditions) before a clear picture of an expansion joint design can emerge. The following is a listing of the basic design conditions that should be supplied to the manufacturer when specifying an expansion joint.

IV-1.4.1   Size.   Size refers to the diameter of the pipeline (or dimensions of the duct in the case of rectangular joints) into which the expansion joint is to be installed. The size of an expansion joint affects its pressure-retaining capabilities, as well as its ability to absorb certain types of movements.

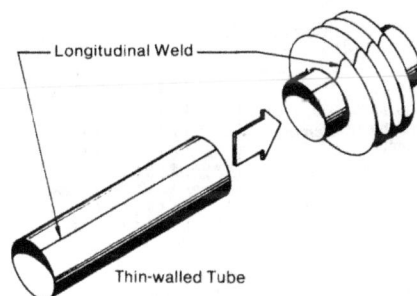

FIG. IV-1.2.1
FORMED BELLOWS

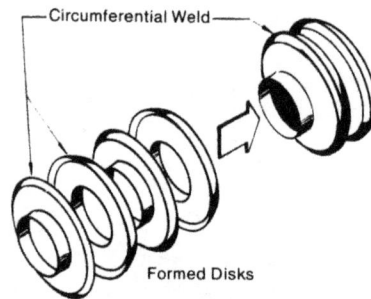

FIG. IV-1.2.2
FABRICATED BELLOWS

Note: When specifying pipe sizes, misunderstandings often result due to the confusing array of nominal pipe sizes (NPS) and different pipe schedules. To eliminate any misconceptions, pipe sizes should be supplied to the expansion joint manufacturer in terms of the actual outside diameter and the actual wall thickness of the pipe. The outside diameters, schedules, and wall thickness of pipe sizes up thru 36 inches can be found in the American Society of Mechanical Engineers (ASME B36.10).

IV-1.4.2   Flowing Medium.   The substances that will come in contact with the expansion joint should be specified. In some cases, due to excessive erosion, or corrosion potential, or in cases of high viscosity, special materials and accessories should be specified. When piping systems containing expansion joints are cleaned periodically, the cleaning solution must be compatible with the bellows materials.

IV-1.4.3   Pressure.   Pressure is possibly the most important factor determining expansion joint design. Minimum and maximum anticipated pressure should be accurately determined. If a pressure test is to be performed, this pressure should be specified as well. While the determination of pressure requirements is important, care should be exercised to insure that these specified pressures are not increased by unreasonable safety factors as this could result in a design which may not adequately satisfy other performance characteristics.

IV-1.4.4   Temperature.   The operating temperature of the expansion joint will affect its pressure capacity, allowable stresses, cycle life, and material requirements. All possible temperature sources should be investigated when determining minimum and maximum temperature requirements. In so

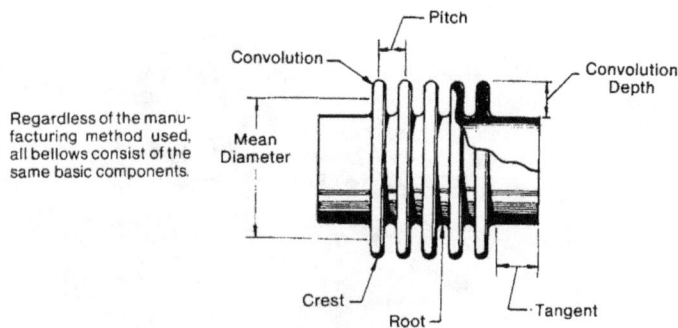

FIG. IV-1.3
BELLOWS COMPONENTS

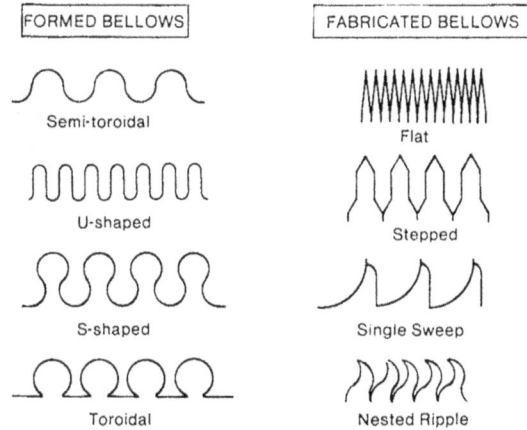

FIG. IV-1.4
CONVOLUTION SHAPES

doing, however, it is important that only those temperatures occurring at the expansion joint location itself be specified. Specifying temperatures remote from the expansion joint may unnecessarily result in the need for special materials at additional expense.

IV-1.4.5   Motion.   Movements due to temperature changes or mechanical motion to which the expansion joint will be subjected must be specified. (Methods for determining movements are presented in Section IV-2.) In addition, other extraneous movements, such as wind loading or installation misalignment must be considered. The various dimensional changes which an expansion joint is required to absorb, such as those resulting from thermal changes in a piping system, are as follows (Fig. IV-1.5):

Axial Motion is motion occurring parallel to the center line of the bellows and can be either extension or compression.

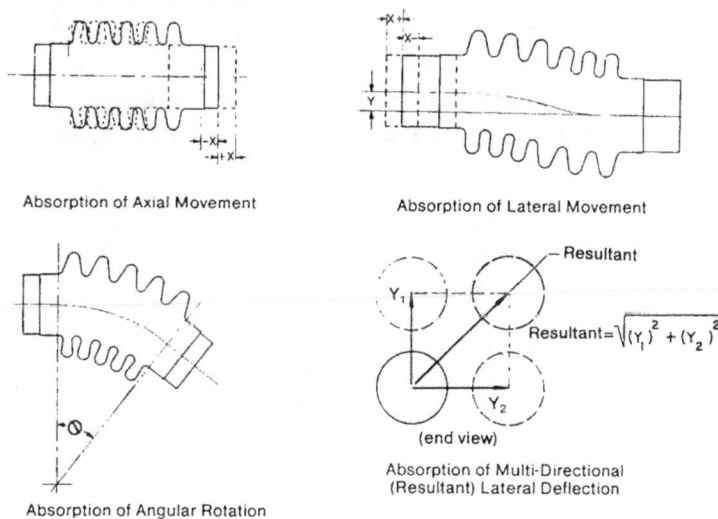

FIG. IV-1.5
TYPES OF EXPANSION JOINT MOVEMENT

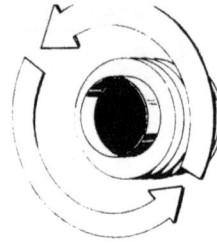

In addition to Axial, Lateral, and Angular Movements, an expansion joint may be subjected to Torsional motion, or twisting. Torsion imposes severe stresses on the expansion joint and all such cases should be referred to the manufacturer.

FIG. IV-1.6
TORSIONAL MOVEMENT

Lateral Deflection, or offset, is motion which occurs perpendicular, or at right angles to the centerline of the bellows. Lateral deflection can occur along one or more axis simultaneously.

Angular Rotation is the bending of an expansion joint along its centerline.

IV-1.5   Expansion Joint Accessories

The basic unit of every expansion joint is the bellows. By adding additional components, expansion joints of increasing complexity and capability are created which are suitable for a wide range of applications.

Refer to Figure IV-1.7 and the following list for a description of the basic expansion joint accessories.

1. BELLOWS – The flexible element of an expansion joint, consisting of one or more convolutions, formed from thin material.

2. LINER – A device which minimizes the effect on the inner surface of the bellows by the fluid flowing through it. Liners are primarily used in high velocity applications to prevent erosion of the inner surface of the bellows and to minimize the likelihood of flow inducted vibration. Liners come in single, tapered, or telescoping configurations according to the application requirements. An expansion joint, if provided with liners, must be installed in the proper orientation with respect to flow direction. Liners are sometimes referred to as internal sleeves.

3. COVER – A device used to provide external protection to the bellows from foreign objects, mechanical damage, and/or external flow. The use of a cover is strongly recommended to all applications. A cover is sometimes referred to as a shroud.

4. WELD END – The ends of an expansion joint equipped with pipe for weld attachment to adjacent equipment or piping. Weld ends are commonly supplied beveled for butt welding.

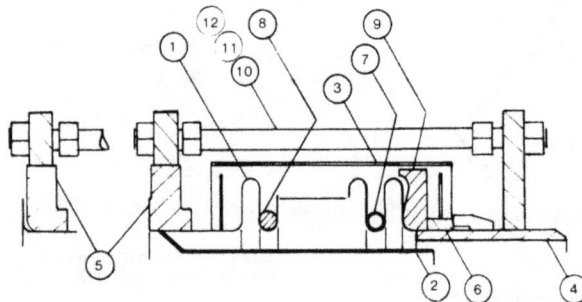

FIG. IV-1.7
EXPANSION JOINT ACCESSORIES

5. FLANGED END – The ends of an expansion joint equipped with flanges for the purpose of bolting the expansion joint to the mating flanges of adjacent equipment or piping.

6. COLLAR – A ring of suitable thickness which is used to reinforce the bellows tangent, or cuff, from bulging due to pressure.

7. HOLLOW REINFORCING RINGS – Devices used on some expansion joints, fitting snugly in the roots of the convolutions. The primary purpose of these devices is to reinforce the bellows against internal pressure. Hollow rings are usually formed from a suitable pipe or tubing section.

8. SOLID ROOT RINGS – Identical in function to hollow root rings, but formed from solid bar stock for greater strength.

9. EQUALIZING RINGS – "T" Shaped in cross-section, these rings are made of cast iron, steel, stainless steel or other suitable alloys. In addition to resisting internal pressure, equalizing rings limit the amount of compression movement per convolution.

10. CONTROL RODS – Devices, usually in the form of rods or bars, attached to the expansion joint assembly whose primary function is to distribute the applied movement between the two bellows of a universal expansion joint. Control rods are NOT designed to restrain bellows pressure thrust.

11. LIMIT RODS – Devices, usually in the form of rods or bars, attached to the expansion joint assembly whose primary function is to restrict the bellows movement range (axial, lateral and angular) during normal operation. In the event of a main anchor failure, limit rods are designed to prevent bellows over-extension or over-compression while restraining the full pressure loading and dynamic forces generated by the anchor failure.

12. TIE RODS – Devices, usually in the form of rods or bars, attached to the expansion joint assembly whose primary function is to continuously restrain the full bellows pressure thrust during normal operation while permitting only lateral deflection. Angular rotation can be accommodated only if two tie rods are used and located 90° from the direction of rotation.

13. PANTOGRAPHIC LINKAGES – A scissors-like device. A special form of control rod attached to the expansion joint assembly whose primary function is to positively distribute the movement equally between the two bellows of the universal joint throughout its full range of movement. Pantographic linkages, like control rods, are NOT designed to restrain pressure thrust.

IV-1.6   Types of Expansion Joints

There are several different types of expansion joints. Each is designed to operate under a specific set of design conditions. The following is a listing of the most basic types of expansion joint designs, along with a brief description of their features and application requirements. A more in-depth discussion of specific expansion joint applications appears in Section IV-2.2.

All of the following expansion joint types are available in both circular and rectangular models

(1) SINGLE EXPANSION JOINT. The simplest form of expansion joint; of single bellows construction, it absorbs all of the movement of the pipe section into which it is installed.

(2) DOUBLE EXPANSION JOINT. A double expansion joint consists of two bellows jointed by a common connector which is anchored to some rigid part of the installation by means of an anchor base. The anchor base may be attached to the common connector either at installation or time of manufacturer. Each bellows of a double expansion joint functions independently as a single unit. Double bellows expansion joints should not be confused with universal expansion joints.

(3) UNIVERSAL EXPANSION JOINT. A universal expansion joint is one containing two bellows joined by a common connector for the purpose of absorbing any combination of the three (3)

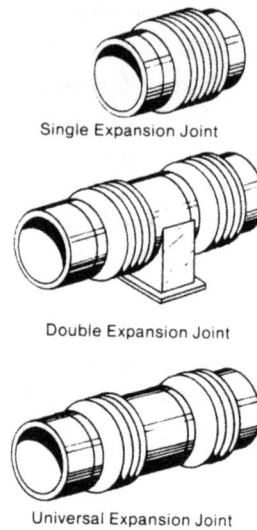

FIG. IV-1.8
EXPANSION JOINT TYPES

basic movements. (Section IV-1.4.5) A universal expansion joint is used in cases to accommodate greater amounts of lateral movement than can be absorbed by a single expansion joint.

(4) UNIVERSAL TIED EXPANSION JOINTS. Tied universal expansion joints are used when it is necessary for the assembly to eliminate pressure thrust forces from the piping system. In this case the expansion joint will absorb lateral movement and will not absorb any axial movement external to the tied length.

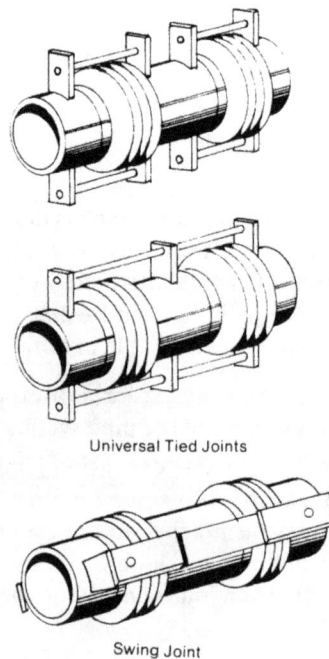

FIG. IV-1.8.1
EXPANSION JOINT TYPES (CONT.)

Hinged Expansion Joint

Gimbal Expansion Joint

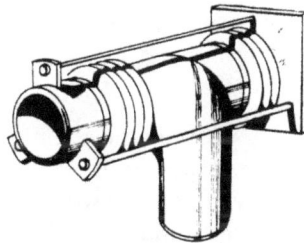

Pressure-Balanced Expansion Joint

FIG. IV-1.8.2
EXPANSION JOINT TYPES (CONT.)

(5) SWING EXPANSION JOINT. A swing expansion joint is designed to absorb lateral deflection and/or angular rotation in one plane only by the use of swing bars, each of which is pinned at or near the ends of the unit.

(6) HINGED EXPANSION JOINT. A hinged expansion joint contains one bellows and is designed to permit angular rotation in one plane only by the use of a pair of pins running through plates attached to the expansion joint ends. Hinged expansion joints should be used in sets of 2 or 3 to function properly.

(7) GIMBAL EXPANSION JOINT. A gimbal expansion joint is designed to permit angular rotation in any plane by the use of 2 pairs of hinges affixed to a common floating gimbal ring.

NOTE: Expansion joints Types 4, 5, 6, and 7 are normally used for restraint of pressure thrust forces. (Section IV-2.1.3.2)

(8) PRESSURE-BALANCED EXPANSION JOINT. A pressure-balanced expansion joint is designed to absorb axial movement and/or lateral deflection while restraining the bellows pressure thrust force (Section IV-2.1.3.2) by means of the devices interconnecting the flow bellows with an opposed bellows also subjected to line pressure. This type of joint is installed where a change of direction occurs in a run of pipe.

IV-2   USING EXPANSION JOINTS

IV-2.1   System Preparation

When thermal growth of a piping system has been determined to be a problem, and the use of an expansion joint is indicated, the most effective expansion joint should be selected as a solution. When using expansion joints it should be kept in mind that movements are not eliminated, but are only

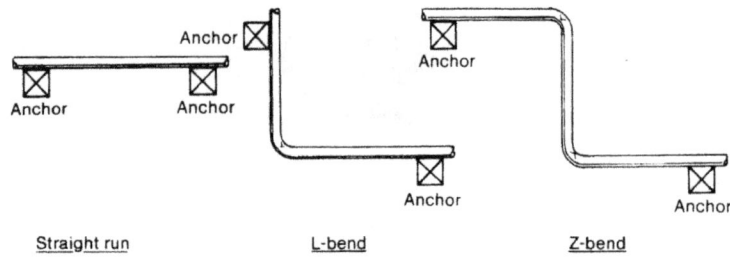

FIG. IV-2.1
STANDARD PIPE SECTIONS

directed to certain points where they can best be absorbed by the expansion joint. The methods used and the steps that must be taken in preparing a piping system for the introduction of an expansion joint are covered in this section. As will be seen, many factors must be considered to insure the proper functioning of an expansion joint.

IV-2.1.1   Simplify the System.   The first step in the selection of expansion joints is to choose the tentative location of the pipe anchors. The purpose of these anchors is to divide a piping system into simple, individual expanding sections. Since thermal growth cannot be restrained, it becomes the function of the pipe anchors to limit and control the amount of movement that the expansion joints, installed between them, must absorb. It is generally advisable to begin with the assumption that using single or double expansion joints in straight axial movement will provide the most simple and economical layout. *Never install more than one 'single' type joint between any two anchors in a straight run of pipe.*

In general, piping systems should be anchored so that a complex system is divided into several sections which conform to one of the configurations shown in Fig. IV-2.1.

Major pieces of equipment such as turbines, pumps, compressors, etc. may function as anchors provided they are designed for this use. Fig. IV-2.2 illustrates a hypothetical piping system in which expansion joints are to be installed. The following sections will describe expansion joint application methods as applied to this system.

IV-2.1.2   Calculating Thermal Growth.   Step 1 in determining thermal movement is to choose tentative anchor positions. The goal here is to divide a system into several straight-line sections exhibiting only axial movement. Referring to Fig. IV-2.3 demonstrates this method.

The pump and two tanks are logical anchor positions. The addition of anchors at pts. B and C result in the desired series of straight-line piping legs. The next step is to determine the actual change in length of each of the piping legs due to changes in temperature.

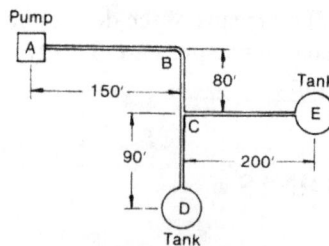

FIG. IV-2.2
TYPICAL PIPING LAYOUT

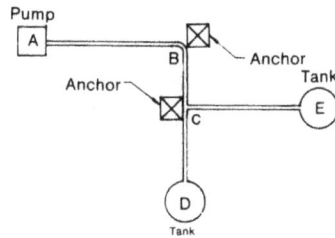

**FIG. IV-2.3
ANCHOR POSITIONS**

In determining thermal movement, all potential sources of temperature must be considered. Usually, the temperature of the flow medium is the major source of dimensional changes, but in certain extreme cases, ambient temperature outside the system can contribute to thermal movement. Movements caused by other sources such as mechanical movements or wind loading, must also be considered.

IV-2.1.3   Pipe Anchors and Forces.   Pipe anchors, their attachments, and the structures to which they are attached, must be designed to withstand all of the forces acting upon them. In addition to the normal pipe system forces, anchors in systems which contain expansion joints are subjected to additional forces. Two significant forces which are unique to expansion joint systems are spring force and pressure thrust force.

IV-2.1.3.1   Spring Force

Spring force is the force required to deflect an expansion joint a specific amount. It is based on operating conditions and manufacturing methods and materials. Spring force imparts a resisting force throughout the system much as a spring would when compressed or otherwise deflected. (Fig. IV-2.4) In order for an expansion joint to operate properly, this spring force must be restrained by anchors.

The magnitude of spring force is determined by the expansion joint's spring rate, and by the amount of movement to which the expansion joint is subjected. The spring rate varies for each expansion joint and is based upon the physical dimensions and material of a specific expansion joint.

A simple example will illustrate the magnitude of spring force for an expansion joint installed in a 24 inch diameter pipe system. The axial spring rate is given as: 1568 lbs./in. If the expansion joint deflects $\frac{1}{2}$ in. axially, the spring force is:

$$\text{Spring Force} = 1568 \times .5 = 784 \text{ lbs.}$$

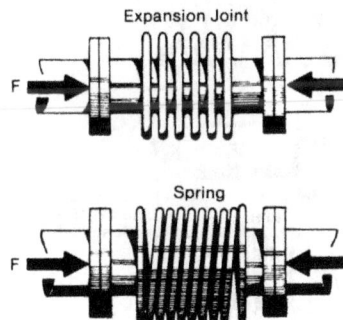

**FIG. IV-2.4
SPRING FORCE**

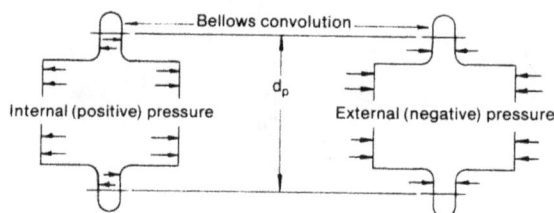

FIG. IV-2.5
PRESSURE THRUST

IV-2.1.3.2   <u>Pressure Thrust Force</u>

Pressure thrust force is often misunderstood. It is a condition created by the installation of a flexible unit, such as an expansion joint, into a rigid piping system which is under pressure. Pressure thrust force is a function of the system pressure and mean diameter (Fig. IV-2.5) of the bellows. Mean diameter is determined by the height, or span, of the bellows and can vary from unit to unit. Mean diameter (d_p) is usually greater than the pipe diameter. Fig IV-2.5 illustrates the effect of pressure on a bellows convolution. In cases of internal, or positive pressure, the convolutions are pushed apart, causing the bellows to extend or increase in length while the opposite is observed in cases of external or negative pressure. The force required to maintain the bellows at it's proper length is equal to this pressure thrust and can be significantly higher than all other system forces combined. See Fig. (IV-2.6) for further details of pressure thrust effects.

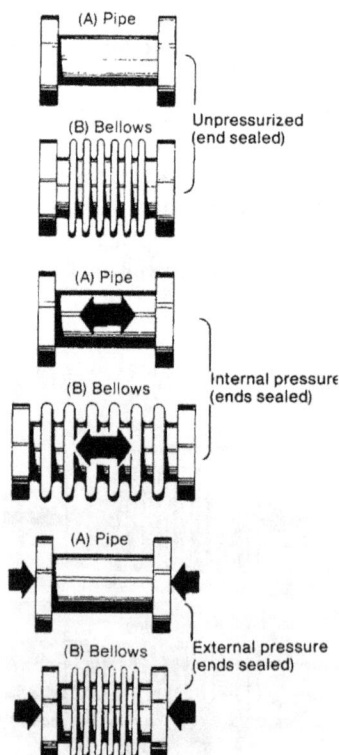

FIG. IV-2.6
PRESSURE THRUST EFFECTS

IV-2.1.3.3   Determining Pressure Thrust Force

The magnitude of pressure thrust force (Fs) in lbs., is determined by the following equation:

$$Fs = Pa$$

where

(P) is the pressure (psig) and
(a) is the effective area of the expansion joint (in^2)

The following example will illustrate the effect of pressure thrust. Suppose that an expansion joint is to be installed in a 24 in. diameter pipe system which operates a 150 psig pressure. From the manufacturer of the expansion joint the effective area is found to be 560 in^2.

$$a = \frac{\pi \, (d_p)^2}{4}$$

The pressure thrust force (Fs) is found to be:

$$Fs = Pa \text{ or } Fs = (150)(560) = 84{,}000 \text{ lbs.}$$

Note that when determining pressure thrust forces, the pressure value used in Fs = Pa should equal the maximum anticipated pressure that the system is likely to experience. For this reason, any pressure test must be considered. Be careful not to specify unrealistic safety factors which could lead to over-design of the anchors at additional cost.

The force is transmitted from the ends of the expansion joint along the pipe. Referring to Fig. IV-2.6, a comparison of the effect of pressure thrust is shown between a section of rigid pipe (A) and a flexible bellows (B).

In an unpressurized state, there is no pressure thrust force transmitted by A or B and their lengths remain the same.

Under internal pressure, a force is acting upon both A and B that is equal to the pressure times the effective areas of A & B. The force of this thrust is totally absorbed by the rigid walls of the pipe (A) and its length remains unchanged. The length of (B) will tend to increase under internal pressure since the flexible expansion joint is unable to resist the force. It therefore yields, or stretches.

In cases of external pressure or vacuum conditions, the effect of internal pressure is reversed and the expansion joint will tend to compress.

To stop the expansion joint from extending or compressing due to pressure thrust, main anchors must be located at some point on each side of the joint to withstand the forces of pressure thrust and keep the expansion joint at its proper length. Tie rods, hinges and/or gimbals may also be used to restrain pressure thrust force by attaching the ends of the expansion joint to each other. In this case, these restraints prevent the bellows from absorbing any axial movement external to the expansion joint. Pressure-balanced expansion joints restrain pressure thrust and absorb axial movement external to the joint.

As can be seen, the forces exerted on a piping system due to the presence of expansion joints can be quite significant. In the preceding example the spring force was found to be 784 lbs., while the pressure thrust force was equal to 84,000 lbs.—a combined force of 84,784 lbs! One can see that proper anchoring of a piping system containing expansion joints is vital.

IV-2.1.3.4   Main Anchors

A main anchor must be designed to withstand the forces and moments imposed upon it by each of the pipe sections to which it is attached. In the case of a pipe section containing one or more unrestrained expansion joints, these will consist of the full bellows thrust due to pressure, media flow, the forces and/or moments required to deflect the expansion joint or joints their full rated movement, and the frictional forces caused by the pipe guides, directional anchors and supports. In certain applications it may be necessary to consider the weight of the pipe and its contents and any other forces and/or moments resulting from wind loading etc.

In systems containing expansion joints, main anchors are installed at any of the following locations:

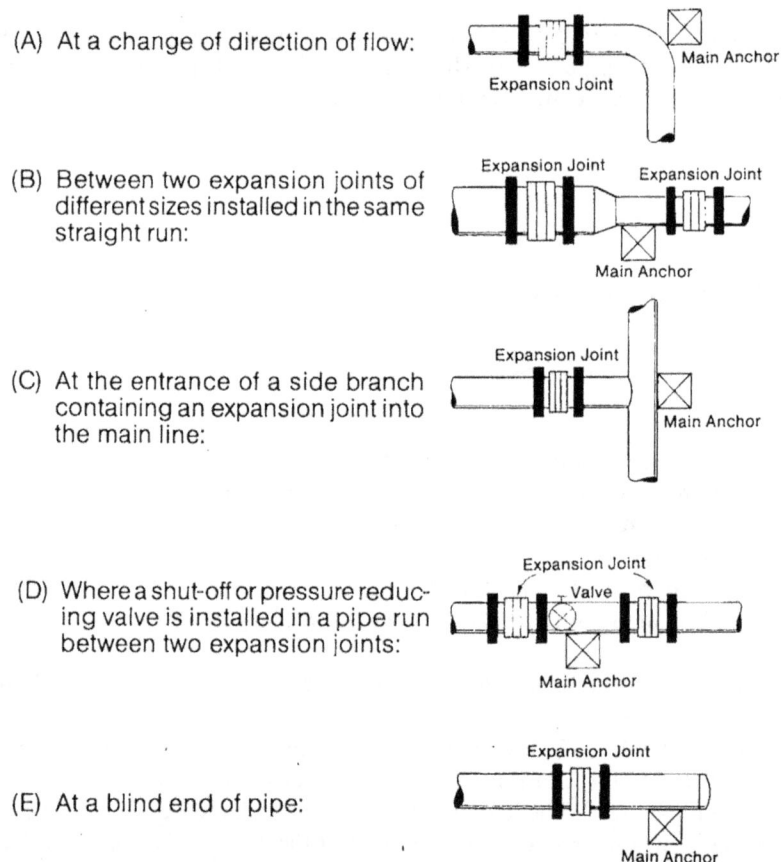

(A) At a change of direction of flow:

(B) Between two expansion joints of different sizes installed in the same straight run:

(C) At the entrance of a side branch containing an expansion joint into the main line:

(D) Where a shut-off or pressure reducing valve is installed in a pipe run between two expansion joints:

(E) At a blind end of pipe:

FIG. IV-2.7

IV-2.1.3.5   Intermediate Anchors

Intermediate anchors are not intended to withstand pressure thrust force. This force is absorbed by main anchors, by devices on the expansion joint such as tie rods, swing bars, hinges, gimbals, etc., or, as in the case of a pressure-balanced or double expansion joint, is balanced by an equal pressure force acting in the opposite direction. An intermediate anchor must withstand all of the non-pressure forces acting upon it by each of the pipe sections to which it is attached. In the case of a pipe section containing an expansion joint, these forces will consist of the force required to move the expansion joint and the frictional forces caused by the pipe guides.

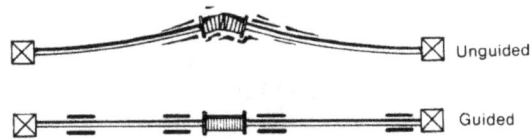

FIG. IV-2.8.1
PIPE GUIDES AND STABILITY

IV-2.1.4   Pipe Guides and Supports.   Correct alignment of the pipe adjoining an expansion joint is of vital importance to its proper function. Although expansion joints are designed and built for a long and satisfactory life, maximum service will be obtained only when the pipeline has the recommended number of guides and is anchored and supported in accordance with good piping practice. Proper supporting of the pipeline is required not only to support the pipe itself, but also to provide support at each end of the expansion joint. Pipe guides are necessary to insure proper alignment of movement to the expansion joint and to prevent buckling of the line. Buckling is caused by a combination of the expansion joint flexibility and the internal pressure loading on the pipe which causes it to act like a column loaded by the pressure thrust of the expansion joint. (Fig. IV-2.8.1).

IV-2.1.4.1   Pipe Guide Application

When locating pipe guides for applications involving axial movement only, it is generally recommended that the expansion joint be located near an anchor, and that the first guide be located a maximum of 4 pipe diameters away from the expansion joint. This arrangement will provide proper movement guiding as well as proper support for each end of the expansion joint. The distance between the first and second guide should not exceed 14 pipe diameters, and the distance between the remaining anchors should be determined from paragraph IV-2.2.4.2. of the EJMA Standards.

Common types of pipe guides are shown in Fig. IV-2.9. For systems subjected to lateral motion or angular rotation, directional anchors or planar guides may be required. Examples of these guides are shown in Section IV-2.2.

The recommendations given for pipe anchors and guides represent the minimum requirements for controlling pipelines which contain expansion joints and are intended to protect the expansion joint and pipe system from abuse and failure. However, additional pipe supports are often required between guides in accordance with standard piping practice.

IV-2.1.4.2   Pipe Support Application

A pipe support is any device which permits free movement of the piping while carrying the dead weight of the pipe and any valves or attachments. Pipe supports must also be capable of carrying the live weight. Pipe supports cannot be substituted for pipe alignment or planar guides. Pipe rings, U-bolts, roller supports, spring hanger etc., are examples of conventional pipe support devices.

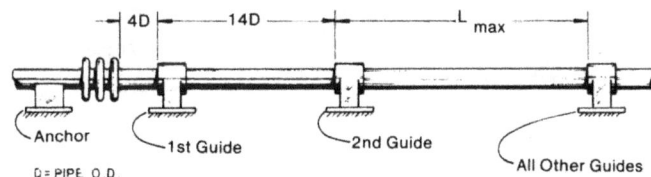

FIG. IV-2.8.2
PROPER PIPE GUIDE LOCATION

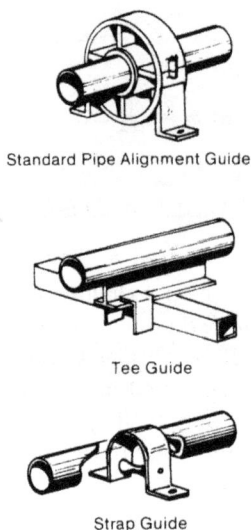

FIG. IV-2.9
COMMON PIPE GUIDES

IV-2.2   Expansion Joint Applications

The following examples demonstrate common types of expansion joint applications, some dealing with simple axial movement and others with more complex and specialized applications.

IV-2.2.1   Axial Movement.   Figure IV-2.10 typifies good practice in the use of a single expansion joint to absorb axial pipeline expansion. Note the use of just one expansion joint between anchors, the nearness of the joint to an anchor and the spacing of the pipe guides.

Fig. IV-2.11 typifies good practice in the use of a double expansion joint to absorb axial pipeline expansion. Note the addition of the intermediate anchor which, in conjunction with the two main anchors effectively divides this section into two distinct expanding segments, so that there is only one expansion joint between anchors.

Fig. IV-2.12 typifies good practice in the use of a pressure-balanced expansion joint (Section IV-1.6, Paragraph 8) to absorb axial movement. Note that the expansion joint is located at a change of direction of the piping and that the elbow and the end of the pipe are secured by intermediate anchors. Since the pressure thrust is absorbed by the expansion joint itself, and only the force required to deflect the expansion joint is imposed on the piping, a minimum of guiding is required. Guiding near the expansion joint as shown will frequently suffice. In long, small-diameter pipelines, additional guiding may be necessary.

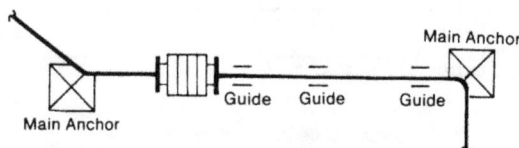

FIG. IV-2.10
SINGLE EXPANSION JOINT APPLICATION

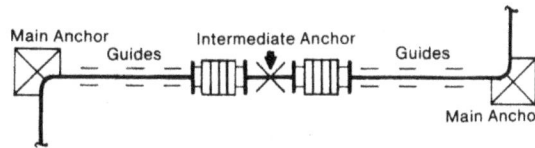

FIG. IV-2.11
DOUBLE EXPANSION JOINT APPLICATION

IV-2.2.2   Combined Movements, Lateral Deflection, and Angular Rotation.   Because it offers the lowest expansion joint cost, the single expansion joint is usually considered first for any application. Fig. IV-2.13 shows a typical application of a single expansion joint used to absorb lateral deflection, as well as axial compression. The joint is located near the end of the long pipe run with main anchors at each end and guides properly spaced for both movement control and protection of the piping against buckling. In this case, as opposed to the similar configuration in Fig. IV-2.10, the anchor at left is a directional main anchor which, while absorbing the main anchor loading in the direction of the expansion joint axis, permits the thermal expansion of the short leg to act upon the expansion joint as lateral deflection. Because the main anchor loading (pressure thrust) exists only in the leg containing the expansion joint, the anchor at the end of the short piping leg is an intermediate anchor.

Figure IV-2.14 represents modifications of Figure IV-2.10 in which the main anchors at either end of the expansion joint are replaced by tie rods. Where the piping configuration permits, the use of the tie rods adjusted to prevent axial movement frequently simplifies and reduces the cost of the installation. Because of these tie rods, the expansion joint is not capable of absorbing any axial movement other than its own thermal expansion (See Section IV-2.1.3.3). The thermal expansion in the short leg is imposed as deflection on the long piping leg.

Fig. IV-2.15 shows a tied universal expansion joint used to absorb lateral deflection in a single plane 'Z' bend. The thermal movement of the horizontal lines is absorbed as lateral deflection by the expansion joint. Both anchors need only to be intermediate anchors since the pressure thrust is restrained by the tie rods. Where possible, the expansion joint should fill the entire offset leg so that its expansion is absorbed within the tie rods as axial movement, and bending in the horizontal lines is minimized.

Figure IV-2.16 shows a typical application of a tied universal expansion joint in a 3-plane 'Z' bend. Since the universal expansion joint can absorb lateral deflection in any direction, the two horizontal legs may lie at any angle in the horizontal plane.

Figure IV-2.17 shows a case of pure angular rotation using a pair of hinge joints in a single-plane 'Z' bend. Since the pressure thrust is restrained by the hinges, only intermediate anchors are required. The axial expansion of the pipe leg between the expansion joints is imposed on the two horizontal legs if there is sufficient flexibility.

If not, a 3 hinge system such as in Fig. IV-2.18 may be required. The thermal expansion of the offset leg is absorbed by joints (A), (B), & (C). Expansion joint (B) must be capable of absorbing the total of the rotation of (A) & (C). Hence, it is frequently necessary that the center expansion joint contain a greater number of convolutions than those at either end.

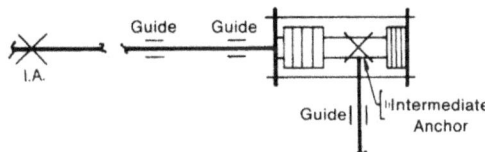

FIG. IV-2.12
PRESSURE-BALANCED APPLICATION

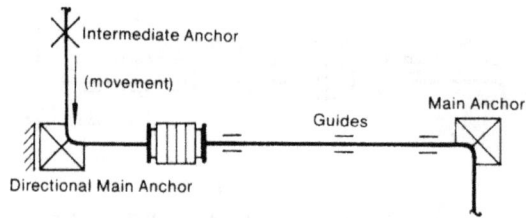

FIG. IV-2.13
COMBINED AXIAL AND LATERAL MOVEMENT

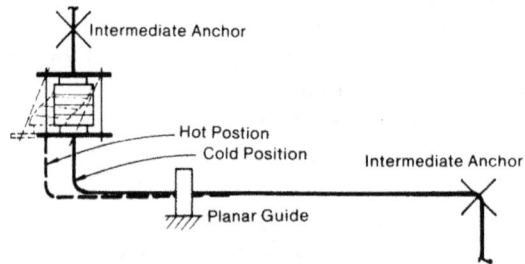

FIG. IV-2.14
SINGLE-TIED APPLICATION

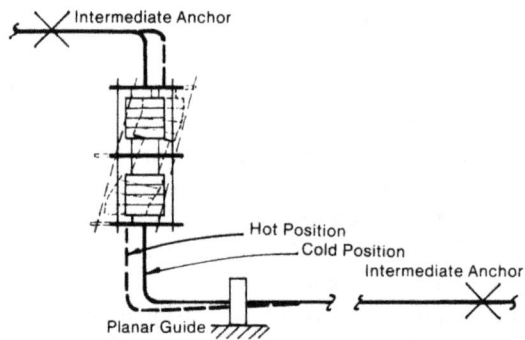

FIG. IV-2.15
UNIVERSAL-TIED APPLICATION

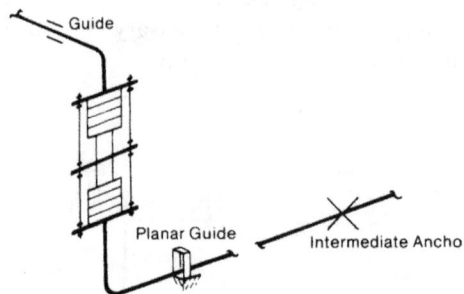

FIG. IV-2.16
UNIVERSAL-TIED APPLICATION

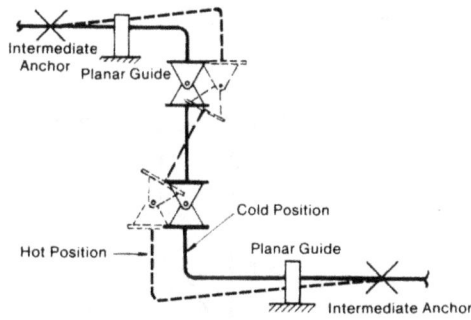

FIG. IV-2.17
TWO-HINGE APPLICATION

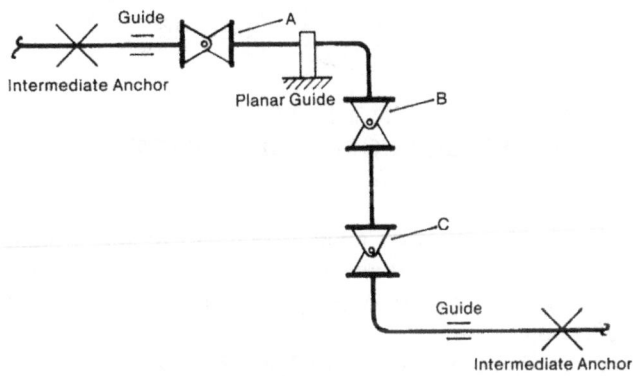

FIG. IV-2.18
THREE-HINGE APPLICATION

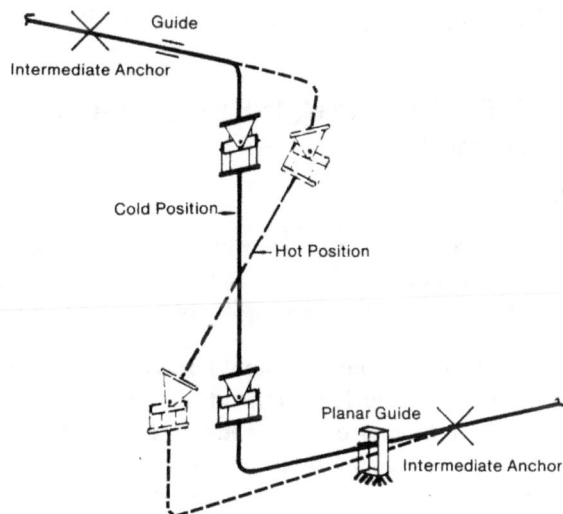

FIG. IV-2.19
TWO-GIMBAL APPLICATION

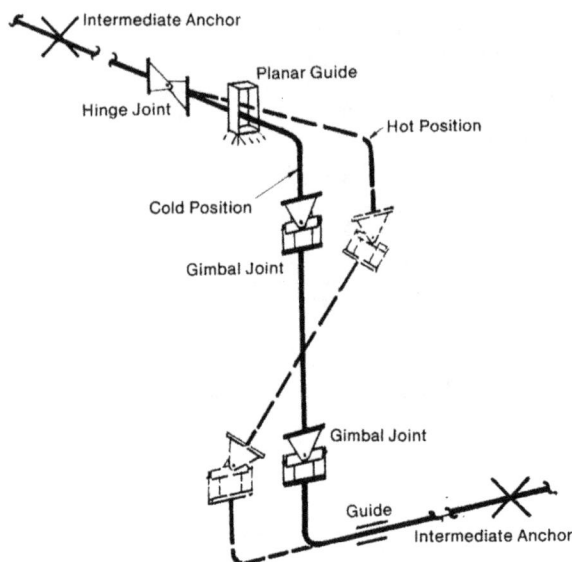

FIG. IV-2.20
ONE HINGE/TWO GIMBAL APPLICATION

Just as hinged expansion joints may offer great advantages in single-plane applications, gimbal expansion joints are designed to offer similar advantages in multi-plane systems. The ability of gimbal expansion joints to absorb angular rotation in any plane is frequently applied by utilizing two such units to absorb lateral deflection.

Figure IV-2.19 shows a case of pure angular rotation using a pair of gimbal expansion joints in a multi-plane 'Z' bend. Since the pressure thrust is restrained by the gimbal assemblies, only intermediate anchors are required.

In applications where the flexibility of the two horizontal piping legs is not sufficient to absorb the axial growth of the pipe section between the two gimbal joints in Figure IV-2.19, the addition of a hinge joint, as shown in Figure IV-2.20, is indicated. Since the expansion of the offset leg takes place in only one plane, the use of the simpler hinge expansion joint is justified.

IV-3   EXPANSION JOINT HANDLING, INSTALLATION, AND SAFETY RECOMMENDATIONS

IV-3.1   Shipping and Handling

Responsible manufacturers of expansion joints take every design and manufacturing precaution to assure the user of a reliable product. The installer and the user have the responsibility to handle, store, install, and apply the expansion joints in a way which will not impair the quality built into them.

IV-3.1.1   Shipping Devices.   All manufacturers should provide some way of maintaining the face-to-face dimension of an expansion joint during shipment and installation. These usually consist of overall bars or angles welded to the flanges or weld ends at the extremities of the expansion joint. Washers or wooden blocks between equalizing rings are also used for this purpose. (Fig. IV-3.1) Do not remove these shipping devices until all expansion joints, anchors, and guides in the system have been installed. Shipping devices manufactured by members of the Expansion Joint Manufacturers Association, Inc., are usually painted yellow or otherwise distinctively marked as an aid to the installer. The shipping devices must be removed prior to start-up or testing of the system. (Ref. Section IV-3.4)

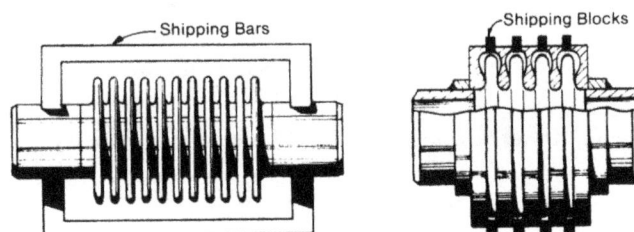

FIG. IV-3.1
SHIPPING DEVICES MAINTAIN OVERALL LENGTH DURING SHIPMENT
AND INSTALLATION

IV-3.1.2   Storage.   Some conditions of outdoor storage may be detrimental, and where possible, should be avoided; preferably, the storage should be in a cool, dry area. Where this cannot be accomplished, the expansion joint manufacturer should be so advised either through the specifications or purchase contract. Care must be taken to avoid mechanical damage such as caused by stacking, bumping or dropping. For this reason, it is strongly suggested that covers be specified on all expansion joints to protect the bellows. Certain industrial and natural atmospheres can be detrimental to some bellows materials. If expansion joints are to be stored or installed in such atmospheric environments, materials should be specified which are compatible with these environments.

IV-3.1.3   Installation Instructions.   Expansion joints are shipped with documents which furnish the installer with instructions covering the installation of the particular expansion joint. These documents should be left with the expansion joint until installation is completed.

IV-3.2   Installation

Metal bellows-type expansion joints have been designed to absorb a specified amount of movement by flexing of the thin-gage bellows. If proper care is not exercised in the installation of the expansion joint, cycle life and pressure capacity could be reduced, leading to premature failure and damage to the piping system.

It is important that the expansion joint be installed at the length specified by the manufacturer. They should never be extended or compressed in order to make up for deficiencies in length, nor should they be offset to accommodate misaligned pipe.

Remember that a bellows is designed to absorb motion by flexing. The bellows, therefore, must be sufficiently thick to withstand the design pressure, while being thin enough to absorb the required flexing. Optimum design will always require a bellows to be of thinner material than virtually any other component in the piping system in which it is installed. The installer must recognize this relative fragility of the bellows and take every possible measure to protect it during installation. Avoid denting, weld spatter, arc strikes or the possibility of allowing foreign matter to interfere with the proper flexing of the bellows. It is highly recommended that a cover be specified for every expansion joint. The small cost of a cover is easily justified when compared to the cost of replacing a damaged bellows element. With reasonable care during storage, handling, and installation, the user will be assured of the reliability designed and built into the expansion joint.

IV-3.3   Do's and Don't's – Installation and Handling

The following recommendations are included to avoid the most common errors that occur during installation. When in doubt about an installation procedure, contact the manufacturer for clarification before attempting to install the expansion joint.

DO'S...

Do ... Inspect for damage during shipment such as: dents, broken hardware, water marks on carton, etc.

Do ... Store in a clean, dry area where it will not be exposed to heavy traffic or damaging environment.

Do ... Use only designated lifting lugs when provided.

Do ... Make the piping system fit the expansion joint. By stretching, compressing, or offsetting the joint to fit the piping, the expansion joint may be overstressed when the system is in service.

Do ... Leave one flange loose on the adjacent piping when possible, until the expansion joint has been fitted into position. Make necessary adjustments of this loose flange before welding.

Do ... Install the joint with the arrow pointing in the direction of the flow.

Do ... Install single vanstone liners pointing in the direction of flow. Be sure also to install a gasket between a vanstone liner and flange.

Do ... In case of telescoping liner, install the smallest I.D. liner pointing in the direction of flow.

Do ... Remove all shipping devices after the installation is complete and before any pressure test of the fully installed system.

Do ... Remove any foreign material that may have become lodged between the convolutions.

Do ... Refer to the proper guide spacing and anchoring recommendations in Section IV-2.1.3 & IV-2.1.4.

DON'T'S...

Don't ... Drop or strike expansion joint.

Don't ... Remove the shipping bars until the installation is complete.

Don't ... Remove any moisture-absorbing desiccant bags or protective coatings until ready for installation.

Don't ... Use hanger lugs or shipping bars as lifting lugs.

Don't ... Use chains or any lifting device directly on the bellows or bellows cover.

Don't ... Allow weld spatter to hit unprotected bellows.

Don't ... Use cleaning agents which contain chlorides.

Don't ... Use steel wool or wire brushes on bellows.

Don't ... Force or rotate one end of an expansion joint for alignment of bolt holes. Bellows are not ordinarily capable of absorbing torsion.

Don't ... Hydrostatic pressure test or evacuate the system before proper installation of all guides and anchors.

Don't ... Use shipping bars to restrain the pressure thrust during testing.

Don't ... Use pipe hangers as guides.

Don't ... Exceed the manufacturers rated test pressure of the expansion joint.

   Caution: The manufacturers warranty may be void if improper installation procedures have been used.

IV-3.4   Safety Recommendations

This section was prepared in order to better inform the user of those factors which many years of experience have shown to be essential for the successful installation and performance of piping systems containing bellows expansion joints.

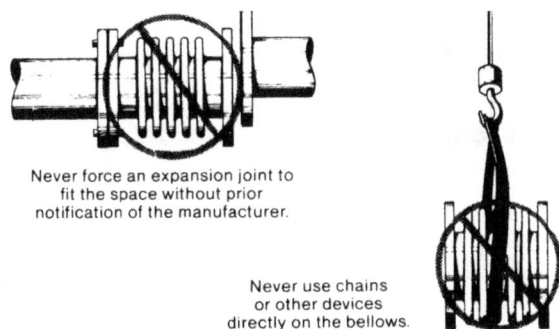

Never force an expansion joint to
fit the space without prior
notification of the manufacturer.

Never use chains
or other devices
directly on the bellows.

FIG. IV-3.2

IV-3.4.1 Inspection Prior to Start-up or Pressure Test. Expansion joints are usually considered to be non-repairable items and generally do not fall into the category for which maintenance procedures are required. However, immediately after the installation is complete a careful visual inspection should be made of the entire piping system to insure that there is no evidence of damage, with particular emphasis on the following:

1. Are anchors, guides, and supports installed in accordance with the system drawings?
2. Is the proper expansion joint in the proper location?
3. Are the expansion joint's flow direction and pre-positioning correct?
4. Have all of the expansion joint shipping devices been removed?
5. If the system has been designed for a gas, and is to be tested with water, has provision been made for proper support of the additional dead weight load on the piping and expansion joint? Some water may remain in the bellows convolutions after the test. If this is detrimental to the bellows or system operation, means should be provided to remove this water.
6. Are all guides, pipe supports and the expansion joints free to permit pipe movement?
7. Has any expansion joint been damaged during handling and installation?
8. Is any expansion joint misaligned? This can be determined by measuring the joint overall length, inspection of the convolution geometry, and checking clearances at critical points on the expansion joint and at other points in the system.
9. Are the bellows and other movable portions of the expansion joint free of foreign material?

IV-3.4.2 Inspection During and Immediately after Pressure Test – Warning: Extreme care must be exercised while inspecting any pressurized system or component.
A visual inspection of the system should include the following:

1. Evidence of leaking or loss of pressure.
2. Distortion or yielding of anchors, expansion joint hardware, the bellows and other piping components.
3. Any unanticipated movement of the piping due to pressure.
4. Evidence of instability in the bellows.
5. The guides, expansion joints, and other movable parts of the system should be inspected for evidence of binding.
6. Any evidence of abnormality or damage should be reviewed and evaluated by competent design authority.

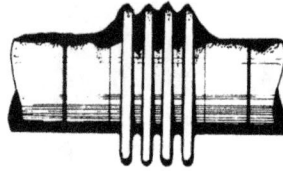

Inspect periodically for a build-up of debris between the bellows convolutions. or any other circumstance which may restrict the free-flexing of the bellows.

FIG. IV-3.3

IV-3.4.3   Periodic Inservice Inspection – Warning: Extreme care must be exercised while inspection any pressurized system or component.

1. Immediately after placing the system in operation, a visual inspection should be conducted to insure that the thermal expansion is being absorbed by the expansion joints in the manner for which they were designed.

2. The bellows should be inspected for evidence of unanticipated vibration.

3. A program of periodic inspection should be planned and conducted throughout the operating life of the system. The frequency of these inspections should be determined by the service and environmental conditions involved. Such inspections can spot the more obvious potential problems such as external corrosion, loosening of threaded fasteners, and deterioration of anchors, guides and other hardware.

IT MUST BE UNDERSTOOD THAT THIS INSPECTION PROGRAM, WITHOUT ANY OTHER BACKUP INFORMATION, CANNOT GIVE EVIDENCE OF DAMAGE DUE TO FATIGUE, STRESS CORROSION OR GENERAL INTERNAL CORROSION. THESE CAN BE THE CAUSE OF SUDDEN FAILURES AND GENERALLY OCCUR WITHOUT ANY VISIBLE OR AUDIBLE WARNING.

4. When any inspection reveals evidence of malfunction, damage, or deterioration, this should be reviewed by competent design authority for resolution. Additionally, any changes in the system operating conditions such as pressure, temperature, movement, flow, velocity, etc. that may adversely affect the expansion joint should be reported to and evaluated by a competent design authority.

IV-3.5   Key to Symbols Used

IV-3.6   Specification and Ordering Guide

This guide has emphasized that the most reliable and safe bellows expansion joint installations are the result of a high degree of understanding between the user and the manufacturer. The prospect for successful long-life performance of piping systems containing expansion joints begins with the submittal of accurate data by the system designer.

When preparing specifications for expansion joints, it is imperative that the system designer completely review the piping system layout and provide the necessary data relating to the selection of an expansion joint: flow medium, pressure, temperature, movements, etc.

To aid the user in the specification and ordering of expansion joints, the Expansion Joint Manufacturers Association has provided the following expansion joint specification sheets. Complete data, when submitted in accordance with these specifications, will help in obtaining an optimum design. Other data that must be considered but which is not covered, may be included in the "remarks" section of the specification sheets.

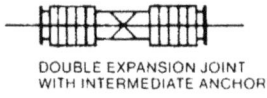 SINGLE EXPANSION JOINT

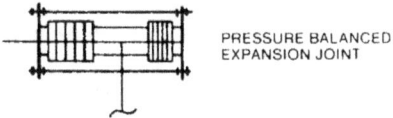

DOUBLE EXPANSION JOINT
WITH INTERMEDIATE ANCHOR

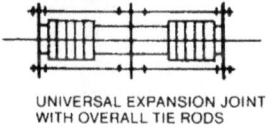 PRESSURE BALANCED
EXPANSION JOINT

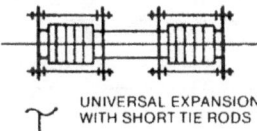 SINGLE EXPANSION JOINT
WITH TIE RODS

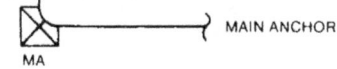

UNIVERSAL EXPANSION JOINT
WITH OVERALL TIE RODS

UNIVERSAL EXPANSION JOINT
WITH SHORT TIE RODS

MAIN ANCHOR

MA

DMA   DIRECTIONAL
MAIN ANCHOR

 INTERMEDIATE
ANCHOR

IA

 PIPE ALIGNMENT
GUIDE

G

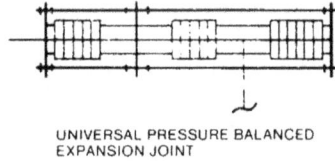

UNIVERSAL PRESSURE BALANCED
EXPANSION JOINT

 HINGED EXPANSION JOINT

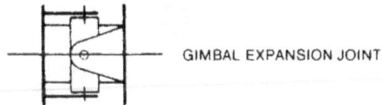 GIMBAL EXPANSION JOINT

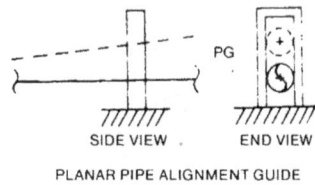

PG

SIDE VIEW   END VIEW

PLANAR PIPE ALIGNMENT GUIDE

PIPE REDUCER

APPENDIX A-1

STANDARD EXPANSION JOINT SPECIFICATION SHEET

COMPANY:		DATE / /			
		SHEET OF			
PROJECT:		INQUIRY NO.			
		JOB NO.			

	ITEM NO./EJ TAG NO.		EJMA PAGE REFERENCE			
1	QUANTITY					
2	NOMINAL SIZE/I.D./O.D. (IN.)					
3	EXPANSION JOINT TYPE		1			
4a	FLUID INFORMATION	MEDIUM GAS/LIQUID	5, 6, 147 77			
4b		VELOCITY (FT./SEC)				
4c		FLOW DIRECTION				
5	DESIGN PRESSURE. PSIG		6, 19, 83, 135			
6	TEST PRESSURE. PSIG					
7a	TEMPERATURE	DESIGN (°F)	6, 13			
7b		MAXIMUM/MINIMUM (°F)				
7c		INSTALLATION (°F)				
8a	MAXIMUM INSTALLATION MOVEMENT	AXIAL COMPRESSION (IN.)	6, 7, 8, 141			
8b		AXIAL EXTENSION (IN.)				
8c		LATERAL (IN.)				
8d		ANGULAR (DEG.)				
9a	MAXIMUM DESIGN MOVEMENTS	AXIAL COMPRESSION (IN.)	6, 7, 13, 47			
9b		AXIAL EXTENSION (IN.)				
9c		LATERAL (IN.)				
9d		ANGULAR (DEG.)				
9e		NO. OF CYCLES				
10a	OPERATING FLUCTUATIONS	AXIAL COMPRESSION (IN.)	84			
10b		AXIAL EXTENSION (IN.)				
10c		LATERAL (IN.)				
10d		ANGULAR (DEG.)				
10e		NO. OF CYCLES				
11a	MATERIALS OF CONSTRUCTION	BELLOWS	5, 6, 45 77, 78			
11b		LINERS				
11c		COVER	3, 7, 72			
11d		PIPE SPECIFICATION				
11e		FLANGE SPECIFICATION	3, 43			
12	RODS (TIE/LIMIT/CONTROL)		3, 4, 41			
13	PANTOGRAPHIC LINKAGE		4			
14	ANCHOR BASE (MAIN/INTERMEDIATE)		1, 2, 17			
15a	DIMENSIONAL LIMITATIONS	OVERALL LENGTH (IN.)				
15b		OUTSIDE DIAMETER (IN.)				
15c		INSIDE DIAMETER (IN.)				
16a	SPRING RATE LIMITATIONS	AXIAL (LBS./IN.)	54			
16b		LATERAL (LBS./IN.)				
16c		ANGULAR (IN.-LBS./DEG.)				
17	INSTALLATION POSITION HORIZ./VERT.		8, 141			
18a	QUALITY ASSURANCE REQUIREMENTS	BELLOWS WELD NDE	LONG. SEAM	133			
18b			ATTACH.				
18c		PIPING NDE				
18d		DESIGN CODE REQRD.				
18e		PARTIAL DATA REQRD.				
18f						
18g						
19	VIBRATION AMPLITUDE/FREQUENCY					

APPENDIX A-2

SUPPLEMENTAL SPECIFICATION SHEET
To be used with Standard Expansion Joint Specification Sheet

Company _____ Date _____

_____ Proposal No._____

Project _____ Inquiry/Job No._____

_____ Sheet _____ of _____

	ITEM NO.				
20.	PURGE, INSTRUMENTATION CONNECTION				
21a.		FACING			
21b.		O.D. (IN.)			
21c.		I.D. (IN.)			
21d.	SPECIAL FLANGE DESIGN	THICKNESS (IN.)			
21e.		B.C. DIAMETER (IN.)			
21f.		NO. HOLES			
21g.		SIZE HOLES			
21h.		HOLE ORIENTATION			

ISOMETRIC PIPING SKETCH:

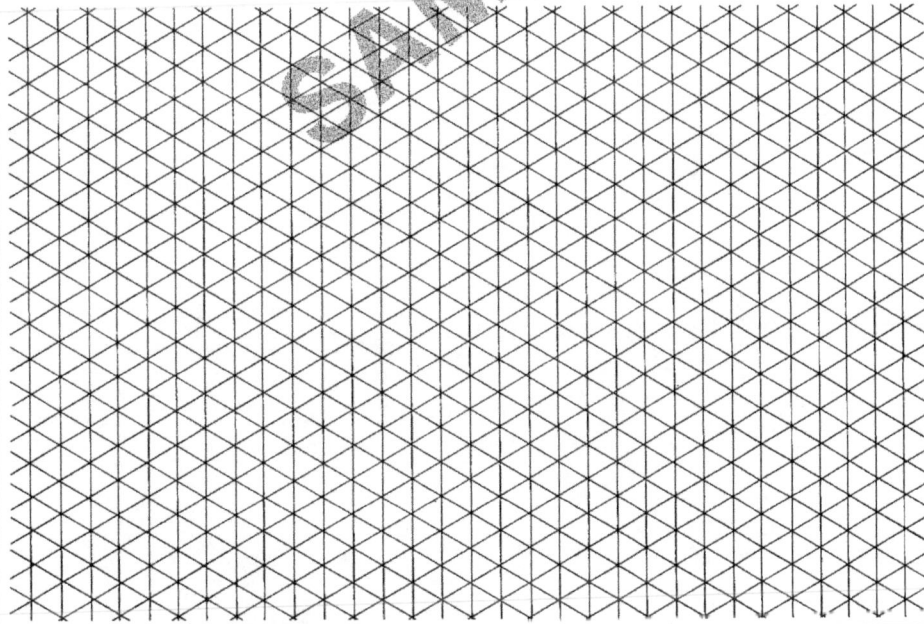

APPENDIX V
CONVERSION FACTORS

CONVERSION FACTORS
(Courtesy of Anvil International)

Multiply	by	To Obtain
Absolute viscosity (poise)	1	Gram/second centimeter
Absolute viscosity (centipoise)	0.01	Poise
Acceleration due to gravity (*g*)	32.174	feet/second²
	980.6	Centimeters/second²
Acres	0.4047	Hectares
	10	Square Chains
	43,560	Square Feet
	4047	Square Meters
	0.001562	Square Miles
	4840	Square Yards
	160	Square Rods
Acre-feet	43,560	Cubic Feet
	325,851	Gallons (US)
	1233.49	Cubic Meters
	1,233,490	Liters
Acre-feet/hour	726	Cubic feet/Minute
	5430.86	Gallons Minute
Angstroms	10⁻¹⁰	Meters
Ares	0.01	Hectares
	1076.39	Square Feet
	0.02471	Acres
Atmospheres	76.0	Cms of Hg at 32°F
	29.921	Inches of Hg at 32° F
	33.94	Feet of Water at 62° F
	10,333	Kgs Square meter
	14.6963	Pounds Square inch
	1.058	Tons/Square foot
	1013.15	Millibars
	235.1408	Ounces/Square inch
Bags of cement	94	Pounds of cement
Barrels of oil	42	Gallons of oil (US)
Barrels of cement	376	Pounds of cement
Barrels (not legal) or	31	Gallons (US)
	31.5	Gallons (US)
Board feet	144 × 1 in.*	Cubic inches
Boiler horse power	33,479	BTU/hour
	9.803	Kilowatts
	34.5	pounds of water evaporated/ hour at 212° F
BTU	252.016	Calories (gm)
	0.252	Calories (Kg)
	778.26	Foot pounds
	0.0003927	Horse power hours
	1055.1	Joules
	107.5	Kilogram meters
	0.0002928	Kilowatt hours
BTU/Cu foot	8.89	Calories (Kg)/Cu meter at 32° F
BTU/Hr/ft²/°F/ft	0.00413	Cal (gm)/Sec/cm²/°C/cm
	1.49	Cal (Kg)/ Hr/M²/°C/Meter
BTU/minute	12.96	Foot pounds/second
	0.02356	Horse Power
	0.01757	Kilowatts

*For thickness less than 1 in. use actual thickness in decimals of an inch.

Multiply	by	To Obtain
BTU/minute	17.57	Watts
BTU/pound	0.556	Calories (Kg)/Kilogram
Bushels	2150.4	Cubic inches
	35.24	Liters
	4	Pecks
	32	Quarts (dry)
Cables	120	Fathoms
Calories (gm)	0.003968	BTU
	0.001	Calories (Kg)
	3.088	Foot pounds
	1.558 × 10⁻⁶	Horse power hours
	4.185	Joules
	0.4265	Kilogram meters
	1.1628 × 10⁻⁶	Kilowatt hours
	0.0011628	Watt hours
Cal (gm)/sec/cm²/°C/cm	242.13	BTU/Hr/ft²/°F/ft
Calories (Kg)	3.968	BTU
	1000	Calories (gm)
	3088	Foot pounds
	0.001558	Horse power hours
	4185	Joules
	426.5	Kilograms meters
	0.0011628	Kilowatt hours
	1.1628	Watt hours
Calories (Kg) Cu meter	0.1124	BTU/Cu foot at 0° C
Cal (Kg)/Hr/M²/°C/M	0.671	BTU Hr/ft²/°F/foot
Calories (Kg)/Kg	1.8	BTU pound
Calories (Kg) minute	51.43	Foot pounds/second
	0.09351	Horse power
	0.06972	Kilowatts
Carats (diamond)	200	Milligram
Centares (Centiares)	1	Square meters
Centigram	0.01	Grams
Centiliters	0.01	Liters
Centimeters	0.3937	Inches
	0.032808	Feet
	0.01	Meters
	10	Millimeters
Centimeters of Hg at 32° F	0.01316	Atmospheres
	0.4461	Feet of water at 62° F
	136	Kgs/Square meter
	27.85	Pounds/Square foot
	0.1934	Pounds/Square inch
Centimeters/second	1.969	Feet/minute
	0.03281	Feet/second
	0.036	Kilometers/hour
	0.6	Meters/minute
	0.02237	Miles/hour
	0.0003728	Miles/minute
Centimeters/second²	0.03281	Feet/second²
Centipoise	0.000672	Pounds/sec foot
	2.42	Pounds/hour foot
	0.01	Poise
Chains (Gunter's)	4	Rods
	66	Feet
	100	Links

CONVERSION FACTORS (CONTINUED)
(Courtesy of Anvil International)

Multiply	by	To Obtain
Cheval-vapeur	1	Metric horse power
	75	Kilogram meters/second
	0.98632	Horse power
Circular inches	10^6	Circular mils
	0.7854	Square inches
	785,400	Square mils
Circular mils	0.7854	Square mils
	10^{-6}	Circular inches
	7.854×10^{-5}	Square inches
Cubic centimeters	3.531×10^{-5}	Cubic feet
	0.06102	Cubic inches
	10^{-6}	Cubic meters
	1.308×10^{-6}	Cubic yards
	0.0002642	Gallons (US)
	0.001	Liters
	0.002113	Pints (liq. US)
	0.001057	Quarts (Liq. US)
	0.0391	Ounces (fluid)
Cubic feet	28.320	Cubic centimeters
	1728	Cubic inches
	0.02832	Cubic meters
	0.03704	Cubic yards
	7.48052	Gallons (US)
	28.32	Liters
	59.84	Pints (liq. US)
	29.92	Quarts (liq. US)
	2.296×10^{-5}	Acre feet
	0.803564	Bushels
Cubic feet of water	62.4266	Pounds at 39.2° F
	62.3352	Pounds at 62° F
Cubic feet/minute	472	Cubic centimeters/sec
	0.1247	Gallons (US)/second
	0.472	Liters/second
	62.34	Pounds water/min at 62°F
	7.4805	Gallons (US)/minute
	10,772	Gallons/24 hours
	0.033058	Acre feet/24 hours
Cubic feet/second	646.317	Gallons (US)/24 hours
	448.831	Gallons/minute
	1.98347	Acre feet/24 hours
Cubic inches	16.387	Cubic centimeters
	0.0005787	Cubic feet
	1.639×10^{-5}	Cubic meters
	2.143×10^{-5}	Cubic yards
	0.004329	Gallons (US)
	0.01639	Liters
	0.03463	Pints (liq. US)
	0.01732	Quarts (liq. US)
Cubic meters	10^6	Cubic centimeters
	35.31	Cubic feet
	61.023	Cubic inches
	1.308	Cubic yards
	264.2	Gallons (US)
	1000	Liters
	2113	Pints (liq. US)
	1057	Quarts (liq. US)
Cubic yards	764,600	Cubic centimeters
	27	Cubic feet
	46,656	Cubic inches
	0.7646	Cubic meters
	202	Gallons (US)
	764.6	Liters
	1616	Pints (liq. US)

Multiply	by	To Obtain
	807.9	Quarts (liq. US)
Cubic yards/minute	0.45	Cubic feet/second
	3.367	Gallons (US)/second
	12.74	Liters/second
Cubit	18	Inches
Days (mean)	1440	Minutes
	24	Hours
	86,400	Seconds
Days (sidereal)	86,164.1	Solar seconds
Decigrams	0.1	Grams
Deciliters	0.1	Liters
Decimeters	0.1	Meters
Degrees (angle)	60	Minutes
	0.01745	Radians
	3600	Seconds
Degrees F [less 32]	0.5556	Degrees C
Degrees F	1 [plus 460]	Degrees F above absolute 0
Degrees C	1.8 [plus 32]	Degrees F
	1 [plus 273]	Degrees C above absolute 0
Degrees/second	0.01745	Radians/second
	0.1667	Revolutions/minute
	0.002778	Revolutions/second
Dekagrams	10	Grams
Dekaliters	10	Liters
Dekameters	10	Meters
Diameter (circle)	3.14159265359	Circumference
(approx)	3.1416	
(approx)	3.14	
(approx)	$\frac{22}{7}$	
Diameter (circle)	0.88623	Side of equal square
	0.7071	Side of inscribed square
Diameter³ (sphere)	0.5236	Volume (sphere)
Diam (major) × diam (minor)	0.7854	Area of ellipse
Diameter² (circle)	0.7854	Area (circle)
Diameter² (sphere)	3.1416	Surface (sphere)
Diam (inches) × RPM	0.262	Belt speed ft/minute
Digits	0.75	Inches
Drams (avoirdupois)	27.34375	Grains
	0.0625	Ounces (avoir.)
	1.771845	Grams
Fathoms	0.16667	Feet
Feet	30.48	Centimeters
	12	Inches
	0.3048	Meters
	1/3	Yards
	0.06061	Rods
Feet of water at 62	0.029465	Atmospheres
	0.88162	Inches of Hg at 32° F
	62.3554	Pounds/square foot
	0.43302	Pounds/square inch
	304.44	Kilogram/sq meter
Feet/minute	0.5080	Centimeters/second
	0.01667	Feet/second
	0.01829	Kilometers/hour

CONVERSION FACTORS (CONTINUED)
(Courtesy of Anvil International)

Multiply	by	To Obtain
Feet/minute	0.3048	Meters/minute
	0.01136	Miles/hour
Feet/Second	30.48	Centimeters/second
	1.097	Kilometers/hour
	0.5921	Knots
	18.29	Meters/minute
	0.6818	Miles/hour
	0.01136	Miles/minute
Feet/second^2	30.48	Centimeters/second^2
	0.3048	Meters/second^2
Flat of a hexagon	1.155	Distance across corners
Flat of a square	1.414	Distance across corners
Foot pounds	0.00128492	BTU
	0.32383	Calories (gm)
	0.0003238	Calories (Kg)
	5.05×10^{-7}	Horse power hours
	1.3558	Joules
	0.13826	Kilogram meters
	3.766×10^{-7}	Kilowatt hours
	0.0003766	Watt hours
Foot pounds/minute	0.001286	BTU/minute
	0.01667	Foot pounds/second
	3.03×10^{-5}	Horse power
	0.0003241	Calories (Kg)/minute
	2.26×10^{-5}	Kilowatts
Foot pounds/second	0.07717	BTU/minute
	0.001818	Horse power
	0.01945	Calories (Kg)/minute
	0.001356	Kilowatts
Furlong	40	Rods
	220	Yards
	660	Feet
	0.125	Miles
	0.2042	Kilometers
Gallons (Imperial)	277.42	Cubic inches
	4.543	Liters
	1.20095	Gallons (US)
Gallons (US)	3785	Cubic centimeters
	0.13368	Cubic feet
	231	Cubic inches
	0.003785	Cubic meters
	0.004951	Cubic yards
	3.785	Liters
	8	Pints (liq. US)
	4	Quarts (liq. US)
	0.83267	Gallons (Imperial)
	3.069×10^{-6}	Acre feet
Gallons (US) of water at 62° F	8.333	Pounds of water
Gallons (US) of water/ minute	6.0086	Tons of water/24 hours
Gallons(US)/minute	0.002228	Cubic feet/second
	0.13368	Cubic feet/minute
	8.0208	Cubic feet/hour
	0.06309	Liters/second
	3.78533	Liters/minute
	0.0044192	Acre feet/24 hours
Grains	1	Grains (avoirdupois)
	1	Grains (apothecary)
	1	Grains (troy)
	0.0648	Grams
	0.0020833	Ounces (troy)

Multiply	by	To Obtain
	0.0022857	Ounces (avoir.)
Grains/gallon (US)	17.118	Parts/million
	142.86	Pounds/million gallons (US)
Grams	980.7	Dynes
	15.43	Grains
	0.001	Kilograms
	1000	Milligrams
	0.03527	Ounces (avoir.)
	0.03215	Ounces (troy)
	0.002205	Pounds
Grams/centimeters	0.00521	Pounds/inch
Grams/cubic centimeter		62.45 Pounds/cubic foot
	0.03613	Pounds/cubic inch
	4.37	Grains/100 cubic ft
Grams/liter	58.405	Grains/gallon (US)
	8.345	Pounds/100 gallons (US)
	0.062427	Pounds/cubic foot
	1000	Parts/million
Gravity (g)	32.174	Feet/second^2
	980.6	Centimeters/second^2
Hand	4	Inches
	10.16	Centimeters
Hectares	2.471	Acres
	107,639	Square feet
	100	Ares
Hectograms	100	Grams
Hectoliters	100	Liters
Hectometers	100	Meters
Hectowatts	100	Watts
Hogshead	63	Gallons (US)
	238.4759	Liters
Horse power	42.44	BTU/minute
	33,000	Foot pounds/minute
	550	Foot pounds/second
	1.014	Metric horse power (Cheval vapeur)
	10.7	Calories (Kg)/min
	0.7457	Kilowatts
	745.7	Watts
Horse power (boiler)	33,479	BTU/hour
	9.803	Kilowatts
	34.5	Pounds of water evaporated/hour at 212° F
Horse power hours	2546.5	BTU
	641,700	Calories (gm)
	641.7	Calories (Kg)
	1980198	Foot pounds
	2688172	Joules
	273,740	Kilogram meters
	0.7455	Kilowatt hours
	745.5	Watt hours
Inches	2.54	Centimeters
	0.08333	Feet
	1000	Mils
	12	Lines
	72	Points
Inches of Hg at 32° F	0.03342	Atmospheres
	345.3	Kilograms/square meter
	70.73	Pounds/square foot
	0.49117	Pounds/square inch
	1.1343	Feet of water at 62° F

CONVERSION FACTORS (CONTINUED)
(Courtesy of Anvil International)

Multiply	by	To Obtain
Inches of Hg at 32° F	13.6114	Inches of water at 62° F
	7.85872	Ounces/square inch
Inches of water at 62° F	0.002455	Atmospheres
	25.37	Kilograms/square meter
	0.5771	Ounces/square inch
	5.1963	Pounds/square foot
	0.03609	Pounds/square inch
	0.07347	Inches of Hg at 32° F
Joules	0.00094869	BTU
	0.239	Calories (gm)
	0.000239	Calories (Kg)
	0.73756	Foot pounds
	3.72×10^{-7}	Horse power hours
	0.10197	Kilogram meters
	2.778×10^{-7}	Kilowatt hours
	0.0002778	Watt hours
	1	Watt second
Kilograms	980,665	Dynes
	2.205	Pounds
	0.001102	Tons (short)
	1000	Grams
	35.274	Ounces (avoir.)
	32.1507	Ounces (troy)
Kilogram meters	0.009302	BTU
	2.344	Calories (gm)
	0.002344	Calories (Kg)
	7.233	Foot pounds
	3.653×10^{-6}	Horse power hours
	9.806	Joules
	2.724×10^{-6}	Kilowatt hours
	0.002724	Watt hours
Kilograms/ cubic meter	0.06243	Pounds/ cubic foot
Kilograms/meter	0.6720	Pounds/foot
Kilograms/sq centimeter	14.223	Pounds/sq inch
	1	Metric atmosphere
Kilogram/sq meter	9.678×10^{-5}	Atmospheres
	0.003285	Feet of water at 62° F
	0.002896	Inches of Hg at 32° F
	0.2048	Pounds/square foot
	0.001422	Pounds/square inch
	0.007356	Centimeters of Hg at 32° F
Kiloliters	1000	Liters
Kilometers	100,000	Centimeters
	1000	Meters
	3281	Feet
	0.6214	Miles
	1094	Yards
Kilometers/hour	27.78	Centimeters/second
	54.68	Feet/minute
	0.9113	Feet/second
	16.67	Meters/minute
	0.6214	Miles/hour
	0.5396	Knots
Kilometers/hr/sec	27.78	Centimeters/sec/sec
	0.9113	Feet/sec/sec
	0.2778	Meters/sec/sec
Kilowatts	56.92	BTU/minute
	44,250	Foot pounds/minute
	737.6	Foot pounds/second
	1.341	Horse power
	14.34	Calories (Kg)/min
	1000 Watts	
Kilowatt hours	3413	BTU

Multiply	by	To Obtain
Kilowatt hours	859,999	Calories (gm)
	858.99	Calories (Kg)
	2,655,200	Foot pounds
	1.341	Horse power hours
	3,600,000	Joules
	367,100	Kilogram meters
	1000	Watt hours
Knots	1	Nautical miles/hour
	1.1516	Miles/hour
	1.8532	Kilometers/hours
Leagues	3	Miles
Lines	0.08333	Inches
Links	7.92	Inches
Liters	1000	Cubic Centimeters
	0.03531	Cubic feet
	61.02	Cubic inches
	0.001	Cubic meters
	0.001308	Cubic yards
	0.2642	Gallons (US)
	0.22	Gallons (Imp)
	2.113	Pints (liq. US)
	1.057	Quarts (liq. US)
	8.107×10^{-7}	Acre Feet
	2.2018	Pounds of water at 62° F
Liter/minute	0.0005886	Cubic feet/second
	0.004403	Gallons (US)/second
	0.26418	Gallons (US)/minute
Meters	100	Centimeters
	3.281	Feet
	39.37	Inches
	1.094	Yards
	0.001	Kilometers
	1000	Millimeters
Meters/minute	1.667	Centimeters/second
	3.281	Feet/minute
	0.05468	Feet/second
	0.06	Kilometers/hour
	0.03728	Miles/hour
Meters/second	196.8	Feet/minute
	3.281	Feet/second
	3.6	Kilometers/hour
	0.06	Kilometers/minute
	2.237	Miles/hour
	0.03728	Miles/minute
Microns	10^{-6}	Meters
	0.001	Millimeters
	0.03937	Mils
Mils	0.001	Inches
	0.0254	Millimeters
	25.4	Microns
Miles	160,934	Centimeters
	5280	Feet
	63,360	Inches
	1.609	Kilometers
	1760	Yards
	80	Chains
	320	Rods
	0.8684	Nautical miles
Miles/hour	44.70	Centimeters/second
	88	Feet/minute
	1.467	Feet/second
	1.609	Kilometers/hours
	0.8684	Knots
	26.82	Meters/minute

CONVERSION FACTORS (CONTINUED)
(Courtesy of Anvil International)

Multiply	by	To Obtain	Multiply	by	To Obtain
Miles/minute	2682	Centimeters/second		1.315	Horse power
	88	Feet/second	Pounds (avoirdupois)	16	Ounces (avoir.)
	1,609	Kilometers/minute		256	Drams (avoir.)
	60	Miles/hour		7000	Grains
Millibars	0.000987	Atmosphere		0.0005	Tons (short)
Milliers	1000	Kilograms		453.5924	Grams
Milligrams	0.01	Grams		1.21528	Pounds (troy)
	0.01543	Grains		14.5833	Ounces (troy)
Milligrams/liter	1	Parts/million	Pounds (troy)	5760	Grains
Milliliters	0.001	Liters		240	Pennyweights (troy)
Million gals/24 hours	1.54723	Cubic feet/second		12	Ounces (troy)
Millimeters	0.1	Centimeters		373.24177	Grams
	0.03937	Inches		0.822857	Pounds (avoir.)
	39.37	Mils		13.1657	Ounces (avoir.)
	1000	Microns		0.00036735	Tons (long)
Miner's inches	1.5	Cubic feet/minute		0.00041143	Tons (short)
Minutes (angle)	0.0002909	Radians		0.00037324	Tons (metric)
Nautical miles	6080.2	Feet	Pounds of water at 62° F		0.01604 Cubic feet
	1.516	Miles		27.72	Cubic inches
Ounces (avoirdupois)	16	Drams (avoir.)		0.120	Gallons (US)
	437.5	Grains	Pounds of water/min. at 62° F	0.0002673	Cubic feet/second
	0.0625	Pounds (avoir.)	Pounds/cubic foot	0.01602	Grams/cubic centimeter
	28.349527	Grams		16.02	Kilograms/cubic meter
	0.9115	Ounces (troy)		0.0005787	Pounds/cubic inch
Ounces (fluid)	1.805	Cubic inches	Pounds/cubic inch	27.68	Grams/cubic centimeter
	0.02957	Liters		27.680	Kilograms/cubic meter
	29.57	Cubic centimeters		1728	Pounds/cubic foot
	0.25	Gills	Pounds/foot	1.488	Kilograms/meter
Ounces (troy)	480	Grains	Pounds/inch	178.6	Grams/centimeter
	20	Pennyweights (troy)	Pounds/hour foot	0.4132	Centipoise
	0.08333	Pounds (troy)		0.004132	Poise grams/sec cm
	31.103481	Grams	Pounds/sec foot	14.881	Poise grams/sec cm
	1.09714	Ounces (avoir.)		1488.1	Centipoise
Ounces/square inch	0.0625	Pounds/square inch	Pounds/square foot	0.016037	Feet of water at 62° F
	1.732	Inches of water at 62° F		4.882	Kilograms/square meter
	4.39	Centimeters of water at 62° F		0.006944	Pounds/square inch
				0.014139	Inches of Hg at 32° F
	0.12725	Inches of Hg at 32° F		0.0004725	Atmospheres
	0.004253	Atmospheres	Pounds/square inch	0.068044	Atmospheres
Palms	3	Inches		2.30934	Feet of water at 62° F
Parts/million	0.0584	Grains/gallon (US)		2.0360	Inches of Hg at 32° F
	0.07016	Grains/gallon (Imp)		703.067	Kilograms/square meter
	8.345	Pounds/million gal (US)		27.912	Inches of water at 62° F
Pennyweights (troy)	24	Grains	Quadrants (angular)	90	Degrees
	1.55517	Grams		5400	Minutes
	0.05	Ounces (troy)		324,000	Seconds
	0.0041667	Pounds (troy)		1.751	Radians
Pints (liq. US)	4	Gills	Quarts (dry)	67.20	Cubic inches
	16	Ounces (fluid)	Quarts (liq.US)	2	Pints (liq. US)
	0.5	Quarts (liq. US)		0.9463	Liters
	28.875	Cubic inches		32	Ounces (fluid)
	473.1	Cubic centimeters		57.75	Cubic inches
Pipe	126	Gallons (US)		946.3	Cubic centimeters
Points	0.01389	Inches	Quintal, Argentine	101.28	Pounds
Poise	0.0672	Pounds/sec foot	Brazil	129.54	Pounds
	242	Pounds/hour foot	Castile, Peru	101.43	Pounds
	100	Centipoise	Chile	101.41	Pounds
Poncelots	100	Kilogram meters/second	Metric	220.46	Pounds
			Mexico	101.47	Pounds

CONVERSION FACTORS (CONTINUED)
(Courtesy of Anvil International)

Multiply	by	To Obtain
Quires	25	Sheets
Radians	57.30	Degrees
	3438	Minutes
	206.186	Seconds
	0.637	Quadrants
Radians/second	57.30	Degrees/second
	0.1592	Revolutions/second
	9.549	Revolutions/minute
Radians/second^2	573.0	Revolutions/minute^2
	0.1592	Revolutions/second^2
Reams	500	Sheets
Revolutions	360	Degrees
	4	Quadrants
	6.283	Radians
Revolutions/minute	6	Degrees/second
	0.1047	Radians/second
	0.01667	Revolutions/second
Revolutions/minute^2	0.001745	Radians/second^2
	0.0002778	Revolutions/second^2
Revolutions/second	360	Degrees/second
	6.283	Radians/second
	60	Revolutions/minute
Revolutions/second^2	6.283	Radians/second^2
	3600	Revolutions/minute^2
Rods	16.5	Feet
	5.5	Yards
Seconds (angle)	4.848 × 10^{-6}	Radians
Sections	1	Square miles
Side of a square	1.4142	Diameter of inscribed circle
	1.1284	Diameter of circle with equal area
Span	9	Inches
Square centimeters	0.001076	Square feet
	0.1550	Square inches
	0.0001	Square meters
	100	Square millimeters
Square feet	2.296 × 10^{-5}	Acres
	929.0	Square centimeters
	144	Square inches
	0.0929	Square meters
	3.587 × 10^{-8}	Square miles
	0.1111	Square yards
Sqaure inches	6.452	Square centimeters
	0.006944	Square feet
	645.2	Square millimeters
	1.27324	Circular inches
	1,273,239	Circular mils
	1,000,000	Square mils
Square kilometers	247.1	Acres
	10,760,000	Square feet
	1,000,000	Square meters
	0.3861	Square miles
	1,196,000	Square yards
Square meters	0.0002471	Acres
	10.764	Square feet
	1.196	Square yards
	1	Centares
Square miles	640	Acres
	27,878,400	Square feet

Multiply	by	To Obtain
Square Miles	2.590	Square kilometers
	259	Hectares
	3,097,600	Square yards
	102,400	Square rods
	1	Sections
Square millimeters	0.01	Square centimeters
	0.00155	Square inches
	1550	Square mils
	1973	Circular mils
Square mils	1.27324	Circular mils
	0.0006452	Square millimeters
	10^{-6}	Square inches
Square yards	0.0002066	Acres
	9	Square feet
	0.8361	Square meters
	3.228 × 10^{-7}	Square miles
Stere	1	Cubic meters
Stone	14	Pounds
	6.35029	Kilograms
Tons (long)	1016	Kilograms
	2240	Pounds
	1.12	Tons (short)
Tons (metric)	1000	Kilograms
	2205	Pounds
	1.1023	Tons (short)
Tons (short)	2000	Pounds
	32,000	Ounces
	907.185	Kilograms
	0.90718	Tons (metric)
	0.89286	Tons (long)
Tons of refrigeration	12,000	BTU/hour
	288,000	BTU/24 hours
Tons of water/24 hours at 62° F	83.33	Pounds of water/hour
	0.16510	Gallons (US)/minute
	1.3263	Cubic feet/hour
Watts	0.05692	BTU/minute
	44.26	Foot pounds/minute
	0.7376	Foot pounds/second
	0.001341	Horse power
	0.01434	Calories (Kg)/minute
	0.001	Kilowatts
	1	Joule/second
Watt hours	3.413	BTU
	860.5	Calories (gm)
	0.8605	Calories (Kg)
	2655	Foot pounds
	0.001341	Horse power hours
	3600	Joules
	367.1	Kilogram meters
	0.001	Kilowatt hours
Watts/square inch	8.2	BTU/square foot/minute
	6373	Foot pounds/sq ft/minute
	0.1931	Horse power/ square foot
Yards	91.44	Centimeters
	3	Feet
	36	Inches
	0.9144	Meters
	0.1818	Rods
Year (365 days)	8760	Hours

REFERENCES

American Lifelines Alliance, *Guidelines for the Design of Buried Steel Pipe*, 2001/2005 addenda.

ANSI B16.1, *Cast Iron Pipe Flanges and Flanged Fittings*, American National Standards Institute.

ANSI B18.2.1, *Square and Hex Bolts and Screws – Inch Series*, American National Standards Institute.

ANSI B18.2.4.6M, *Metric Heavy Hex Nuts*, American National Standards Institute.

ANSI B18.22M, *Metric Plain Washers*, American National Standards Institute.

ANSI B18.22.1, *Plain Washers*, American National Standards Institute.

API 579/ASME FFS-1, *Fitness for Service*, American Petroleum Institute and American Society of Mechanical Engineers.

ASME B1.1, *Unified Inch Screw Threads*, American Society of Mechanical Engineers.

ASME B1.13M, *Metric Screw Threads – M Profile*, American Society of Mechanical Engineers.

ASME B1.20.1, *Pipe Threads, General Purpose (Inch)*, American Society of Mechanical Engineers.

ASME B16.1, *Cast Iron Pipe Flanges and Flanged Fittings – 25, 125, 250 & 800 Class*, American Society of Mechanical Engineers.

ASME B16.3, *Malleable Iron Threaded Fittings*, American Society of Mechanical Engineers.

ASME B16.4, *Gray Iron Threaded Fittings*, American Society of Mechanical Engineers.

ASME B16.5, *Pipe Flanges and Flanged Fittings*, American Society of Mechanical Engineers.

ASME B16.9, *Factory-Made Wrought Steel Buttwelding Fittings*, American Society of Mechanical Engineers.

ASME B16.10, *Face-to-Face and End-to-End Dimensions of Valves*, American Society of Mechanical Engineers.

ASME B16.11, *Forged Steel Fittings, Socket-Welding and Threaded*, American Society of Mechanical Engineers.

ASME B16.14, *Ferrous Pipe Plugs, Bushings, and Locknuts with Pipe Threads*, American Society of Mechanical Engineers.

ASME B16.15, *Cast Bronze Threaded Fittings, Classes 125 and 250*, American Society of Mechanical Engineers.

ASME B16.18, *Cast Copper Alloy Solder Joint Pressure Fittings*, American Society of Mechanical Engineers.

ASME B16.22, *Wrought Copper and Copper Alloy Solder Joint Pressure Fittings*, American Society of Mechanical Engineers.

ASME B16.24, *Cast Copper Alloy Pipe Flanges and Flanged Fittings: Classes 150, 300, 600, 900, 1500, and 2500*, American Society of Mechanical Engineers.

ASME B16.34, *Valves—Flanged, Threaded, and Welding End*, American Society of Mechanical Engineers.

ASME B16.42, *Ductile Iron Pipe Flanges and Flanged Fittings, Classes 150 and 300*, American Society of Mechanical Engineers.

ASME B16.47, *Large Diameter Steel Flanges, NPS 26 Through NPS 60*, American Society of Mechanical Engineers.

ASME B16.50, *Wrought Copper and Copper Alloy Braze-Joint Pressure Fittings*, American Society of Mechanical Engineers.

ASME B18.2.1, *Square and Hex Bolts and Screws (inch series)*, American Society of Mechanical Engineers.

ASME B18.2.2, *Square and Hex Bolts (inch series)*, American Society of Mechanical Engineers.

ASME B18.2.3.5M, *Metric Hex Bolts*, American Society of Mechanical Engineers.

ASME B18.2.3.6M, *Metric Heavy Hex Bolts*, American Society of Mechanical Engineers.

ASME B18.21.1, *Lock Washers (Inch Series)*, American Society of Mechanical Engineers.

ASME B16.48, *Steel Line Blanks*, American Society of Mechanical Engineers.

ASME B31.1, *Power Piping*, American Society of Mechanical Engineers.

ASME B31.3, *Process Piping*, American Society of Mechanical Engineers.

ASME B31.4, *Pipeline Transportation Systems for Liquid Hydrocarbons and Other Liquids*, American Society of Mechanical Engineers.

ASME B31.5, *Refrigeration Piping*, American Society of Mechanical Engineers.

ASME B31.8, *Gas Transmission and Distribution Piping Systems*, American Society of Mechanical Engineers.

ASME B31.9, *Building Services Piping*, American Society of Mechanical Engineers.

ASME B31.11, *Slurry Piping*, American Society of Mechanical Engineers.

ASME B21.12, *Hydrogen Piping and Pipelines*, American Society of Mechanical Engineers.

ASME B31E, *Standard for Seismic Design and Retrofit of Above-Ground Piping*, American Society of Mechanical Engineers.

ASME B31H, *Standard Method to Establish Maximum Allowable Design Pressures for Piping Components* (under development, not published), American Society of Mechanical Engineers.

ASME B31J, *Standard Method to Determining Stress Intensification Factors (i-Factors) and Flexibility Factors for Metallic Piping Components*, American Society of Mechanical Engineers.

ASME B31T, *Standard Toughness Requirements for Piping*, American Society of Mechanical Engineers.

ASME Boiler and Pressure Vessel Code, Section I, *Power Boilers*, American Society of Mechanical Engineers.

ASME Boiler and Pressure Vessel Code, Section II, *Materials*, Part A, *Ferrous Material Specifications*, American Society of Mechanical Engineers.

ASME Boiler and Pressure Vessel Code, Section II, *Materials*, Part B, *Nonferrous Material Specifications*, American Society of Mechanical Engineers.

ASME Boiler and Pressure Vessel Code, Section II, *Materials*, Part D, *Properties*, American Society of Mechanical Engineers.

ASME Boiler and Pressure Vessel Code, Section III, *Rules for Construction of Nuclear Power Plant Components*, American Society of Mechanical Engineers.

ASME Boiler and Pressure Vessel Code, Section V, *Nondestructive Examination*, American Society of Mechanical Engineers.

ASME Boiler and Pressure Vessel Code, Section VIII, Division 1, *Pressure Vessels*, American Society of Mechanical Engineers.

ASME Boiler and Pressure Vessel Code, Section VIII, Division 2, *Pressure Vessels, Alternative Rules*, American Society of Mechanical Engineers.

ASME Boiler and Pressure Vessel Code, Section VIII, Divisions 1 and 2, Code Case 2286, *Alternative Rules for Determining Allowable Compressive Stresses for Cylinders, Cones, Spheres, and Formed Heads*, American Society of Mechanical Engineers.

ASME Boiler and Pressure Vessel Code, Section IX, *Welding and Brazing Qualifications*, American Society of Mechanical Engineers.

ASME PCC-1, *Guidelines for Pressure Boundary Bolted Flange Joint Assembly*, American Society of Mechanical Engineers.

ASME PCC-2, *Repair of Pressure Equipment and Piping*, American Society of Mechanical Engineers.

ASME PCC-3, *Inspection Planning Using Risk Based Methods*, American Society of Mechanical Engineers.

ASME Standard OM-S/G, *Standards and Guides for Operation and Maintenance of Nuclear Power Plants*, American Society of Mechanical Engineers.

ASTM A47, *Standard Specification for Ferritic Malleable Iron Castings,* American Society for Testing and Materials.

ASTM A53, *Standard Specification for Pipe, Steel, Black and Hot-Dipped, Zinc-Coated, Welded and Seamless,* American Society for Testing and Materials.

ASTM A106, *Standard Specification for Seamless Carbon Steel Pipe for High-Temperature Service,* American Society for Testing and Materials.

ASTM A278, *Standard Specification for Gray Iron Castings for Pressure-Containing Parts for Temperatures Up to 650°F (350°C),* American Society for Testing and Materials.

ASTM C582, *Standard Specification for Contact-Molded Reinforced Thermosetting Plastic (RTP) Laminates for Corrosion Resistant Equipment,* American Society for Testing and Materials.

ASTM D1599, *Standard Test Method for Short-Time Hydraulic Failure Pressure of Plastic Pipe, Tubing, and Fittings,* American Society for Testing and Materials.

ASTM D2837, *Test Method for Obtaining Hydrostatic Design Basis for Thermoplastic Pipe Materials or Pressure Design Basis for Thermoplastic Pipe Products,* American Society for Testing and Materials.

ASTM D2992, *Standard Practice for Obtaining Hydrostatic or Pressure Design Basis for "Fiberglass" (Glass-Fiber-Reinforced Thermosetting-Resin) Pipe and Fittings,* American Society for Testing and Materials.

AWS D10.10, *Recommended Practices for Local Heating of Welds in Piping and Tubing,* American Welding Society.

AWWA C110, *Ductile-Iron and Gray-Iron Fittings, 3 Inch Through 48 Inch (75 mm Through 1200 mm), for Water and Other Liquids,* American Water Works Association.

AWWA C115, *Flanged Ductile-Iron Pipe with Threaded Flanges,* American Water Works Association.

AWWA C207, *Steel Pipe Flanges for Water Works Service, Sizes 4 Inch Through 144 Inch (100 mm Through 3,600 mm),* American Water Works Association.

AWWA C208, *Dimensions for Fabricated Steel Water Pipe Fittings,* American Water Works Association.

AWWA C500, *Metal-Seated Gate Valves for Water Supply Service,* American Water Works Association.

AWWA C504, *Rubber-Seated Butterfly Valves,* American Water Works Association.

AWWA C509, *Resilient Seated Gate Valves for Water Supply Service,* American Water Works Association.

Becht IV, C., 1988, "Elastic Follow-up Evaluation of a Piping System with a Hot Wall Slide Valve," *Design and Analysis of Piping, Pressure Vessels, and Components—1988,* PVP-Vol. 139, American Society of Mechanical Engineers.

Becht IV, C., 1989, "Considerations in Bellows Thrust Forces and Proof Testing," *Metallic Bellows and Expansion Joints—1989,* PVP-Vol. 168, American Society of Mechanical Engineers.

Becht IV, C., Chen, Y., and Benteftifa, C., 1992, "Effect of Pipe Insertion on Slip-On Flange Performance," *Design and Analysis of Pressure Vessels, Piping, and Components—1992,* PVP-Vol. 235, American Society of Mechanical Engineers.

Becht IV, C., and Chen, Y., 2000, Span Limits for Elevated Temperature Piping," *Journal of Pressure Vessel Technology,* **122**(2), pp. 121–124.

Bednar, H., 1986, *Pressure Vessel Design Handbook,* Van Nostrand Reinhold, New York.

Bergman, E. O., 1960, " The New-Type Code Chart for the Design of Vessels Under External Pressure," *Pressure Vessel and Piping Design, Collected Papers 1927–1959,* American Society of Mechanical Engineers, pp. 647–654.

Bernstein, M D. and Yoder, L. W., 1998, *Power Boilers, A Guide to Section I of the ASME Boiler and Pressure Vessel Code,* American Society of Mechanical Engineers.

Biersteker, M., Dietemann, C., Sareshwala, S., and Haupt, R. W., 1991, "Qualification of Nonstandard Piping Product Form for ASME Code for Pressure Piping, B31 Applications," *Codes and Standards and*

Applications for Design and Analysis of Pressure Vessels and Piping Components, PVP-Vol. 210–1, American Society of Mechanical Engineers.

Bijlaard, P. P., "Stresses from Local Loadings in Cylindrical Pressure Vessels," *Trans. A.S.M.E.,* **77,** 802–816 (1955).

Bijlaard, P. P., "Stresses from Radial Loads in Cylindrical Pressure Vessels," *Welding Jnl.,* **33** (12), Research supplement, 615-s to 623-s (1954).

Bijlaard, P. P., "Stresses from Radial Loads and External Moments in Cylindrical Pressure Vessel," *Ibid.,* **34** (12). Research supplement, 608-s to 617-s (1955).

Bijlaard, P. P., "Computation of the Stresses from Local Loads in Spherical Pressure Vessels or Pressure Vessel Heads," *Welding Research Council Bulletin* No. 34, (March 1957).

Bijlaard, P. P., "Local Stresses in Spherical Shells from Radial or Moment Loadings," *Welding Jnl.,* **36** (5), Research Supplement, 240-s to 243-s (1957).

Bijlaard, P. P., "Stresses in a Spherical Vessel from Radial Loads Acting on a Pipe," *Welding Research Council Bulletin* No. 49, 1–30 (April 1959).

Bijlaard, P. P., "Stresses in a Spherical Vessel from External Moments Acting on a Pipe," *Ibid.,* No. 49, 31–62 (April 1959).

Bijlaard, P. P., "Influence of a Reinforcing Pad on the Stresses in a Spherical Vessel Under Local Loading," *Ibid.,* No. 49, 63–73 (April 1959).

Bijlaard, P. P., "Stresses in Spherical Vessels from Local Loads Transferred by a Pipe," *Ibid.,* No. 50, 1–9 (May 1959).

Bijlaard, P. P., "Additional Data on Stresses in Cylindrical Shells Under Local Loading," *Ibid.,* No. 50, 10–50 (May 1959).

Boardman, H. C., 1943, "Formulas for the Design of Cylindrical and Spherical Shells to Withstand Uniform Internal Pressure," *Water Tower,* **30,** pp. 14–15.

Copper Development Association, 1995, *The Copper Tube Handbook,* Copper Development Association, New York.

EJMA, *Standards of the Expansion Joint Manufacturer's Association,* EJMA, Tarrytown, New York.

Hinnant, C., and Paulin, T., 2008, "Experimental Evaluation of the Markl Fatigue Methods and ASME Piping Stress Intensification Factors," *Proceedings of PVP 2008,* The American Society of Mechanical Engineers.

Holt, M., 1960, "A Procedure for Determining the Allowable Out-of-Roundness for Vessels Under External Pressure," *Pressure Vessel and Piping Design, Collected Papers 1927–1959,* American Society of Mechanical Engineers, pp. 655–660.

ISO 15649, 2001, Petroleum and natural gas industries-piping.

MacKay, J. R. and Pillow, J. T., 2011, *Power Boilers, A Guide to Section I of the ASME Boiler and Pressure Vessel Code, Second Edition,* American Society of Mechanical Engineers.

Markl, A., 1960a, " Fatigue Tests of Piping Components," *Pressure Vessel and Piping Design, Collected Papers, 1927–1959,* American Society of Mechanical Engineers, pp. 402–418.

Markl, A., 1960b, " Fatigue Tests of Welding Elbows and Comparable Double-Mitre Bends," *Pressure Vessel and Piping Design, Collected Papers, 1927–1959,* American Society of Mechanical Engineers, pp. 371–393.

Markl, A., 1960c, " Fatigue Tests on Flanged Assemblies," *Pressure Vessel and Piping Design, Collected Papers, 1927–1959,* American Society of Mechanical Engineers, pp. 91–101.

Markl, A., 1960d, Piping-Flexibility Analysis, *Pressure Vessel and Piping Design, Collected Papers, 1927–1959,* American Society of Mechanical Engineers, pp. 419–441.

Mershon, J., Mokhtarian, K., Ranjan G., and Rodabaugh, E., 1984, *Local Stresses in Cylindrical Shells Due To External Loadings on Nozzles—Supplement to WRC Bulletin No. 107,* Bulletin 297, Welding Research Council, New York.

MSS SP-42, *Corrosion-Resistant Gate, Globe, Angle and Check Valves with Flanged and Buttweld Ends (Classes 150, 300 & 600),* Manufacturers Standardization Society of the Valve and Fittings Industry.

MSS SP-43, *Wrought and Fabricated Butt Welding Fittings for Low Pressure, Corrosion Resistant Applications,* Manufacturers Standardization Society of the Valve and Fittings Industry.

MSS SP-51, *Class 150LW Corrosion Resistant Cast Flanges and Flanged Fittings,* Manufacturers Standardization Society of the Valve and Fittings Industry.

MSS SP-58, *Pipe Hangers and Supports—Materials, Design, and Manufacture,* Manufacturers Standardization Society of the Valve and Fittings Industry.

MSS SP-67, *Butterfly Valves,* Manufacturers Standardization Society of the Valve and Fittings Industry.

MSS SP-75, *Specifications for High Test Wrought Butt-Welding Fittings,* Manufacturers Standardization Society of the Valve and Fittings Industry.

MSS SP-79, *Socket-Welding Reducer Inserts,* Manufacturers Standardization Society of the Valve and Fittings Industry.

MSS SP-80, *Bronze Gate, Globe, Angle and Check Valves,* Manufacturers Standardization Society of the Valve and Fittings Industry.

MSS SP-83, *Class 3000 Steel Pipe Unions, Socket-Welding and Threaded,* Manufacturers Standardization Society of the Valve and Fittings Industry.

MSS SP-88, *Diaphragm Valves,* Manufacturers Standardization Society of the Valve and Fittings Industry.

MSS SP-95, *Swage(d) Nipples and Bull Plugs,* Manufacturers Standardization Society of the Valve and Fittings Industry.

MSS SP-97, *Integrally Reinforced Forged Branch Outlet Fittings—Socket Welding, Threaded, and Buttwelding Ends,* Manufacturers Standardization Society of the Valve and Fittings Industry.

MSS SP-105, *Instrument Valves for Code Applications,* Manufacturers Standardization Society of the Valve and Fittings Industry.

MSS SP-106, *Cast Copper Alloy Flanges and Flanged Fittings, Class 125, 150 and 300,* Manufacturers Standardization Society of the Valve and Fittings Industry.

Paulin, T., 2012, *SIF and K-factor Alignment Project,* ASME ST-LLC 07-02, American Society of Mechanical Engineers.

PFI ES-24, *Pipe Bending Methods, Tolerances, Process and Material Requirements,* Pipe Fabrication Institute.

Robinson, E., 1960, "Steam-Piping Design to Minimize Creep Concentrations," *Pressure Vessel and Piping Design, Collected Papers 1927–1959,* American Society of Mechanical Engineers, pp. 451–466.

Rodabaugh, E., and George, H., 1960, "Effect of Internal Pressure on Flexibility and Stress-Intensification Factors of Curved Pipe or Welding Elbows," *Pressure Vessel and Piping Design, Collected Papers 1927–1959,* American Society of Mechanical Engineers, pp. 467–477.

Rossheim, D. B., and Markl, A.R.C., (1960), "The Significance of, and Suggested Limits for, the Stress in Pipelines Due to the Combined Effects of Pressure and Expansion," *Pressure Vessels and Piping, Collected Papers, 1927–1959,* American Society of Mechanical Engineers, pp. 362–370.

SAE J513, *Refrigeration Tube Fittings—General Specifications,* Society of Automotive Engineers.

SAE J514, *Hydraulic Tube Fittings,* Society of Automotive Engineers.

SAE J518, *Hydraulic Flange Tube, Pipe, and Hose Connections, Four-Bolt Split Flanged Type,* Society of Automotive Engineers.

Saunders, H. E., and Windenburg, D., 1960, "Strength of Thin Cylindrical Shells Under External Pressure," *Pressure Vessel and Piping Design, Collected Papers 1927–1959,* American Society of Mechanical Engineers, pp. 600–611.

Short II, W. E., Leon, G. F., Widera, G. E. O., and Zui, C. G., 1996, *Literature Survey and Interpretive Study on Thermoplastic and Reinforced-Thermosetting-Resin Piping and Component Standards,* WRC 415, Welding Research Council.

Wichman, K., Hopper A., and Mershon J., 1979, *Local Stresses in Spherical and Cylindrical Shells due to External Loadings,* Bulletin 107, Welding Research Council.

Windenburg, D., 1960, "Vessels Under External Pressure: Theoretical and Empirical Equations Represented in Rules for the Construction of Unfired Pressure Vessels Subjected to External Pressure," *Pressure Vessel and Piping Design, Collected Papers 1927–1959,* American Society of Mechanical Engineers, pp. 625–632.

Windenburg, D., and Trilling, C., 1960, "Collapse by Instability of Thin Cylindrical Shells Under External Pressure," *Pressure Vessel and Piping Design, Collected Papers 1927–1959,* American Society of Mechanical Engineers, pp. 612–624.

INDEX